TRANSPORT SYSTEMS
Modelling, Planning, and Evaluation

"For a man to achieve all that is demanded of him, he must regard himself as greater than he is."

— Johann Wolfgang von Goethe

"We can't solve problems by using the same kind of thinking we used when we created them."

"As far as the laws of mathematics refer to reality, they are not certain; as far as they are certain, they do not refer to reality."

— Albert Einstein

TRANSPORT SYSTEMS
Modelling, Planning, and Evaluation

Milan Janić

Department of Transport & Planning, Faculty of
Civil Engineering and Geosciences
&
Department of Air Transport and Operations, Faculty of
Aerospace Engineering, Delft University of Technology
Delft, The Netherlands

CRC Press is an imprint of the
Taylor & Francis Group, an **informa** business

A SCIENCE PUBLISHERS BOOK

CRC Press
Taylor & Francis Group
6000 Broken Sound Parkway NW, Suite 300
Boca Raton, FL 33487-2742

First issued in paperback 2021

© 2017 by Taylor & Francis Group, LLC
CRC Press is an imprint of Taylor & Francis Group, an Informa business

No claim to original U.S. Government works

ISBN-13: 978-0-367-78278-8 (pbk)
ISBN-13: 978-1-4987-1908-7 (hbk)

Library of Congress Cataloging-in-Publication Data

Names: Janiâc, Milan, author.
Title: Transport systems : modelling, planning and evaluation / Milan Janiâc, Department of Transport & Planning, Faculty of Civil Engineering and Geosciences & Department of Air Transport and Operations, Faculty of Aerospace Engineering, Delft University of Technology, Delft, The Netherlands.
Description: First Edition. | Boca Raton, FL : Taylor & Francis, 2016. | Includes bibliographical references and index.
Identifiers: LCCN 2016028261| ISBN 9781498719087 (hardback) | ISBN 9781498719094 (e-book)
Subjects: LCSH: Transportation. | Transportation--Planning. | Intelligent transportation systems.
Classification: LCC HE151 .J356 2016 | DDC 388--dc23
LC record available at https://lccn.loc.gov/2016028261

**Visit the Taylor & Francis Web site at
http://www.taylorandfrancis.com**

**and the CRC Press Web site at
http://www.crcpress.com**

To my wife Vesna
for her continuous inspiration and support

PREFACE

The transport sector consists of different transport modes, each with systems serving generally growing demand for mobility of persons and transport of goods/freight shipments. Under such conditions, the need for expanding the capacity of infrastructure and rolling stock of particular systems operated by particular modes has been increasing, while on the other hand, the needs regarding economic efficiency, effectiveness and environmental and social friendliness have been strengthening. Consequently, this has implied development of a 'greener', i.e. more sustainable transport sector, which would, in order to meet the growing demand, at the same time increase its overall contribution to social-economic welfare and reduce its negative impacts on the environment and society.

The above-mentioned development of the transport sector, particular modes and their systems has increased the complexity of dealing with them by academics-researchers, planners, the transport industry in general and policy makers at different geographical and institutional levels. As far as the academics-researchers and planners are concerned, there has been an increasing need for a problem-solving approach. This has implied comprehensive analyzing of the systems in order to identify the problem(s) and then modelling them in order to obtain solution(s) more easily, systematically and transparently. In addition, modelling has contributed to easier generating and planning of alternative solutions to problems, which have then been evaluated by means of different MCDM (Multi-Criteria Decision Making) methods as complements or counterparts to the well-known single-criterion NPV (Net Present Value) or BAU (Business As Usual) method. The main objective has been to select the preferred, i.e. the best among the offered alternative solutions, usually addressing and respecting the individual interests/preferences of particular actors/stakeholders involved under the given conditions.

This book aims to contribute to the above-mentioned problem-solving approach by generally analyzing, modelling and planning the performances of different transport systems and then using them as criteria for evaluation of the selected cases of these systems using MCDM methods. The book is organized in chapters addressing modelling, planning, or evaluation of these selected cases of transport systems. As such the book possesses three elements of added value: (i) modelling, planning and evaluation elaborated in the general sense and as such can be applied not only to the presented cases but also to any other similar ones without the need for substantial (if any) modification; (ii) the data for the application of modelling, planning and evaluation to particular cases originate from the time periods when the research had been carried out; as such they try to guide similar considerations at present and in the future; and (iii) the results from particular cases are generic in the relative and specific sense, at least in the medium- to long-term period of time considering the dynamism of changing particular transport modes and their systems.

Therefore, the book would be particularly useful to readers from the academia and profession/transport sector with some pre-knowledge and familiarity with transport system components, operations and related problems and to those who are interested in dealing with these through modelling, planning and/or evaluation. Finally, the book can be an inspiring material for upgrading and further development of the above-mentioned approach, particularly if such developments are in line with the further increasing needs for a 'greener', i.e. more sustainable transport sector.

Milan Janić

ACKNOWLEDGEMENTS

The author gratefully acknowledges the support of Transport & Planning Departnment of the Faculty of Civil Engineering and Geosciences at the Delft Unversity of Technology (Delft, The Netherlands). In particular, I would like to thank very much Professor Bart van Arem, Head of Department, for providing suitable conditions for writing and for providing financial support for finalization of the book. In addition, I express my special gratitude to Mr Andrej Grah Whatmough for his excellent help in polishing the language.

On the personal front, the great effort of writing this book was continuously supported and inspired by my wife Vesna and my son Miodrag, Doctor of Medicine, who is engaged in finalizing his PhD thesis.

Milan Janić

CONTENTS

Preface　　vii

Acknowledgements　　ix

Abbreviations　　xiii

About the Author　　xvii

1. Introduction: Setting the Scene　　1
2. Transport Systems: Components & Concept of Performances　　4
3. Modelling Transport Systems—I: Operational, Economic, Environmental and Social Performances　　76
4. Modelling Transport Systems—II: Influence of New Technologies on Performances　　137
5. Modelling Transport Systems—III: Resilience　　195
6. Planning Transport Systems: Infrastructure, Rolling Stock & Planning Process　　249
7. Evaluation of Transport Systems: Methodology & Cases　　285
8. Conclusions: Summary & Lessons Learnt　　389

Index　　393

ABBREVIATIONS

AA	American Airlines
AC	Alternating Current
ACM	Aircraft Movement
ADS-B	Automatic Dependent Surveillance – Broadcast
AHP	Analytical Hierarchy Process
AMS	Amsterdam (Schiphol)
APT	Air Passenger Transport
ASDE X	Airport Surface Detection Equipment – Model X
ATC	Air Traffic Control
ATC	Automated Train Control
ATFM	Air Traffic Flow Management
ATM	Air Traffic Management
atm	air transport movement
ATP	Automatic Train Protection
BACC	BloccoAutomatico a CorrentiCodificate
BAU	Business As Usual
BCR	Benefit-Cost Ratio
BTS	Base Transceiver Station
CAA	Civil Aviation Authority
CDTI	Cockpit Display Traffic Information
CI	Consistency Index
COST	(European) COoperation in Science and Technology
CREAM	Customer-driven Rail-freight services on a European mega-corridor based on Advanced business and operating Models
CRH	China Railway High (speed)
dB	decibel
DC	Direct Current
DM	Decision-Maker

DME	Distance Measuring Equipment
DSS	Decision Support System
dwt	dead-weight-ton
EAT	Economic Analysis Technique
EBICAB	Electrique Bureau CABine
EC	European Commission
ECMT	European Conference of Ministers of Transport
€-ct	Euro cent
EEC	EUROCONTROL
ELECTRE	ELimination Et Choix Traduisant la REalité (ELimination and Choice Expressing Reality)
ETCS	European Train Control and Command System
FAA	Federal Aviation Administration
FAG	Final Approach Gate
FFP	Frequent Flyer Programmed
FMP	Framework Program
ft	feet
GDP	Gross Domestic Product
GHP	Ground Holding Program
GHG	Green House Gases
GS	Glide Slope
GSM-R	Global System for Mobile Communications – Railway
hr	hour
HS	High Speed
H-S	Hub-and-Spoke
HSR	High Speed Rail
IATA	International Air Transport Association
ICAO	International Civil Aviation Organization
ICE	The Intercity-Express
ICT	Information/Communication Technologies
IDACS	Integrated Departure and Arrival Coordination System
IFR	Instrument Flight Rules
ILS	Instrument Landing System
IRR	Internal Rate of Return
JIT	Just-in-Time
kg	kilogram
kp	kilopond
kt	knot
kV	kilovolt
kW	kilowatt
kWh	kilowatt-hour

l	liter
LCC	Low Cost Carrier
LEU	Lineside Equipment Unit
LHR	London Heathrow
LTO	Landing-and-Take-Off (cycle)
LZB	LinienZugBeeinflussung
LZZ	Localizer
m	meter
MADM	Multi-Attribute Decision Making
MCA	Multi-Criteria Analysis
MCDM	Multi-Criteria Decision Making
MLS	Microwave Landing System
mm	millimeter
MTOW	Maximum Take-Off Weight
mi	mile
mph	mile per hr
N	Newton
NextGen	Next Generation
nm	nautical mile
NPV	Net Present Value
O-D	Origin-Destination
OECD	Organization for Economic Co-operation and Development
pass	Passenger
PCI	Per Capita Income
p-km	passenger-kilometer
PROMETEE	Preference Ranking Organization Method for Enrichment of Evaluations
RAMS	Reliability, Availability, Maintainability and Safety
RBC	Radio Block Centre
RFC	Rail Freight Corridor
RETRACK	REorganization of Transport Networks by Advanced RAil Freight Concepts
RI	Random Index
RNE	Rail Net Europe
RWY	Runway
SAW	Simple Additive Weighting
SCM	Supply Chain Management
SCP	Supply Chain Performances
SESAR	Single European Sky ATM Research
s	second
s-km	seat kilometer

STAR	Standard Terminal Arrival Route
SWIM	System Wide Information Management
TCC	Trackside Control Center
TFDM	Terminal Flight Data Manager
TFMS	Traffic Flow Management System
TEN	Trans-European Transport Network
TEU	Twelve Equivalent Unit
TGV	Train à Grande Vitesse
t-km	ton-kilometer
TOPSIS	Technique for Order Preference by Similarity to Ideal Solution
TU	Transport Unit
TVM	Transmission Vole Machine
U.S.	United States
USA	United States of America
UK	United Kingdom
UIC	International Union of Railways
UNCTAD	United Nations Conference on Trade And Development
USDOT	United States Department of Transportation
VFR	Visual Flight Rules
VHF	Very High Frequency
VOR	VHF Omni Directional Radio Range
WP	Way-Point
WTMA	Wake Turbulence Mitigation for Arrivals
WTMD	Wake Turbulence Mitigation for Departures

ABOUT THE AUTHOR

Milan Janić (PhD) is a transport and traffic engineer and planner. At present, he is a Senior Researcher in the Department of Transport & Planning of the Faculty of Civil Engineering and Geosciences and Visiting Researcher at the Department of Air Transport and Operations of Faculty of Aerospace Engineering, both of the Delft University of Technology (Delft, The Netherlands). He is also Research Professor at the Faculty of Traffic and Transport Engineering in the University of Belgrade (Belgrade, Serbia). Previously he used to be a Leader of the Research Program of the Transport & Infrastructure Section at the OTB Research Institute for the Built Environment of Delft University of Technology (Delft, The Netherlands), Senior Researcher at Manchester Metropolitan University (Manchester, UK) and Loughborough University (Transport Studies Group) (Loughborough, UK) and the leader of Research Program at the Institute of Transport of the Slovenian Railways (Ljubljana, Slovenia).

Milan has been involved in many transport-related research and planning projects on both the national and international scale for almost thirty years. He has also published many scientific and professional transport-related papers and presented many of them at national and international transport conferences. In addition to contributing to many edited books, he has published five books: *Advanced Transport Systems: Analysis, Modelling and Evaluation of Performances; Greening Airports: Advanced Technology and Operations; Airport Analysis, Planning and Design: Demand, Capacity and Congestion; The Sustainability of Air Transportation: A Quantitative Analysis and Assessment;* and *Air Transport System Analysis and Modelling: Capacity, Quality of Services and Economics.*

Milan has also been a member of international scientific and professional organizations, such as NECTAR (Network on European Communications and Transport Activity Research), Delft Aviation Centre (Delft University of Technology, Delft, The Netherlands), GARS (German Aviation Research Society, Cologne, Germany), ATRS (Air Transport Research Society) and Airfield and Airspace Capacity and Delay Committee of TRB (Transportation Research Board) (Washington DC, USA).

CHAPTER 1

INTRODUCTION
Setting the Scene

The transport sector consists of different transport modes—land, sea or air—each serving the need for mobility of people and transport of goods/freight shipments. For a long time, transport systems subsisted on the basis of demand and supply, i.e. capacity, which has been under comprehensive consideration by the different actors/stakeholders involved. On the demand side were the users of transport services, like passengers and goods/freight shippers and receivers; on the supply/capacity side were the providers of transport infrastructure and services. In addition, the transport planners and policy makers at different institutional levels—local, regional, national and international—have been involved with planning and implementing solutions on development of transport systems and their particular components, besides the regulation of their operations, both internally and externally. In general, the actors/stakeholders have had specific, sometimes conflicting, interests and expectations, from the particular transport system. The users of transport services usually expect to receive safe, efficient, quick and effective door-to-door services at reasonable prices. The providers of transport infrastructure and services try to satisfy the expectations of users under conditions characterized by a growing transport demand on the one side and voicing operational, economic, environmental and social constraints on the other. The latter aim at mitigating the impact of overall transport-related activities on the society and environment.

Transport planners are engaged in planning adequate transport capacity in the wider social, economic and environmental context. Transport policy makers create institutional/regulative conditions for maintaining the transport capacity as per the demand by setting up internal and external market, social and environmental framework for implementation and operation of particular transport systems and their components according to the regulations based on various constraints.

The main objectives of both transport policy makers and the transport sector itself are 'greening' or developing more sustainable transport systems, which contribute to social welfare while reducing their impact on the society and environment in both absolute and relative terms in order to keep space with the growing demand. Fulfilment of such an objective influences changes in dealing with transport systems and their particular problems by all actors/stakeholders involved directly and/or indirectly. This is mainly carried out by widening the context and content, i.e. the increasing complexity of addressing these systems and related problems.

This book aims to illustrate some aspects of the transport system as seen from the analyzing, modelling, planning and evaluation perspective. Thus, *widening the context and content* is addressed through a multi-dimensional examination of the performances of transport systems, i.e. by considering them simultaneously. These performances are classified as infrastructural, technical/technological, operational, economic, environmental and social. The *increasing complexity* is addressed by considering the dependability of particular performances more explicitly or implicitly. Consequently, the modelling, planning and evaluation of transport system performances follow the above-mentioned features of both the context and content. In addition, despite being written from the research and planning perspective, this book can prove useful to actors/stakeholders who deal directly or indirectly with different problems in the transport sector. Therefore, in addition to this introductory chapter, the book consists of seven other chapters which cover aspects like modelling, planning and evaluating transport systems.

Chapter 2 describes the general characteristics of transport systems illustrated by analyzing and modelling of performances of the HSR (High Speed Rail) system.

Chapter 3 deals with modelling the operational economic, environmental and social performances of transport systems illustrated under three heads—utilization of the runway system capacity at a large hub airport; the full (internal and external) costs of intermodal rail/road and road freight transport networks competing with each other under given conditions and the effects/impacts in terms of savings of externalities of rail/road substitution in the given freight transport corridor(s).

Chapter 4 deals with modelling the effects and impacts of new technologies and related innovative procedures on the performances of transport systems by elaborating two cases: the performances of supply chain(s) operated by mega vehicles and the capacity of the system of two closely-spaced parallel airport runways where the aircraft landings and taking-offs are supported by new ATC/ATM (Air Traffic Control/Air Traffic Management) technologies and related innovative procedures.

Chapter 5 presents modelling of the resilience of transport systems illustrated by two cases: a logistics networks operating under regular and irregular (disturbing) conditions and an air transport network affected by a large-scale disruptive event.

Chapter 6, without referring to any specific cases, describes the main characteristics of planning the transport systems, such as the procedure of long-term planning of transport infrastructure and rolling stock, the prospective effects and impacts of such planning and the main components of the planning process of both transport infrastructure and service network(s).

Chapter 7 deals with evaluation of the transport system alternatives by using different MCDM (Multi-Criteria Decision Making) methods. Three cases are elaborated: (i) selection of the new hub airport for an airline, (ii) location of the new runway at an airport of a given airport system, and (iii) evaluation of the feasibility of alternative rail freight transport corridors.

The last chapter summarizes the conclusions and lessons learnt.

Despite the chapters covering diverse transport modes and their systems and therefore appearing rather wide and heterogeneous in terms of context and specific topics dealt with, each is actually very coherent regarding the activity carried out—modelling, planning and/or evaluation. In addition, each chapter is organized into the following sections: an introduction, presentation, elucidation, conclusion and references. Each section of the chapters except the concluding one is organized into sub-sections dealing with the particular cases as follows: background, description of the system and problem, the basic structure of the proposed methodology—some related research, objectives and assumptions, structure of the methodology/models and their application to either real-life or hypothetical case(s) and an interim summary. As such, some sections and sub-sections look similar to the modified papers published in transport-related scientific and professional journals. This is simply because most of the presented material originates from the author's research carried out over the past decade-and-a-half and which has been partially published in the above-mentioned journals.

TRANSPORT SYSTEMS
Components & Concept
of Performances

2.1 Introduction

Transport systems enable mobility of persons and transport of goods/freight shipments between their origins and destinations. These represent the demand served by the supply capacity of transport systems under given conditions. The transport systems are operated by different modes, such as road, rail, inland waterways, air, sea, and their intermodal or multimodal combinations. In the latter case, the systems of two or more modes are combined in the sequential order for serving given passenger and/or goods/freight transport demand between origin(s) and destination(s), i.e. 'door-to-door'.

In general, the transport systems operated by each mode consist of physical components such as: (i) transport infrastructure; (ii) rolling stock, i.e. vehicles; (iii) supporting facilities and equipment; (iv) operating rules and procedures; (v) staff; and (vi) fuel/energy.

(i) The transport infrastructure of each mode consists of links and nodes, which, as mutually connected, constitute the infrastructure networks. These can be considered at different spatial scales such as urban, sub-urban, regional, national, international between countries, and intercontinental global. As far as the individual transport systems are concerned (i.e. use of private cars), the nodes are usually (regulated or non-regulated) intersections of urban and sub-urban streets, regional roads and interurban roads and highways. The segments of streets, roads and highways spreading between them are considered the links. In case of passenger mass transport systems, the nodes are the passenger bus, rail, stations, ports and airports, while the segments of roads and highways, rail lines, inland waterways, sea routes and airways, respectively, connecting them, are considered the links. In case of goods/freight mass transport systems, the nodes include freight road, rail, port, airport and intermodal terminals and the links are segments of the corresponding infrastructure lines connecting them. The main physical characteristics of the infrastructure nodes and links of transport systems operated by all modes are their specific design standards, size and spatial layout and position, i.e. location in the wider geographical area. The design standards generally provide compatibility of their use by both users and suppliers of transport services, including interoperability. The size and spatial layout depend on the current and prospective volumes and structure of demand to be accommodated on the one hand and the available land for settling down the given nodes and links, on the other. The position, i.e. location in the wider geographical area mainly refers to the maximum convenience of accessibility for users and at the same time compromising the existing non-transport activities as little as possible. All the above-mentioned characteristics are particularly relevant for the passenger and goods/ freight terminals located within or very close to densely populated urban and sub-urban areas, as well as for the road and rail lines passing through them.

(ii) Rolling stock, i.e. transport vehicles, represent the mobile component of transport systems. They carry out transport services and thus facilitate mobility of persons and transport of goods/freight shipments. The vehicles are both demand- and mode-specific. For the user-passenger demand, these include individual cars and buses (road), urban, sub-urban and interurban trains (rail), passenger barges (inland waterways) and cruiser ships (sea), and aircraft (air), all of different sizes and payload capacity. For the goods/freight demand, these include trucks (road), freight trains (rail), barges (inland waterways), freight ships (sea), and freight/cargo aircraft (air), again all of different sizes and payload capacity. The vehicle size influences the area of land/space needed for maneuvering and parking. It also influences the required engine power. In addition, the engine power influences the maximum and operating speed, which together with payload capacity, influences productivity and energy/fuel consumption. The latter particularly influences the vehicle operating cost, i.e. economic efficiency on the one hand and impacts the environment through emission of GHG (Green House Gases), on the other.

(iii) The supporting facilities and equipment of transport systems operated by particular transport modes have two basic functions. The first is to manage the flow and control the individual vehicles operating on the infrastructure in order to provide safe, efficient and effective means of transport. The typical components are different control, signaling, overhead line interface and information/communications systems, all with the components, the infrastructure and onboard the rolling stock/vehicles. The latter two are particularly relevant for mass passenger and goods/freight transport. The other function of the supportive facilities and equipment is to provide users with transport means/services. The typical components are different information/communications systems, elevators and moving stairs/belts for passengers, consolidation units (boxes, containers of different sizes) and loading/unloading devices (usually cranes) for goods/freight shipments, both mainly located in the corresponding nodes/terminals.

(iv) The facilitation of users and transport services on the one hand and movement of vehicles, on the other are carried out according to the specified operating rules and procedures aimed at providing safe, efficient and effective transport operations and related services. In this context, safety implies operations without traffic incidents/accidents with related consequences (damage to property, injuries and loss of life) due to already known reasons. Efficiency implies providing transport services at reasonable prices by the operators. Effectiveness implies carrying out transport services punctually on time or with minimum delays, say up to 15 minutes and reliably i.e. without cancellations.

(v) Staff is engaged by all transport operators to carry out direct and indirect tasks of providing transport services. In general, direct tasks include operation of infrastructure, rolling stock, i.e. transport vehicles and supportive facilities and equipment. Indirect tasks include those related to maintenance of the above-mentioned components, regulating energy/fuel supply, and administrative and managerial activities at different hierarchal/organizational level(s).

(vi) Fuel/energy consumption is an additional specific feature of transport systems operated by particular modes. For example, most of them currently consume electric energy and derivatives of crude oil—the former is usually obtained from a combination of different primary sources, such as water, coal, crude oil, atomic energy and the sun; the latter by refining crude oil as the primary source. At the road transport mode, mostly derivatives of crude oil are consumed. More recently, hybrid and pure electric cars are in use. At railways, both passenger and freight trains consume electric energy, while diesel-powered trains are being gradually phased out. Both passenger and goods/freight inland waterways and sea transportation usually depend on derivatives of crude oil. Air transport still uses kerosene as a derivative of crude oil. Nevertheless, in future it can be expected that solar and wind energy would become more intensively used as primary sources for generating electricity. LH_2 (Liquid Hydrogen) could also become (at least partially) a fuel for commercial air transport.

In addition to the above-mentioned introduction, this chapter analyzes the components and performances of the HSR (High Speed Rail) system. This is preferred as it is one of the most illustrative cases of contemporary transport systems. Therefore, Section 2.2 contains an analysis of its main components. Section 2.3 deals with analyzing and modelling of its performances dependent and influenced by the characteristics of particular components. The last Section 2.4 contains some concluding remarks.

2.2 Components

2.2.1 Background

The main criterion for choosing the HSR (High Speed Rail) system for presenting an analysis of the components and modelling performances of transport systems is its fast development over the past two-and-a-half decades worldwide (Europe, Far East and USA) as a rather innovative system within the railway transport mode, particularly as compared to its conventional rail passenger counterpart. Despite the common name, different definitions of HSR systems are used in particular global regions as follows:

2.2.1.1 Japan

The country's HSR system called 'Shinkansen' (i.e. 'new trunk line') is defined as the main line. Along almost its entire length (i.e. route), trains can run at speeds of at least 200 km/hr and above. The HSR system's network is built with specific technical standards (i.e. dedicated tracks without level crossings and a standardized and special loading gauge). This HSR system represents a part of the overall Japanese Shinkansen transportation system (UIC 2010a).

2.2.1.2 Europe

The European definition of HSR system includes: infrastructure; rolling stock; and compatibility of infrastructure and rolling stock (EC 1996).

(*a*) *Infrastructure*
- Infrastructure of the trans-European HSR system is considered to be a part of the trans-European rail transport system/network. It is

specially built and/or upgraded for HS (High Speed) travel. This may include connecting lines and junctions of new lines upgraded for HS and stations located on them, where speeds must take into account the local conditions.

- The HSR lines include those specially built for speeds equal to or greater than 250 km/h (*Category I*), those specially upgraded for speeds to the tune of 200 km/h (*Category II*), and those upgraded with particular features resulting from topographical relief or town-planning constraints (*Category III*).

Therefore, in the given context, *Category I* lines are exclusively considered as true HSR lines.

(b) Rolling stock

HS trains are designed to guarantee safe and uninterrupted travel at speeds of at least 250 km/h on *Category I* lines, at 300 km/h under appropriate circumstances, about 200 km/h on the specially upgraded *Category II* lines and at the highest possible speed on other *Category III* lines.

(c) Compatibility of infrastructure and rolling stock

HS trains are designed to be fully compatible with the characteristics of the infrastructure and vice versa. This compatibility influences the performances in terms of safety, quality and cost of services.

2.1.2.3 China

According to Order No. 34, 2013 of China's Ministry of Railways, the HSR system refers to the newly built passenger-dedicated lines with (actual or reserved) speed equal and/or greater than 250 km/h. Its specific acronym is CRH (China Railway High) speed. In addition, a number of new 200 km/h express passenger and 200 km/h mixed (passenger and freight) lines are being built as components of the country's entire HSR network (Ollivier et al. 2014).

2.2.1.4 USA

The country's HSR system consists of frequent express services between the major population centers at distances of 200 to 600 miles with few intermediate stops, at the speeds of at least 150 mph on completely grade-separated, dedicated rights-of-way lines (one mile = 1.609 kms). Regional, relatively frequent services between the major and moderate population centers at distances between 100 and 500 miles with some intermediate stops at speeds from 110 to 150 mph, grade-separated with some dedicated and some shared tracks using positive train control technology (USDOT 2009) are also considered to belong to the HSR system. In both cases, the HSR system is expected to relieve congestion at highways and airports; in the latter case particularly by competing with the short- to medium-haul flights.

The above-mentioned definitions mean that the HSR system is generally characterized by specially designed dedicated and/or upgraded conventional lines and tracks that enable operating speeds of HS trains of over 200 km/h. This also implies that the system possesses completely new above-mentioned physical components compared to its conventional rail counterpart. Since much higher operating speeds are the main distinction, all components are designed to enable such speeds.

Consequently, the HSR system is considered to consist of the physical components, such as infrastructure (lines/tracks and stations constituting the infrastructure network), rolling stock (trains of given technology, design and comfort), supporting facilities and equipment, i.e. power supply and signaling system(s) and the corresponding maintenance systems and related policies. The operational rules and procedures aimed at enabling safe operations of the above-mentioned physical components are also an additional (non-physical) component.

2.2.2 Infrastructure

The main elements of the HSR system infrastructure are the rail lines with tracks connecting the stations. In the given context, both are considered exclusively as the above-mentioned *Category I* of HSR lines. The lines and stations constitute the HSR network spreading over a given region, country and/or continent. Table 2.1 gives an illustration of the progress in developing the HSR networks around the world.

Table 2.1: Development of the HSR network around the world
(CSP 2014, UIC 2014, Yanase 2010)

Status	Continent			
	Europe	**Asia**	**Others**[1]	**World**
In operation (km)	7351	15241	362	22954
Under construction (km)	2929	9625	200	12754
Total (km)	10280	24866	562	35708

[1]Latin America, USA, Africa

As can be seen, the longest HSR network currently operating and under construction is in Asia, due to the fast development of the Chinese HSR network, while the shortest are in both the Americas and Africa. The European network is in between.

2.2.2.1 HSR Lines

The lines as links connecting particular stations as nodes of the HSR network are mainly characterized by a three-dimensional layout.

(a) Layout

The most relevant parameters of geometry of the HSR line's tracks are the distance between their centers, gauge, the maximum axle load, gradient, the minimum horizontal and vertical radius of curvature, the maximum cant, the cant gradient and the length of transition curves corresponding to the minimum curve radius. For example, in Europe, except for the track gauge (1435 mm), all the other above-mentioned parameters are dependent on the maximum design speed. Figure 2.1 is an example of the relationship between the minimum radius of track curvature and the maximum design speed of HSR lines in certain European countries.

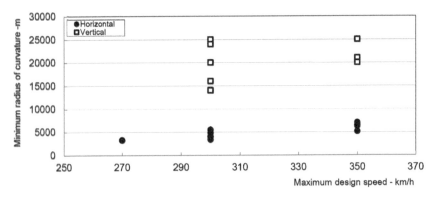

Fig. 2.1: Relationship between the minimum radius of track curvature and the maximum design speed of HSR lines (France, Germany, Italy, Spain, Belgium) (UIC 2002, 2010b, Profillidis 2006).

As can be seen, the minimum radius of horizontal curvature linearly increases with increase in the design speed with variations throughout the selected countries. For the speeds of 300 km/h, it varies between 3,000 and 5,500 m, and for the speeds of 350 km/h and above from 5,000 to 7,000 m. The minimum radius of the vertical curvature also increases with increase of the maximum design speed. It is more diverse (14-25 kms) at <300 km/h than at >300 km/h; 20-25 kms).

The maximum cant, i.e. difference between the inner and outer rail in the curved track(s), is generally constant or increases with increase of the maximum speed. Typically, it varies from 105 mm in Italy for a speed of 300 km/h to 180 mm in France for a speed of 300-350 km/h. The maximum cant gradient as the rate of change (increasing/decreasing) along a given length of track also generally increases or is constant with increase in the maximum speed. It is the lowest in Italy (12 mm/m) and the highest in Germany (40 mm/m). In France it is constant at 35 mm/m.

The length of the transition curves, i.e. those connecting the track segments of constant non-zero curvature to other segments with constant curvature corresponding to the minimum radius is also constant or increases with increase in the maximum design speed as shown in the example in Fig. 2.2.

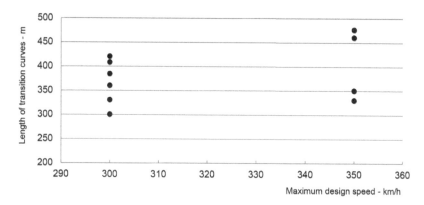

Fig. 2.2: Relationship between the length of transition curves and the maximum design speed of HSR lines (France, Germany, Italy, Spain, Belgium) (UIC 2002, 2010b, Profillidis 2006).

As can be seen, the variations are again greater at lower speeds (300 km/h) (between 300 m and 420 m), but lower and with greater differences at higher speeds (350 km/h) (between 330 m and 475 m). In addition, in European countries—France, Germany, Italy and Spain—the distance between the track centers is typically 4.0-4.2 m for speeds of 250 km/h and 4.5-5.0 m for speeds of 300 km/h and higher. The track gauge is standardized to 1453 mm. The maximum longitudinal track gradient is 35.0, 12.5, 18.0 and 30.0 per cent, respectively. In China and Korea, it is 20.0 and 25.0 per cent, respectively (Profillidis and Botzoris 2013). In addition, in the above-mentioned European countries, the maximum axle load is typically 17 ton/axe with some exceptions in Germany (<16 ton/axe) and Spain (18 ton/axe), both for speeds of 350 km/h and higher.

(b) Tracks

The rails of HSR lines are made of welded steel whose weight is typically 60 kg/m. The rails are elastically fastened to ties or sleepers, which are made as concrete mono- or bi-blocks, each positioned with a longitudinal density of 1666 units/km. The length of tie typically varies from 2.4 to 2.6 m/unit, width from 29 to 33 cm and height from 18.0 to 24.2 cm. The surface area of a tie varies from 0.244 to 0.390 m². Consequently, the weight of a tie also varies between 245 and 450 kg/unit (UIC 2002, 2010b).

The HSR tracks can be broadly categorized as ballasted and ballastless tracks. The former are present at already built HSR lines while the later are considered particularly for lines with long segments of tunnels and/or bridges, such as those in Japan[1]. In addition, they are expected to increase the capacity of the HSR lines, operating speed, reduce maintenance costs also by reducing the frequency of maintenance operations and consequently increase the level of safety. Both categories of tracks are laid on few foundation sublayers made of different materials with a bearing capacity typically greater than 120 N/mm² and protected against frost. In the bottom-up direction, for ballasted tracks the thickness of the first sublayer is 70 cm; above it is a sublayer of about 20 cm of gravel-type material; above this is a ballast layer about 30 cm thick as required for the concrete ties. At ballastless tracks, the thickness of the gravel sublayer above the first layer of 70 cm is typically 30 cm; the thickness of the concrete sublayer under the ties is 25 cm; in some cases, the latter two sublayers can be replaced by concrete slabs and wedging concrete under the ties. Consequently, at both categories of tracks, the total thickness of the structure amounts to about 1.3-1.4 m.

In general, both ballasted and ballastless tracks are designed to last 30 and 60 years, respectively. In particular, the ballastless tracks are flexible to tolerate and adapt to changes of the soil support under them. In addition, they do not need systematic corrections of geometry during their life-cycle.

Both categories of tracks need to take into account interfaces with the overhead contacting lines and signaling systems, mainly in terms of reserving sufficient space for electrical equipment and facilities providing connections to the rail and required insulation. Ballastless tracks are noisier and generate greater vibration than their ballasted counterparts. Noise is greater as it is emitted and partially reflected by the track, as well as due to the rolling stock. At ballasted tracks, ballast has shown to be a better absorber of noise than concrete used in their ballastless counterpart. Greater vibration of ballastless tracks is mainly due to using softer rail fasteners, allowing the rails to vibrate over a greater length. A substantial reduction of both noise and vibration can be achieved by the so-called 'floating ballastless track' concept or by supporting rails with an elastic material in combination with a shorter rail section(s).

Ballastless tracks are generally more expensive than ballasted tracks. Some estimates indicate that their construction costs (suppliers, working and

---

[1] In 1972, ballastless 'slab track' was developed and applied to the Sanyo Shinkansen line; in 2007, 'slab tracks' were used for 1,244 km of line, accounting for about 57 per cent of the total length of the Shinkansen network (Takai 2013).

overhead) are higher by the factor of about 1.3 for earth work and the factor 1.1-1.5 for tunnel work. At the same time, the maintenance costs are generally lower by about 25-50 per cent. According to Japan's 40-year experience, the total (construction and maintenance) costs of ballastless tracks are about 30 per cent higher than those of their ballast counterparts, but the full balance in terms of equalizing these costs is established after nine years of implementation and exploitation (Takai 2013, UIC 2002, 2010b). In China, both ballast and ballastless slab tracks have been used (Takagi 2011).

2.2.2.2 The HSR Stations

The HSR stations mainly characterized by their location and design enable facilitation of the HSR system with its users-passengers.

(a) Location

The main aspects of location of the HSR stations as nodes of the corresponding network is their number along the given rail lines. Then, it is their micro location in cities and at airports, which should enable safe, efficient and effective accessibility by individual (car) and mass urban and sub-urban public transit systems (bus, tram, light rail, metro, etc.). For example, the new CRH South Guangzhou station on the Hangzhou-Shenzhen line (China) has 15 platforms with 28 tracks and is the largest in Asia at the moment (Takagi 2011).

The additional aspects include the capacity of particular facilities, construction and maintenance costs and related revenues from both primary (traveling) and secondary (non-traveling-commercial) activities.

(b) Design

In addition to location, a functional design of the HSR stations is of crucial relevance. This includes: (i) the track and platform spatial and technical aspects (number, arrangement, dimension, safety and electrical signaling and communication systems); (ii) the user-passenger service and comfort aspects (accessibility, intermodal transfer, security, ticketing, and travel information and station facilities); and (iii) the environmental aspects (choice of building/construction materials and noise affecting the local environment) (Anderson and Lindvert 2013, Kido 2005):

- The track and platform spatial and technical aspects generally include the number of parallel tracks and platforms and their mutual arrangement(s), the curvature and gradients of tracks, the number and layout of additional tracks, the dimensions (length, width and height) of platforms, the required distances between the tracks and between the tracks and other objects, designs for safely separating users/passengers from passing HS trains, and signaling, power supply and communication systems;

- The user-passenger service and comfort aspects generally include inside conditions (light, air quality) and the size and content of waiting areas, shopping and eating facilities, toilets, possibilities and easiness for interchange to/from individual and mass transit systems, conveniences for disabled persons, ticket purchase offices and machines, live, audible and visual information, etc.; and
- The environmental aspects embrace materials for construction/building the station and other constructions, and inside and outside noise from the passing HS trains.

An additional important design aspect of the HSR stations is the mutual arrangement of tracks and platforms for users-passengers. In general, this can be carried out according to two main concepts: island platforms with tracks on both sides and side platforms with only one side faced to the track. Both concepts have advantages compared to each other. Island platforms are advantageous due to enabling train interchanges over the platform, while side platforms are advantageous considering the safety of users-passengers while at the platform. However, both types of arrangements are in place at most HSR lines (Anderson and Lindvert 2013). Figure 2.3 (a, b) shows a simplified layout of the arrangement of a station along the line and the begin/end station, i.e. terminus.

The former station is arranged with two side platforms and four tracks. The latter is arranged with two island platforms and four tracks.

The safety aspect of HSR station design is important for users-passengers located on the platforms while non-stopping trains pass by at relatively high speeds. These people could be affected (sucked towards a passing train if standing too close to the platform edge) by air streams generated by HS trains. In addition, some objects could be thrown up on to the platforms. For example, some research indicates that people standing on the platform at a distance of 2 m from a passing HS train at a speed of 240 km/h could be at real risk (USDOT 1999).

A common way to avoid the above-mentioned safety risk is to locate separate passing tracks sufficiently far from the platform. This could be additionally strengthened by installing some kind of fixed barriers between the passing and stopping tracks. These tracks are now located close to the platform(s) where the trains approach at very low speeds before stopping as shown in Fig. 2.3 (a, b) (Anderson and Lindvert 2013). Additional measures for maintaining the specific level of safety and protecting people on platforms prevent them from coming too close to the trains passing by. For such a purpose, safety zones are clearly marked on platforms. This can also include setting up safety barriers with movable gates (in Europe at least 1.6 m wide), which close when a HS train passes through and automatically open when it

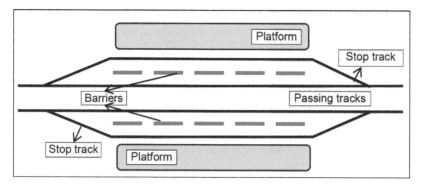

(a) Line station with side platforms and two passing and two stopping tracks (Anderson and Lindvert 2013)

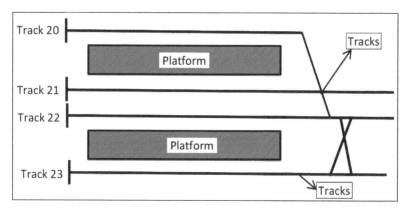

(b) Begin/end station with inland platforms (Tokyo Shinkansen) (Nishiyama 2010)

Fig. 2.3: Simplified scheme of arrangement of the tracks and platforms at the HSR station(s).

stops. In order to avoid delays of trains, functioning of these gates/doors has to be highly reliable with supporting personnel present nearby in cases of failures.

The platforms are dimensioned according to specified standards keeping in mind the minimization of land use and related environmental impacts. The standards embrace the platform length, width and height. According to these standards, the platform heights are specified to be 55 cm or 76 cm. According to European standards, the minimum length of a HSR station platform is 400 m, i.e. approximately equal or slightly longer than HS train(s).

The width of HS rail station platforms generally depends on the number of people simultaneously being present while waiting for arrival of an HS train and the standard area of space allocated to each of them. This should include areas for safety zones, for circulation of passengers including access for disabled persons, and an area for lifts, stairs, etc. For example, the minimum width of the safety zones is from 1.5-2.0 m. The minimum width of the paths for circulation of passengers is 1.6 m. A width of 0.5 m can be added for each 100 passengers and 1.0 m for vehicular traffic. The minimum distance between the particular obstacles and the safety zone varies between 0.8 and 2.5 m. In addition, the total width of 2.5 m for side and 3.3 m for island platforms is specified without including the width of the path for passenger circulation and obstacles.

As mentioned above, in Europe, the distance between the centers of the HSR tracks varies, depending on the design speed from 4.0-4.5 m. This distance is the same at the HS rail stations. In addition, the distance between the track and the platform is standardized at 1.65 m. In the configuration shown in Fig. 2.3 (a, b), by considering the width of each platform of 5 m, the distance to the track of 1.65 m, the distance between the stopping and passing track separated by the barrier of 6.3 m and the distance between two passing tracks of 4.5 m, the total functional width of the station would be about 30.4 m (Anderson and Lindvert 2013).

2.2.2.3 The HSR Network

The above-mentioned lines and stations constitute the HSR infrastructure network, which commonly spreads over the territory of a given country. The lines-links connect any two neighboring stations-nodes. The routes consist of several successive links connecting the stations along them. A country's specific layout/topology of HSR networks mainly depends on their design to connect cities as potentially larger generators and attractors of sufficient user-passenger demand. Figure 2.4 shows simplified generic schemes of the layout/topology of these networks.

As can be seen, three types of spatial layout of HSR networks have generally been developed in different countries as follows: line (for example, in Italy), star (in France) and polygon (for example in Germany). In addition, in some countries, the HSR network consists of different above-mentioned *categories* of lines, which make them rather heterogeneous in terms of the maximum design and operating speed. Specifically, Fig. 2.5 shows a simplified but more detailed scheme of the HSR network in Germany.

As can be seen, the HSR rail lines connect stations located in large cities/ urban areas. Two above-mentioned *Categories I* and *II* of lines are currently in place, i.e. those for the design speed of 300 km/h, 250 km/h and more and 200 km/h. These lines are incorporated into the rest of the network consisting of upgraded lines for the speed of 160 km/h (the above-mentioned *Category III* of lines). This configuration usually implies running HS train services on

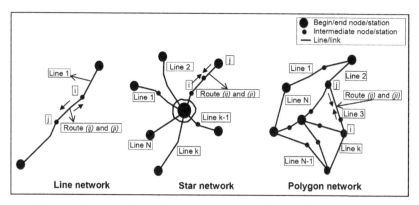

Fig. 2.4: The simplified schemes of layout/topology of HSR networks.
(Crozet 2013, http://www.johomaps.com/eu/europehighspeed.html)

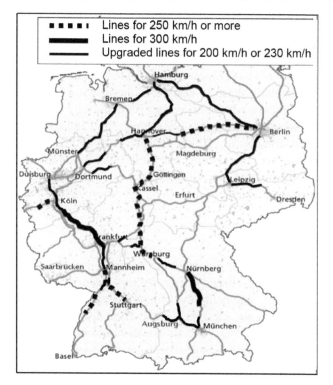

Fig. 2.5: Simplified scheme of the HSR infrastructure network.
(Germany) (http://en.wikipedia.org/wiki/High-speed_rail_in_Germany)

particular links and routes at different speeds and likely mixing with other conventional trains. In addition, Table 2.2 shows some characteristics of the main grid (eight national backbone lines) of the HSR network in China.

Table 2.2: Some characteristics of the main grid of CRH (Chinese Rail High) speed network (Fu et al. 2015, Takagi 2011, https://en.wikipedia.org/wiki/High speed_rail_in_China/)

Relation	Orientation	Length of line (km)	Design speed (km/h)
Beijing-Harbin	N-S[1]	1800	350
Beijing-Shanghai	N-S	1318	350
Beijing-Hong Kong	N-S	2383	350
Hangzhou-Shenzhen	N-S	1499	250/350
Sub-length		7000	
Qingdao-Taiyuan	E-W	940	200/250
Xuzhou-Lanzhou	E-W	1434	250/350
Chengdu-Shanghai	E-W	2066	200/250
Kunming-Shanghai	E-W	2056	350
Sub-length		6496	
Total length		13469	

[1]N-S (North-South); E-W (East-West)

The specificity of this (Chinese) compared to the other HSR rail networks worldwide, particularly those in Europe, is the length of lines between the end stations/terminuses, which varies from 1000 to 2400 km. In Europe, these lengths are much shorter and vary, for example, from 280 km between Berlin and Hamburg (Germany) to 770 km between Paris and Marseilles (France) (UIC 2014). However, the experience so far has shown that the average travel distances on some of these long Chinese lines was about 560-620 km, which appears comparable to some of their (long) European counterparts (Fu et al. 2015).

2.2.3 Rolling Stock

The HSR rolling stock possesses some common characteristics, such as optimized aerodynamic shape, self-propelling, fixed composition and bi-directional train sets, compatibility with infrastructure (track and loading gage, platforms, catenary, etc.), concentrated or distribution power, inside signaling system(s), braking systems, power electronic equipment, control circuits, computer network, automatic diagnostic system, particularly high level of RAMS (Reliability, Availability, Maintainability and Safety) and maintenance characterized by inspections at fixed time intervals and preventive maintenance (UIC 2010a). Table 2.3 gives select technical/technological specifications for different HS trains. As can be seen, their maximum design speed varies from 250-350 km/h. The locomotives are powered by electric energy. They are so-called multi-system locomotives interoperable for at least two different electric supply systems. The traction power varies from 5500 to

Table 2.3: Technical/technological characteristics of different HS trains (Siemens 2014, http://en.wikipedia.org/wiki/ICE_3, http://en.wikipedia.org/wiki/Siemens_Velaro; http://www.trainweb.org/tgvpages/tgvindex.html, http://en.wikipedia.org/wiki/New_Pendolino)

Type of HS train	Build date (yr)	Max. speed[1] (km/h)	Supply voltage[2]	Traction (kW, kV)	Length/ Weight (m, ton)	Configuration[3] (-, seats)	Performance metrics (kW/seat)
TGV PSE (Paris Sud-East)	1978-1985	300/270	25 kV 50 Hz AC 1.5 kV DC	6450, 25 3100, 1.5	200, 385	1+8+1, 385	18.34
TGV LA Poste	1981-1984	270	25 kV 50 Hz AC 1.5 kV DC	6450, 25 3100, 1.5	200, 345	-	-
TGV Atlantique	1989-1992	300	25 kV 50 Hz AC 1.5 kV DC	8800, 25	238, 484	1+10+1, 485	18.14
TGV Resau	1992-1996	300	25 kV 50 Hz AC 1.5 kV DC	8800, 25	200, 386	1+8+1, 377	23.34
Eurostar	1993-1995	300	25 kV 50 Hz AC 3 kV DC 1.5 kV DC, 750 V DC	12200, 25	394, 752	1+18+1, 794	15.90
TGV Duplex	1995-1997	320	25 kV 50 Hz AC 1.5 kV DC	8800, 25	200, 380	1+8+1, 545	16.15
Thalys	1995-1998	300	25 kV 50 Hz AC 1.5 kV DC	8800, 25	200, 385	1+8+1, 377	23.14

(Contd.)

Table 2.3: (*Contd.*)

Type of HS train	Build date (yr)	Max. speed[1] (km/h)	Supply voltage[2]	Traction (kW, kV)	Length/Weight (m, ton)	Configuration[3] (-, seats)	Performance metrics (kW/seat)
ICE 3	1998-1999	330	15 kV 16.7 Hz AC 1.5 kV DC	8000, 1.5	201, 435	1+8+1, 441	18.14
ICE 3M	2000	330	25 kV 50 Hz AC 3 kV DC	8000, 25	201, 435	1+8+1, 430	18.60
ICE Velaro CN	2004	350	25 kV 50 Hz AC	9200, 25	200, 447	1+8+1, 601	15.31
AVE	1991-1992	300	25 kV 50 Hz AC 3 kV DC	8800, 25	200, 392	1+8+1, 320	26.75
KTX (TGV Korea)	1997-2002	300	25 kV 50 Hz AC	13200, 25	381, 701	1+18+1, 935	14.12
N700-I (Japan)	2007	330	25 kV 60 Hz AC	9760, 25	204.7, 365	8, 636	15.35
ETR 600	2008	250	25 kV 50 Hz AC 3 kV DC	5500, 25	187, 387	2+3+2, 430	12.79

[1]Design speed; [2]AC – Alternating current; DC – Direct current; [3]Power car(s) – Trailers – Power car(s)

13200 kW/train set. The length of the train is predominantly about 200 m and the corresponding weight between 350 and 450 tons. A typical configuration of an HS train set is 1 power car + 8 trailers + 1 power car. The performance metrics vary across the considered set of HS trains between 12 and 23 kW/seat.

In addition, Fig. 2.6 shows the relationship between the traction and weight for select HS trains.

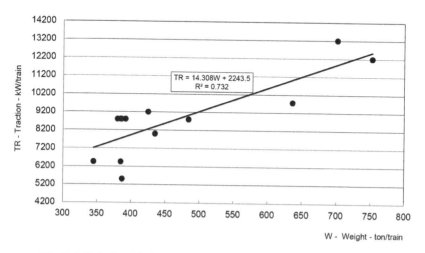

Fig. 2.6: Relationship between the traction and weight of the selected HS trains (Table 2.3).

As can be seen, generally, the required traction linearly increases with increase in the weight of an HS train. Figure 2.7 shows the relationship between the seat capacity and weight of HS trains. Table 2.3 shows the relationship between the seat capacity and weight of select HS trains.

As can be seen, again, the increase in the train's seat capacity is linear with increase in its weight. In this case, the average gross weight per seat is about 1.3 ton.

Figure 2.8 shows the relationship between the HS train's performance metrics and seating capacity.

As can be seen, the performance metrics expressed by the installed traction per seat decreases more than proportionally with increase in the number of seats, thus indicating economies of train size in terms of the installed (and required) traction. This implies that HS trains with higher seating capacity do not need proportionally stronger traction. In addition, Fig. 2.9 shows the relationship between the maximum design and the maximum operating speed of HS trains.

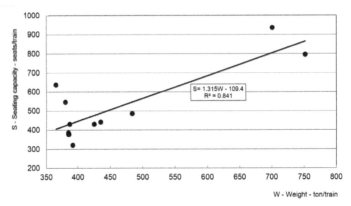

Fig. 2.7: Relationship between the seat capacity and weight of the selected HS trains (Table 2.3).

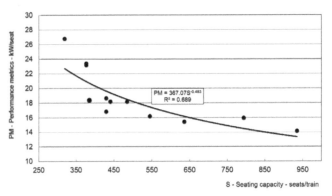

Fig. 2.8: Relationship between the performance metrics and seat capacity of the selected HS trains (Table 2.3).

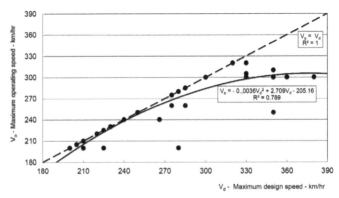

Fig. 2.9: Relationship between the maximum design and the maximum operating speed of the selected HS trains (http://en.wikipedia.org/wiki/List_of_high-speed_trains).

As can be seen, these two speeds between 200 km/h and 320 km/h coincide with each other at many HS trains. Nevertheless, generally, with increase in the maximum design speed, the positive difference between this and the maximum operating speed tends to increase, which particularly occurs at speeds between 270 km/h and 380 km/h. Consequently, at particularly high maximum design speeds (above 300 km/h), lower maximum operating speeds should be expected; in this case, by about 10-20 per cent.

2.2.4 Supportive Facilities and Equipment

The main supportive facilities and equipment of the HSR system in the given context are power supply, signaling system and traffic control/management system.

2.2.4.1 Power Supply System

The power supply constitutes an integrated system including electric high voltage power lines, substations, contact lines, HS trains and a remote command and control system ensuring efficient, reliable and safe supply of electric power to HSR lines and trains, and consequently the operations. The electrified networks for HSR lines generally use alternate current (AC) or direct current (DC). As given in Table 2.3, the typical voltage and frequencies are 25 kV 50Hz AC, 1.5kV DC and 15kV 16.7Hz AC. The latter is installed in Germany and supplied from a dedicated high voltage network called 'Railway Frequency'. The above-mentioned general system components can further be divided into two main components: the HSR electrical infrastructure and the HS rolling stock traction equipment (ABB 2014).

(a) The HSR Electrical Infrastructure

The HSR electrical infrastructure consists of the following main components— traction substations, which feed and distribute power to the lines; static frequency converter stations, which convert the power to the correct frequency and power; power quality systems protecting the network and the grid from voltage disturbances; transformers, which enable traction power supply or substation distribution of power and the network management systems, which monitor, control and manage functioning of the grid and rail distribution networks (ABB 2014).

(b) The HS Rolling Stock Traction Equipment

The HS rolling stock traction equipment includes the main sub-component called traction package, which consists of traction transformers, traction converters and traction motors. In addition, the auxiliary converters distribute power to the train's applications on board (ABB 2014).

2.2.4.2 Signaling Systems

(a) General

Different HSR signaling systems are applied in different countries. For example, in Europe, each country has its own HSR signaling system: in France it is TVM (Transmission Vole Machine), in Germany LZB (LinienZugBeeinflussung), in Spain Germany's LZB (for speeds up to 300 km/h) and EBICAB (Electrique Bureau CABine) (for speeds up to 220 km/h) are used, and in Italy BACC (Blocco Automatico a Correnti Codificate) (for speeds up to 250 km/h). In addition, ERTMS (European Rail Traffic Management System – Level 1 and/or 2) has been introduced in particular countries on particular lines as an alternative and/or in combination with the existing national systems (ABB 2014).

The type of signaling system influences the length of a block of the track, which can be occupied exclusively by a single train. The number of such successive empty blocks determining the (braking) distance between any pair of trains moving in the same direction depends on their operating cruising speed and braking/deceleration rate. In general, this distance can be estimated as follows. Let us assume that a HS train decelerates at a constant rate during the braking phase of a journey starting from the operating cruising speed. Since braking is carried out along an approximately straight line where the train moves as a particle linearly, the relationships between the speed, deceleration rate and braking distance can be expressed as follows:

$$v(t) = v_0 - a^{-}*t \qquad (2.1a)$$

and

$$S(t) = \int_0^t v(t)dt = v_0 * t - \frac{a^{-}*t^2}{2} \qquad (2.1b)$$

where

t is braking time (min.);

$v(t)$ is the train's speed at the end of braking phase (km/h);

v_0 is the train's operating cruising speed at which deceleration and braking starts ($t = 0$) (km/h);

a^{-} is the constant deceleration rate during the braking (m/s^2); and

$S(t)$ is the braking distance (m).

Assuming that the speed at the end of the braking phase will be zero, i.e. $v(t) = 0$ in Equation 2.1a, and by inserting the time $t = v_0/a^{-}$ in Equation 2.1b, the braking distance can be estimated as follows:

$$S(v_0, a^{-}) = \frac{v_0^2}{2a^{-}} \qquad (2.1c)$$

where all the symbols are as in the previous equations.

Figure 2.10 shows some examples of the above-mentioned relationships assuming that the deceleration rate (a) does not depend on the operating cruising speed but remains constant during the entire braking phase.

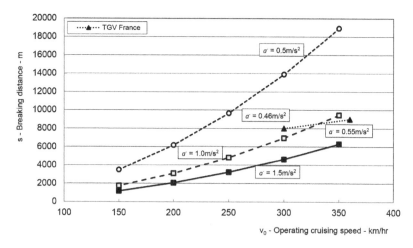

Fig. 2.10: Relationship between the HS train's operating cruising speed, constant deceleration rate and required braking distance.

As can be seen, the braking distance increases with increase in the operating cruising speed more than proportionally (at a square rate) and decreases with increase in the deceleration rate.

(b) TVM (Transmission Vole Machine)

The TVM (Transmission Vole Machine) signaling system is described as a representative example in the given context. Used exclusively on HSR lines in France, the system is based on the ATP (Automatic Train Protection) system, which distributes information on the train speeds depending on the version from 270 to 360 km/h. The ground-based components of the system are TCCs (Trackside Control Center(s)) located approximately every 15 km along each track of the line. They are linked to the line's centralized traffic control center while directly controlling about 10 blocks of track, each equipped with its own track circuit. In addition, the TVM system exclusively relies on cab-signaling, which implies that it operates without trackside signaling. The main characteristic of the cab-signaling system is that the signaling information is transmitted through the tracks as electrical signals, which are picked up by antennas under the train, i.e. continuously transmitted through the track circuits as track-to-train transmissions. Four such antennas, two on each end, are mounted underneath the train, but the two situated in the direction of travel are used. The track circuits in both the tracks are used to transmit the signal information to the train's on-board computers, as well as fixed inductive loop beacons. In addition, the TVM is a fixed block system. This means that the track is subdivided into fixed segments, each of which has a particular state. Only one train may occupy any block at one time under regular operating conditions. The length of a block is about 2,100 m for the speed of 270 km/h and decreases to a length of 1,500 m for speeds between 300 and 360 km/h. The blocks are shorter than the HS train's braking distance, so a braking safe separation interval spreads over several blocks whose number depends on the maximum operating speed and the maximum train deceleration rate. For example, it is usually four blocks for speeds of 270 km/h, five blocks for speeds of 300 km/h, and six blocks for speeds of 320-360 km/h (*see* also Fig. 2.10). Each block possesses the constant relevant properties for the train occupying it, such as length, gradient and the maximum operational safe speed. In addition, the train's target speed as the speed at which the train should exit the current and enter the next block is a variable value. Figure 2.11 shows the simplified generic scheme of the braking pattern of a HS train along several blocks based on the TVM signaling system.

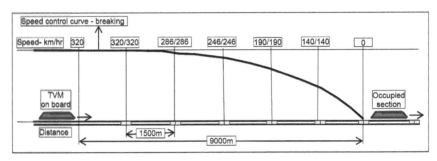

Fig. 2.11: Simplified scheme of the braking pattern of a HS train—Speed control curve by the TVM signaling system.

As can be seen, in the given case, the HS train will cover a distance of about 9000 m (9 km) to stop if starting breaking at the operating cruising speed of about 320 km/h at a constant deceleration rate of 0.46 m/s². All the above-mentioned information is transmitted by the TVM system to the train's computers and the cab displays where the driver can monitor them. Specifically, he/she monitors the target speeds for the current and subsequent blocks (displayed in km/h), full line speed and the speedometer continuously indicates the varying target and current speed (with precision of about 2 per cent). In addition, in order to mitigate the driver's workload, the required (safe) speed is displayed over several blocks ahead of the train. In addition, since the system itself cannot adapt to irregular operating conditions, the human operator-driver is kept in the control loop. For example, if the train exceeds the specified maximum speed of 300 km/h, the computer will undertake an action to reduce it and establish regular operations again. This implies that TGV trains are operated manually, but safety is provided by the automated signaling system. Last but not least, the digital recording system based on a desktop computer system monitors and records every action of the driver, the electro motors' operating regime, activation/deactivation of brakes, pantographs, etc., as well as the above-mentioned signaling information (ABB 2014).

(c) ERTMS (European Rail Traffic Management System)

The ERTMS (European Rail Traffic Management System) was recommended by the EC (European Commission) and gradually implemented in particular countries in addition to the already existing national signaling systems. The main objectives of the ERTMS system are expected as improvement of technical interoperability, safety, train operating performances and availability/reliability.

Improving technical interoperability by using a unified signaling system/equipment is expected to stimulate opening of the rail transport markets for more rail operators. Safety is provided by designing the system according to the given/specified standards. The train-operating performances are expected to imp-rove by enabling safe operations of HS trains at very high speeds in the same direction and separated at much shorter time intervals than those specified by the national signaling systems. Availability/reliability is expected to be achieved by reducing the quantity of equipment along the HSR lines and the probability of component failures and by improving the system's reliability.

The ERTMS consists of two primary components: (i) ETCCS (European Train Control and Command System), which is an automatic control system that controls the speed limits of a train by communicating with the driver, and (ii) GSMCR-R (Global System for Mobile Communications—Railway), i.e. a radio communications system to enable exchange of information between the train (driver) and the traffic management center.

The ERTMS is designed at three levels:

- *The ERTMS Level 1* uses the Eurobalises installed under the tracks as well as the existing trackside signals and track circuits. The Eurobalises are electronic beacons or transponders installed usually below the ties of tracks at a distance of 3 m and represent a part of the ATP (Automatic Train Protection) system. The track signals are the same as those for conventional railways. The track circuits as devices enable collection of information on the train integrity and position. As such, the system can continuously supervise current and generate prospective safe HS train speed(s). The LEU (Lineside Equipment Unit) located by the side of the tracks generates movement authorities [(i.e. safe path(s)] and track description data based on the information received from the trackside signals and track circuits (the latter on the train integrity and position). The movement authorities are transmitted to the train through balises. Then, the on-board computer system calculates the dynamic speed profile ahead (the actual speed, the maximum allowed speed) by taking into account the train's braking characteristics. In addition, it also monitors and controls the indicators in front of the driver. In this case, use of the trackside signals is necessary.

- *The ERTMS Level 2* is based on the radio-based ATC (Automatic Train Control) system, which provides continuous information and supervision of the train speed towards fixed points of the line (ends of block sections, restrictions of speed, etc.). In addition, the RBC (Radio Block Center) generates messages on the movement authorities, state of tracks, current speed, eventual restrictions and emergencies based on the information transmitted from the HS train(s), external interlocking system and the track circuits. The RBS usually covers and manages about 100 km of double track line. The messages constituting the movement authorities are transmitted between RBS and the train(s) means by GSM-R system. Its basis is BTSs (Base Transceiver Station(s)) positioned at approximate distances of about 3-4 km. The GSM-R operates as follows: The moment an HS train passes over a Eurobalise, it transmits its new position and its speed to the RBC. Then it receives back consent (or prohibition) to enter the next section of the track at its new maximum speed. Then, the on-board computer system calculates the train's dynamic speed profile by taking into account the braking characteristics and other commands to be eventually used. In this case, side track signals can also be optionally used.

- *The ERTMS Level 3* has very similar characteristics as the ERTMS Level 2 with some technical and functional differences. The technical differences are the necessity for having on-board equipment for checking the train's integrity, which is then used by the RBC for generating movement authorities. No track circuits are needed for train detection. The functional difference is that the preceding train is considered as the moving block and target while specifying the minimum time/distance separation intervals between the trains (UNIFE 2014a, b, c, d, http://demo.oxalis.be/unife/ertms/?page_id=42, http://en.wikipedia.org/wiki/European_Rail_Traffic_Management_System).

The ERTMS Level 2 has been already installed at railways of different continents. Figure 2.12 shows some statistics related to the rail lines and rolling stock (vehicles).

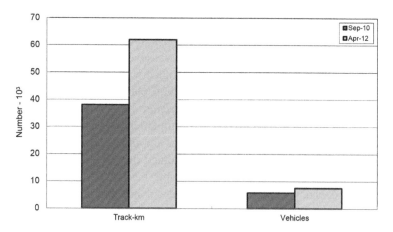

Fig. 2.12: Some statistics of implementing ERTMS around the world. (UNIFE 2014d, http://demo.oxalis.be/unife/ertms/?page_id=42)

As can be seen, the number of both track km and on-board units have increased over the observed period, by about 63 per cent and 32 per cent, respectively.

2.2.4.3 Traffic Control/Management System

In general, at HSR rail lines/networks, the rail traffic control/management system is fully computer-supported and can include the following main components: TOC – Train Operation Controller, PC – Power Controller, STC – Signal and Telecommunication Controller, CCC – Crew and Car Utilization Controller, PSC – Passenger Service Controller, and TSMC – Track and Structure Maintenance Controller. These components are operated by staff accommodated in the same room (JR 2012).

In particular, the main activity of any TOC component is to control and monitor realization of the timetable and provide instructions for mitigating deviations from it due to any reason. Typically, the component is operated by the train dispatcher who carries out control and monitoring activities by using the control panel(s) (usually computer screen(s)) showing a graphical representation of the part of the rail line/network under his/her jurisdiction. The size of the part of the line/network depends on many factors, such as the type of rail line (single, double, multi-track), the number and configuration of stations on the line/network, traffic intensity in terms of the number of trains during the specified period of time, the number and time for executing necessary tasks by the dispatcher and the required productivity of the dispatcher. Consequently, the given HSR rail network can have several traffic control units usually located at large stations or rail yards.

The panels (computer screens) generally display rather simplified schemes of the tracks, including signals and powered switches, the latter usually located at the end of sidings and at crossovers between the main track along the line and at the stations. The occupied tracks are usually displayed by bold or colored lines overlaying the track display. The trains are displayed as tags with the relevant identification code/sign. Under such conditions, the dispatcher monitors their progress along automatically generated conflict-free paths. In addition, the dispatcher possesses the planned timetable enabling him/her to compare the actual with the planned timetable of each particular train. In case of deviations due to any reason, the intervening messages are created and exchanged between the dispatcher and the train drivers either automatically or via voice communication link. In addition, communications are provided between the central and the local dispatchers at the stations included in the part of the centrally controlled/managed line/network.

2.2.5 Operational Rules and Procedures

The operational rules and procedures of each HSR system specify the minimum separation between trains operating in the same direction on a given line and at the stations in the same and different directions. In particular, they specify the minimum time separation between successive trains occupying the same section of the line(s) and station(s). These mainly influence their corresponding 'ultimate' capacity, which is elaborated in the sub-section dealing with the operational performances of HSR system(s).

2.2.6 Staff

As do their conventional counterparts, the staff is also needed to operate each HSR system. However, this staff is usually incorporated into the total staff of particular railway companies, making it rather difficult to clearly extract both absolute and relative numbers. Table 2.4 provides data on the staffing of select railway companies. It contains the total length of rail network, the HSR network share in the total network length, the total staff/employment including both passenger and freight operations and infrastructure provision and the derived measure—the average number of employees per km of the total network length.

Table 2.4: Some characteristics of the networks and staff/employment at particular railway companies (period 2011-2013) (JR 2012, EC 2014)

Railway company	Length of the network (km)	Share of HSR network (%)	Total number of employees	Specific employment (employee/km)
Japan Railways	20,000	8.7	95,706	4.8
SNCF (France)	29,903	6.9	1,58,488	5.3
DB AG (Germany)	33,714	3.8	2,39,888	7.1
ATOC (UK)	16,272	0.7	37,153	2.0
Amtrak (US)	34,080	1.0	19,203	1.4

As can be seen, on the one hand, specific employment generally increases with increase in the total length of the railway network; on the other, it is not particularly correlated with the length of HSR network, i.e. its relative share in the total length of the network. As such this figure could be used just to get an idea of the specific employment by the HSR system if the allocation of staff were in proportion to the length of the corresponding network(s). In any case, the derived numbers should be considered with reservation and just as initial data to be further investigated (JR 2012, EC 2014). Nevertheless, the required number of staff to be assigned to the HSR service on the given line during a specified period of time can be estimated as follows:

$$n_s(T) = k * f(T) * m \qquad (2.2a)$$

where

k is the coefficient of need for the train's staff for one train pair (the 'train pair' is defined by the given train's set return trip along the given line);

$f(T)$ is the train frequency along the given line during time (T) (dep/T);

T is the period of time for which the number of staff is estimated (hr); and

m is the number of persons per pair of train (personnel/train).

In Equation 2.2a, the coefficient *(k)* can be estimated as follows:

$$k = \frac{1}{T} * \left[2\left(\frac{L}{v} + t_1 + t_2\right)\right] * \varphi \qquad (2.2b)$$

where

L is the length of the given HSR line (km);

v is the commercial speed of the HSR services along the given line (km/h);

t_1, t_2 is the time of taking over and delivering the train at the origin and destination station; respectively (hr); and

φ is the coefficient of the train crew workload.

The coefficient (φ) in Equation 2.2b can be estimated as follows:

$$\varphi = \frac{T * d_m}{d_{w/m} * t_{h/d}} \tag{2.2c}$$

where

d_m is the number of days per month (d/m);
$d_{w/m}$ is the number of working days per month (d/m); and
$t_{h/d}$ is the number of working hours per day (hr/d).

For example, if $L = 500$ km, $v = 250$ km/h, $T = 24$ hr, $f(T) = 24$ dep/T, $t_1 = t_2 = 1$ hr, $d_m = 30$ d/m, $d_{w/m} = 22$ d/m, $t_{h/d} = 8$ hr/d, and $m =$ two persons/train pair, the required number of staff will be 65 during the daily shift for the given line. This needs to be increased by about 30 per cent for holidays, sickness and other leave during the year (Kovacevic 1988).

2.3 Performances

2.3.1 The Concept

Performances of transport systems can generally be defined as the latter's ability to fulfill needs and expectations of particular actors/stakeholders involved. Those of the HSR system can generally be classified as infrastructural, technical/technological, operational, economic, social, environmental and policy performances (Janić 2014a):

- *Infrastructural and technical/technological performances* imply physical, constructive, technological and technical characteristics of their infrastructure, vehicles and supporting facilities and equipment;
- *Operational performances* reflect their capabilities to serve the given volumes of passenger and goods/freight demand under the given conditions;
- *Economic performances* express their costs and revenues, the latter based on the charges (prices) to users—passenger and goods/freight shippers/receivers;
- *Social and environmental performances* reflect their effects and impacts on the society and environment, respectively; if monetized, impacts are considered as external costs, i.e. externalities; and
- *Policy performances* reflect compliance with the current and future medium- to long-term policy regulations and specified targets.

The above-mentioned performances are frequently considered individually despite being inherently dependent and influential on each other as shown in Fig. 2.13.

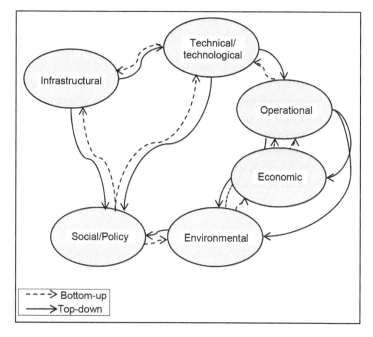

Fig. 2.13: Potential interrelationships of the performances of
transport systems (Janić 2014a).

As can be seen, according to the 'top-down' approach, the infrastructural
performances directly influence the technical/technological performances,
thus causing their and the mutual influence of these and all other performances.
According to the 'bottom-up' approach, the social/policy performances can
directly influence the infrastructural and technical/technological performances,
thus creating mutual influence of these and all other performances.

In the case of an HSR system, its infrastructural and technical/
technological performances have been implicitly analyzed together with
the corresponding system's components in Sub-section 2.2. As such, they
have provided a basis for analyzing the other performances—operational,
economic social and environmental. The policy performances are assumed
to inherently influence all the above-mentioned performances and as such are
not particularly analyzed.

2.3.2 Operational Performances of the HSR System

The operational performances of an HSR system can be considered for an
individual line/route, or the entire network serving a given region, i.e. country.
In the given context, the main operational performances are demand, capacity
and quality of service—the latter as the outcome from the dynamic interaction

between the former two. Demand is represented by the volume(s) of users-passengers requesting service during a specified period of time. Capacity is represented by (i) the maximum number of trains and/or train seats, which can be handled on the particular line(s)/route(s) as link(s) and station(s) as node(s) of the HSR network; (ii) the size of HSR rolling stock operating on particular line(s)/route(s); and (iii) technical productivity. These are estimated for the given conditions mainly determined by constant demand for service ('ultimate' capacity), the average delay per service ('practical' capacity) and the specified period of time (usually one hour). The quality of service is represented by the schedule delay, trip time, reliability, punctuality and price of services, quality of accessibility of the HSR stations and comfort on-board the HS trains.

2.3.2.1 Demand

In general, the demand for HSR services comprises self-generated demand and the demand expected on the competitive routes from other transport modes, such as an individual car, conventional railways and APT (Air Passenger Transport) (Janić 2016).

The self-generated demand for the HSR services is stimulated by expansion of the HSR network and increase in welfare in terms of the national GDP (Gross Domestic Product). Figure 2.14 shows the relationship between the served passenger demand and the length of HSR network in Europe and China.

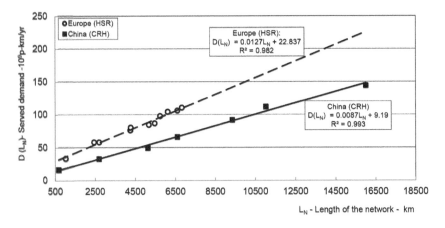

Fig. 2.14: Relationship between the annual passenger demand and the length of HSR networks in Europe and China (Period 1995-2014) (CSP 2014, Janić 2016, EC 2014, Ollivier et al. 2014).

As can be seen, in both the regions, the served passenger demand has grown linearly with increase in the length of HSR networks. In terms of absolute values, the served passenger demand in China exceeded that in Europe during the relatively short period of time (seven years), which indicates a very strong user/passenger preference for the new CRH speed system as shown in Fig. 2.15.

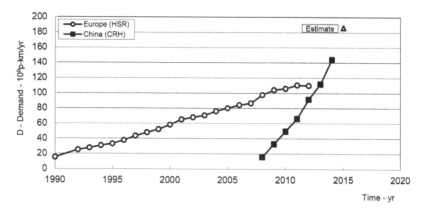

Fig. 2.15: Development of the satisfied passenger demand in the European HSR and Chinese CRH speed network (Period: 1990-2014) (CSP 2014, EC 2014, Janić 2016).

In Europe, the served passenger demand has continuously been growing during the specified period of time. In China, since the start of implementing the CRH speed network, the corresponding passenger demand has been growing tremendously and thus very quickly exceeded that in Europe. In both cases, this has been possible primarily due to the expanding HSR network as shown in Fig. 2.14 and other above-mentioned demand-stimulating factors. Figure 2.16 shows the relationship between GDP and the satisfied HSR passenger demand in Japan during the observed period (Janić 2016, JR 2015).

As can be seen, the passenger demand has increased more than proportionally with rising of GDP, thus indicating that GDP has generally been, is and will continue to be a strong generator of demand in the given context.

In most cases, if appropriate, these numbers on the estimates of passenger demand at the HSR systems are obtained as the outcome after analyzing the potential competition between HSR and other transport modes, such as usually the conventional rail and APT (Air Passenger Transport). An example of the market share of HSR compared to APT in dependence of the line travel time is shown in Fig. 2.17.

As can be seen, the relative market share of HSR (that of APT is complement to 100 per cent) has decreased almost linearly (Europe, Japan) and more than

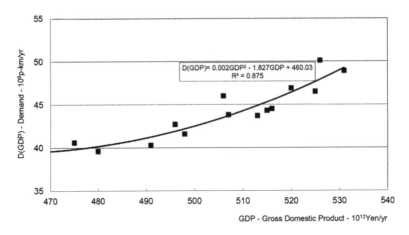

Fig. 2.16: Relationship between the satisfied passenger demand by Japanese Tokaido Shinkansen HSR system and the national GDP (Gross Domestic Product) (Period: 2001-2015) (Janić 2016, JR 2015).

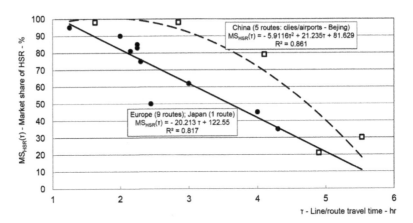

Fig. 2.17: Relationship between the market shares of HSR and APT depending on the line/route travel time (CSP 2014, De Rus 2008, IR 2012, Janić 2016, UIC 2010b, Wu 2013).

linearly (China) with increase in the line/route travel time within the given range. In this case, multiplying the number(s) of passengers and the route length(s), i.e. corresponding travel distances, the prospective HSR passenger demand in terms of the potential volumes of p-km for individual lines/routes or the entire network during the specified period of time can be obtained.

The demand satisfied by the HSR system at a given line/route and the systems is usually expressed by the number of passengers and/or the volumes of p-km carried on the particular lines/routes or the entire network during the specified period of time (hour, day, month, year). Figure 2.18 shows examples of development of user-passenger (served) demand at the Eurostar (Europe) and Shinkansen Tokaido (Japan) HSR line/route.

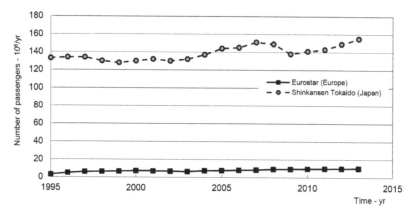

Fig. 2.18: Development of the annual (served) passenger demand on the HSR Eurostar (Europe) and Shinkansen Tokaido (Japan) HSR line/route (JR 2012, 2014; http://en.wikipedia.org/wiki/Eurostar).

The Eurostar system connects the main beginning and ending stations/ terminuses at London St. Pancras (UK), Paris Gare du Nord (France) and Brussels Midi/Zuid (Belgium). All these services (at a frequency of approximately one dep/hr in both directions) pass through the Channel tunnel between the UK and France. The length of this line is 333 km, taking the HSR services two hours and 16 min. in a single direction, which gives an average travel speed of about 147 km/h (including the intermediate stops). The maximum operating speed is 300-350 km/h. As can be seen, this demand, despite a decline during some years, has generally grown at a rather decreasing rate during the observed period. In addition, the annual passenger demand at the Shinkansen Tokaido HSR line connecting Tokyo and Shin-Osaka (552.6 km) is about 15-20 times greater than that of Eurostar.

The volume of passengers has been served by four categories of train services with a frequency of 13 dep/hr in each direction. The particular categories of service are distinguished on the basis of the number of stops along the line. The fastest category takes two hours and 25 min. between the beginning and ending station/terminus, thus indicating an average travel speed of 228 km/h. Both cases illustrate the fluctuation of demand over time on the one hand but also the possible volume of demand at different HSR systems, on the other. In addition, Figure 2.19 shows the relationship between the user-passenger demand density and length of line/route in different world's regions. As can be seen, with two exceptions, this density generally tends to increase with increase in the length of line/route, thus indicating in some sense the current utilization of the HSR infrastructure.

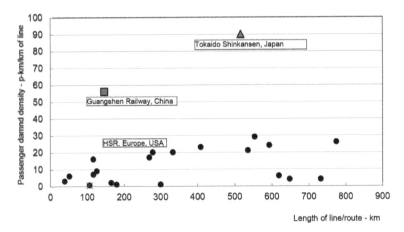

Fig. 2.19: Relationship between the length of HSR line/route and the density of passenger demand (Europe, Japan, China, US) (Period: 2007-2009) (Wendell and Vranich 2008).

2.3.2.2 Capacity

The capacity of the HSR system can generally be calculated for the components of infrastructure, like the stations and lines/routes and the rolling stock. In addition, transport work and productivity are considered as measures integrating in some way the capacities of infrastructure components and that of rolling stock. In general, the capacity of HSR infrastructure components can be defined as their ability to handle the maximum number of trains in a single direction during a specified period of time under given conditions. If these conditions are defined by a constant demand for service, the line/route capacity is called 'ultimate'. If they are defined by the maximum or average delay per train's service, the capacity is called 'practical'. If there are some other influencing conditions, the capacity is referred accordingly (Janić 1988).

(a) Infrastructure—Line(s)/Route(s) and Station(s)

Station(s): The number of required tracks in the given HSR station depends on the intensity of traffic (trains per unit of time) and their average stopping time. Small stations located on double track lines can have only two tracks. The incoming HS trains stop directly on the railway line in which case the train's stopping time should generally be shorter than the inter-arrival time of the incoming trains. The larger stations can have one or more extra tracks in addition to the main tracks in both directions, thus enabling the HS trains to switch and stop there. The main tracks can then be used by passing or other stopping trains simultaneously. In any case, the required number of tracks N_s (τ) for passing and stopping trains can be estimated as follows:

$$N_s(\tau) = 2 + \lambda(\tau) * t_s \tag{2.3a}$$

where

τ is the time period (1/4 or one hour);

$\lambda(\tau)$ is the intensity of arriving HS trains at the station (trains/hr); and

t_s is the average dwell time of a HS train at a given station/terminus (hr).

The first term in Equation 2.3a indicates the minimum required number of the main (usually passing) tracks at the stations along the line(s) and the second term stands for the required number of tracks for stopping trains at the stations along the line(s) and at the beginning/ending stations/terminus(es) where HS trains turn around during a certain time, which is about 20 min. at most HSR terminuses. At the Japanese HSR system (Shinkansen), it is about 12 min. (Nishiyama 2010). This time is used for unloading passengers and their baggage, cleaning the interior, replenishing water, restocking victuals, changing the crew and loading ongoing passengers.

In addition, if the number of tracks at the station N_s (τ) is given, its 'ultimate' capacity will be based on Equation 2.3a as follows:

$$\lambda_s \equiv \mu_s = \frac{N_s - 2}{t_s} \tag{2.3b}$$

where all the symbols are analogous to those in the previous equations.

For example, if the number of tracks is N_s = 16 and the average train's turn around time: t_s = 12 min. the station/terminus capacity from Equation 2.3b will be μ_s = 16/(12/60) = 80 trains/hr.

In some cases, due to safety reasons for users-passengers, the passing tracks at line stations should be spatially separated from the stopping tracks. The number of required platforms depends on the number of simultaneously stopping trains to enable simultaneous exchange of users-passengers. It can be estimated analogously as in the second term of Equation 2.3a.

Line(s)/route(s): The *'ultimate'* capacity of a given HSR line/route is mainly dependent on the minimum 'headway', i.e. the minimum time interval at which successive trains pass in the same direction through the 'reference location' selected for their counting. This 'reference location' can be any location along the open line/route. The minimum time interval is influenced by the HS train's maximum operating speed, deceleration and braking performances, length, the way of its control and also by the station/terminus spacing and design, gradients along the line/route and the type of traffic control (signaling) system (Connor 2011). Therefore, this time interval between a pair of successive HS trains (i) and (j) operating in the same direction of a given line/route can be estimated as follows:

$$t_{ij/min} = \frac{v_j}{a_j^-(v_j)} + \frac{S_{b/j} + L_i}{v_j} \qquad (2.4a)$$

where

i, j is the leading and trailing HS train of the pair of successive trains $(i; j)$;

v_j is the maximum operating speed of the trailing train (j) (km/h);

$a_j^-(v_j)$ is the average deceleration rate of the trailing train (j) during the maximal braking (m/s²);

$S_{b/j}$ is the buffer zone for the trailing train (j) (m); and

L_j is the length of the leading train (i) (m).

The maximum operating speed of HS trains is usually about 250-350 km/h. The deceleration rate varies, i.e. it generally increases with decrease in speed during the breaking phase of trip (*see* Fig. 2.11). For example, it can be 0.30 m/s² for speed between 350 and 300 km/h (first 1,000 m of the breaking distance), 0.35 m/s² for speeds between 300 and 230 km/h (second 1,000 m of the breaking distance) and 0.6 m/s² for speeds 230-0 km/h (the rest of 6000-7000 m of the breaking distance). Consequently, the average deceleration rate of 0.5 m/s² is usually used in such calculations (*see also* Fig. 2.11). The buffer zone is the distance added to the braking distance of the trailing HS train to allow a margin for safe separation from the leading train (i). It is typically 100 m (Hunyadi 2011, Connor 2011). The train length is typically 200 m or 400 m. The latter is the length of the Eurostar and the 2-unit German-designed Velaro train operating in China (Table 2.3). After simple manipulation with the kinematics of the HS train's movement along the given line, similarly as in Equations 2.1 (a, b, c), the breaking distance of the trailing train (j) can be estimated as:

$$S_{br/j} = \frac{v_j^2}{2a_j^-(v_j)} \qquad (2.4b)$$

where all symbols are as in the previous equations. Consequently, the minimum distance between the HS trains *(i)* and *(j)* is equal to:

$$S_{ij/min} = S_{br/j} + S_{b/j} + L_i \qquad (2.4c)$$

If the leading train (i) is to stop and the trailing train (j) is to pass through a station along the line, the 'reference location' for counting the trains, i.e. calculating the capacity, can be at the exit signal of the station. In such a case, the leading train (i) after being dispatched from the station should be at least at the minimum breaking distance of the trailing train (j) at the moment when it arrives at the exit signal of the station, which in this case will allow it to proceed. In such a case, the minimum time interval between two trains can generally is extended by the dwell time of the train (i) as follows:

$$t_{ij/s/min} = \tau_i + \left[\frac{S_{b/j}}{2a_j^-(v_j)}\right]^{1/2} + \frac{v_j}{a_j^-(v_j)} \qquad (2.5a)$$

where τ_i is the dwell time of the leading train (i) at the station (min.).

The other symbols are analogous to those in the previous equations.

At most HSR systems, the dwell time is typically 2-3 min. at the stations located along the lines/routes and 5 min. for those located at airports, the latter mainly due to enabling users-passengers to handle their baggage. This also includes the time for closing the doors, setting up conflict-free exit paths and dispatching the leading train (i).

The minimum inter-arrival time between trains (i) and (j) at the end station/terminus where the 'reference location' for calculating capacity is located can be determined as follows:

$$t_{ij/min/arr} = \frac{v_j}{a_j^-(v_j)} + \left[\frac{S_{b/j}}{2a_j^-(v_j)}\right]^{1/2} \tau_{ij} + \tau_b \qquad (2.5b)$$

where

τ_{ij} is the time for changing the route of trains *(i)* and *(j)* arriving at the end station/terminus of the given line/route (typically 10 s); and

τ_b is the time of blocking the entrance of the end station/terminus by other trains(s) (typically 25 s).

The other symbols are analogous to those in the previous equations.

The minimum inter-departure time between trains (i) and (j) from the beginning station/terminus is estimated considering the 'reference location' as the exit signal as follows:

$$t_{ij/min/dep} = \max\left\{\left[\frac{S_{b/i} + L_i}{2a_i^+(v_i)}\right]^{1/2} + \frac{v_i}{a_i^-(v_i)}; \tau_{j/r} + \tau_{j/gl} + \tau_{j/cf} + \tau_{j/d}\right\} \qquad (2.5c)$$

where

$\tau_{j/r}$ is the time for setting the exit path for the trailing train *(j)* in a given departing sequence *(ij)* (usually 10 s);

$\tau_{j/gl}$ is the time for setting the green light for trailing train *(j)* in a given departing sequence *(ij)* (usually 25 s);

$\tau_{j/cf}$ is the time of blocking exit of the station/terminus for departing trailing train *(j)* by other incoming and outgoing trains (usually 60-75 s); and

$\tau_{j/d}$ is the dispatching time of the trailing train *(j)* in a given departing sequence *(ij)* (usually 30 s).

The other symbols are analogous to those in the previous equations.

Equation 2.5c indicates that the minimum time between departures of the successive trains (i) and (j) from the begin station/terminus should be set up as the maximum of two times: the time the leading train (i) needs to reach the minimum breaking distance from the train (j) and the time for setting up a safe departure for the trailing train (j).

Equations 2.4-2.5 are based on the assumption that all trains operate at approximately the same acceleration/deceleration rate, the maximum operating speed and length, which imply homogeneity of the HS train fleet. In addition, these trains are assumed to appear at the particular 'reference location(s)' without substantive deviations from the prescribed/planned time(s), i.e. they are free of time errors. Then, using Equations 2.4-2.5, the 'ultimate' capacity of a given line/route $\mu_l(\tau)$, the station along the line/route $\mu_{s/l}(\tau)$ and the ending and beginning station/terminus for the incoming and outgoing traffic $\mu_{s/arr}(\tau)$ and $\mu_{s/dep}(\tau)$ respectively, in terms of the number of trains per unit of time can be estimated, respectively, as follows:

$$\mu_l(\tau) = \frac{\tau}{t_{ij\,/\,min}} \; ; \text{ and } \mu_{s/l}(\tau) = \frac{\tau}{t_{ij\,/\,s\,/\,min}} \; ; \text{ and}$$

$$\mu_{s/arr}(\tau) = \frac{\tau}{t_{ij\,/\,min/\,arr}} \; ; \text{ and } \mu_{s/dep}(T) = \frac{\tau}{t_{ij\,/\,min/\,depr}} \qquad (2.6)$$

where τ is the period of time for calculating the 'ultimate' capacity of particular infrastructure component (hr).

The other symbols are analogous to those in the previous equations.

Figure 2.20 shows examples of the calculated 'ultimate' capacity of the HSR line-route and beginning/ending station/terminus independent of the train's maximum operating speed.

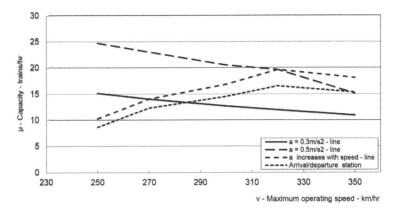

Fig. 2.20: Relationship between the ultimate capacity of the HSR line/route and the begin/end station/terminus, and the maximum train operating speed (Janić 2016).

As can be seen, the line/route capacity decreases with increase in the operating speed if the same average deceleration/acceleration rate is applied (0.5 m/s² for speeds 250-250 km/h). However, if this rate increases with increase in the speed, the capacity generally tends to increase (0.5 m/s² for speeds of 250 km/h, 0.3 m/s² for speeds of 270 km/h, 0.4 m/s² for speeds of 300 km/h and 0.5 m/s² for speeds of 320 and 350 km/h). In the latter case, the capacity again decreases due to applying the same deceleration/acceleration rate to the increasing speed. Similar is the case with the arrival and departure capacities of the beginning and ending station/terminus, respectively. In all cases, the train length is assumed to be 400 m and the buffer distance 100 m (Connor 2011). In practice, typical values of the 'ultimate' capacity for HSR lines/routes and stations is 13-15 trains/hr.

The *'practical'* capacity of a given line/route is defined as the maximum number of HS trains accommodated during a specified period of time under conditions when each of them is imposed the maximum average delay (Janić 1988). However, in this case, the mutual interferences between HSR services of equal priority operating on the *Category I* lines causing their delays are prevented by the so-called stability of the timetable. This implies that the maximum permissible delay of the leading train in the sequence of two trains is defined in a way as not to cause an additional delay of the following train. As such, this delay indicates some kind of the system's margin allowing delays of the HS trains anyway, which under such circumstances implies equivalence of the 'ultimate' and 'practical' capacity. Longer delays causing disruption in the timetable occur generally due to other causes.

(b) Rolling Stock

The capacity of HSR rolling stock reflects its size expressed by the number of trains required to operate under given conditions specified by the timetable. These conditions are usually characterized by the service frequency during the given period of time (hr, d) and the train's turn around time along the given line/route. Consequently, the required number of trains $n_l(\tau)$ can be estimated as follows:

$$n_l(\tau) = f_l(\tau) * \tau_l \tag{2.7a}$$

where

$f_l(\tau)$ is the transport service frequency on the line/route (l) during time (τ) (max $f_l(\tau) = \mu_l(\tau)$) (trains/hr); and

τ_l is the average turn around time of a train along the line/route (l) (hr).

The service frequency of a given line/route $f_l(T)$ can be considered to be either equal to the line/route 'ultimate' capacity determined by Equation 2.6 or be set up to satisfy the expected demand as follows:

$$f_l(\tau) = \frac{Q_l(\tau)}{\theta_l(\tau) * s_l} \tag{2.7b}$$

where

$Q_l(\tau)$ is the expected user-passenger demand on the line/route (l) during time (τ) (pass);

$\theta_l(\tau)$ is the average load factor per service frequency on the line/route (l) during time (τ)(≤ 1.0); and

s_l is the number of seats per frequency (seats).

The other symbols are analogous to those in the previous equations.

The train's turn around time (τ_l) increases with increase in the operating time along the line/route (the ratio between the length of line/route and operating speed), the number and duration of intermediate stops, all in both directions, including those at the beginning and ending station/terminus, and vice versa. The train's seating capacity is usually constant per service frequency, indicating the above-mentioned homogeneous HS train fleet on a given line/route. For example, if the given line/route operates at the service frequency of $f_l(\tau) = 15$ trains/hr, and if the average turn around time per train is $\tau_l = 4$ hr, the required number of trains will be $N_l(\tau) = 60$. In addition, if the average train's seating capacity is $s_l = 485$ (TGV Atlantique, see Table 2.3), the total number of required seats will be $n_l(\tau) = 29,100$.

(c) Transport Work and Productivity

Transport work: The transport work on a given HSR line/route during the specified period of time (p-km) can be calculated based on Equation 2.7 (a, b) as follows:

$$TP_l(\tau) = f_l(\tau) * S_l * d_l = \frac{Q_l(\tau)}{\theta_l(\tau)} * d_l \tag{2.8a}$$

where d_l is the length of route (l) (km).

Productivity: The productivity of a given HSR line/route during a specified period of time (pax-km/hr) can be calculated based on Equations 2.7 (a, b) as follows:

$$TP_l(\tau) = f_l(\tau) * S_l * v_l = \frac{Q_l(\tau)}{\theta_l(\tau)} * v_l \tag{2.8b}$$

where v_l is the operating speed of HS trains on the line/route (l) (km/h).

The other symbols are analogous to those in the previous equations.

As is seen, both transport work and productivity of a given line/route increase with increase in the number of passengers, the length of route and the average operating speed, respectively, and vice versa. For example, for serving passenger demand of $Q(\tau) = 3000$ pax/hr traveling on the route of the length of $d_l = 500$ km, with the average load factor of each train service of $\theta_l(\tau)$ $= 0.85$ operating at the average speed of $v_l = 300$ km/hr, the required service frequency by the trains of the seating capacity $s_l = 485$ seats/train, will be $f_l(\tau)$ $= 3000/(485*0.85) \approx 7$ trains/hr, the transport work $TW_l(\tau) = (3000*500)/0.85$ $= 1.765 *10^6$ seat-km, and productivity $TP_l(\tau) = (3000*300)/0.85 = 1.058 *10^6$ seat-km/h.

2.3.2.3 Quality of Service

In most cases, the most important elements of the service quality influencing the choice of HSR system as one among several available, mass passenger transport modes connecting given origin(s) and destination(s) are transport service frequency and schedule delay, trip time (in combination with punctuality and reliability of services), accessibility, i.e. ease in access by the public and private transport, availability of parking space and the comfort inside of HSR stations and on board the HS trains. The comfort inside of HSR stations includes the feeling of comfort, aesthetics, on-site amenities, shops, restrooms, signing, safety and security, information and announcements, the kindness of personnel, and cleanliness, convenience of platforms attributed by the walking distance, ease in moving baggage, lighting, safety and air quality. The comfort on-board the HS trains refers to the class and quality of services, seat reservations, cleanliness, friendliness of personnel, information, entertainment, baggage storage, safety, noise and temperature control. In addition, particularly for HSR services, these can be trip planning (websites, reservations/information phone lines, printed materials, advertising) and ticketing (on-line, mail, at the station, others) (Carol 2011).

(a) Service Frequency and Schedule Delay

Schedule delay is the difference between the desired and available time of boarding a given HSR service. Under the assumption that users-passengers arrive uniformly during the time interval between two successive HS trains' departures due to being familiar with the timetable at their origin station(s), this delay can be roughly estimated as follow:

$$SD_l = \frac{\tau}{4f_l(\tau)} \tag{2.9a}$$

where all the symbols are analogous to those in the previous equations. For example, for the service frequency of $f(\tau) = 1$ train/hr, the schedule delay will be $SD_l = 15$ min; for the service frequency of $f_l(\tau) = 15$ trains/hr, the schedule delay will be $SD_l = 1$ min. ($\tau =$ one hour or 60 min.).

(b) Trip Time, Punctuality and Reliability

Trip time by HSR is much shorter than that by the conventional rail at the same distance/route. In general, the time savings can be estimated as follows:

$$\Delta_l = d_l/(1/v_{CON} - 1/v_{HSR}) \qquad (2.9b)$$

where

d_l is the length of line/route (l) (km);

v_{CON} is the speed of conventional rail (km/h); and

v_{HSR} is the speed of HSR (km/h).

Equation 2.9b indicates that savings in the travel time increase with increase in differences of operational speeds of both categories of trains and the length of the line/route while all other factors such as the number and duration of stops, punctuality and reliability of services remain the same. Figure 2.21 shows an example for Italy.

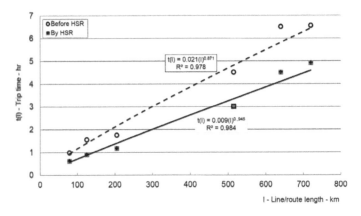

Fig. 2.21: An example of the relationship between the trip time by HS and conventional rail, and line/route length in Italy (Cascetta and Coppola 2011, Janić 2016).

As can be seen, the difference in the trip time by conventional and HSR increases with increase in the line/route length, in the given case from 33 per cent to 42 per cent.

Punctuality of the HSR system is expressed as the ratio of the number of transport services carried out on time, i.e. according to the timetable, or with the specified maximum or average delay and the total number of transport services realized during a given period of time. The experience so far shows that these services in general and on the particular lines/routes have been highly punctual as shown in Fig. 2.22 (UIC 2011).

As can be seen, the Japanese HSR system is generally the most and that in the UK the least punctual. In addition, Fig. 2.23 shows an example of punctuality of the HSR system in Japan over time expressed by an average delay per service.

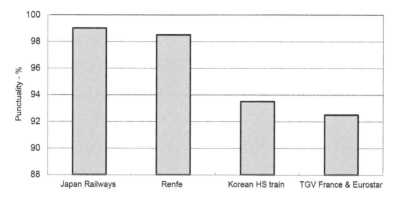

Fig. 2.22: Punctuality of services of the selected HSR systems
(Janić 2016, UIC 2011).

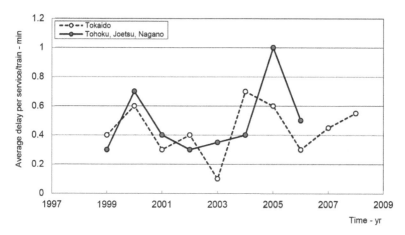

Fig. 2.23: Punctuality of the HSR system in Japan
(Janić 2016, Nishiyama 2010, Tomii 2010).

As can be seen, in the given case, the average delay per HSR service varies from 0.3 to 0.5 min. In addition, the average delay per service over the period of a decade in the Shinkansen HSR system has been about 0.6 min. (JR 2012, Nishiyama 2010).

Figure 2.24 shows the potential influence of punctuality of transport services on the HSR system's market share.

As can be seen, in the given case, the market share of HSR has increased more than proportionally with increase in the punctuality of services, thus indicating the importance of punctuality as a competitive attribute of the HSR system.

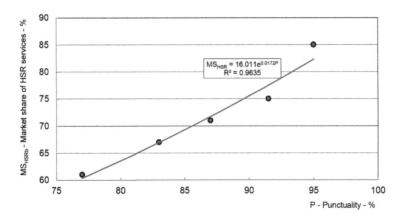

Fig. 2.24: Relationship between punctuality and market share of the HSR Eurostar services (Period: 2002-2007) (Janić 2016, http://en.wikipedia.org/wiki/Eurostar).

Reliability of the HSR system's services can be defined by the proportion of realized transport services as compared to the planned on the given line/route or in the network during the specified period of time (d, mon, yr). This is dependent on the rate of failure of rolling stock due to internal or external reasons causing the cancellation or long delays of the affected transport services. Figure 2.25 shows as an example, the Japanese HSR system.

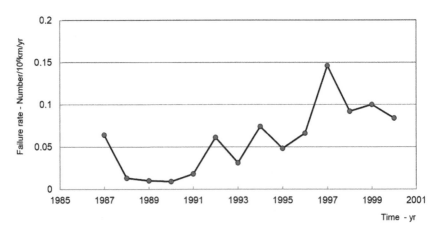

Fig. 2.25: Reliability of the HSR rolling stock (East Japan Railways) (Period: 1987-2000) (Janić 2016, Nishiyama 2010).

As can be seen, this rather very low failure rate has fluctuated during the observed period with an average of 0.084 failures/million km. This has been achieved through maintenance of rolling stock managed by train operators.

In particular, it is carried out at four levels: (i) daily inspection (every 2 d) [inspection of worn out parts (pantograph strip, refreshing water/waste)]; (ii) regular inspection (every 30 d or $30*10^3$ km) (test of conditions and function, inspection of important parts/components without decomposition); (iii) bogie inspection every one-and-a-half year or $600*10^3$ km (inspection of bogie parts by decomposition); and (iv) overall inspection (every three years or $1.2*10^6$ km (inspection of overall rolling stock by decomposition) (Yanase 2010).

(c) Accessibility

Accessibility of the HSR stations represents an important attribute of the HSR system's overall quality of services provided to its users-passengers. In most cases, the new dedicated HSR stations are usually located and designed to fit as well as possible within the surrounding urban and/or sub-urban environment on the one hand and enable a satisfactory quality of accessibility, on the other. In some cases, parts of conventional railway stations have been appropriately upgraded and adapted to serve the HSR system's services. In both cases, the quality of accessibility is expected to be efficient, effective and safe. This implies that the stations need to be accessible at reasonable (acceptable) time and cost by different urban and sub-urban transit modes (car, taxi and frequent, punctual and reliable bus, tram, metro, regional rail, etc.) and safely, i.e. without incidents/accidents due to known reasons, respectively.

(d) Comfort on Board the HS Trains

Comfort on board the HS train services offered to users/passengers usually includes booked seats and a very limited number of stops compared to those of their conventional train counterparts. As far as comparison with APT (Air Passenger Transport) as the main competitor on the short- and medium-haul routes is concerned, the attributes for comparison are distance between seats and mobility within coaches, diversity and type of services and noise on board and potential impact on health.

- In general, the distance between the seats is greater in HS trains than in aircraft, particularly as compared to those operated by LCCs [(Low Cost Carrier(s)] (87-97 cm vs 78-85 cm, respectively) (UIC 2011). Mobility within coaches is also better and higher than that in aircraft mainly due to greater freedom of movement. This is due not to fastening of seat-belt as in an aircraft, particularly during take-off, landing, on-board services and as a result of air turbulence.
- The services on board HS train(s) and aircraft are on the one hand similar and on the other diverse. They are mainly influenced by the construction options and commercial offers by the HSR operators and airlines. In general, these services at both modes consist of internet access, catering (requiring extra charges on HS trains and LCC and free of charge on

the other airlines) and video on board (mostly provided during long-haul flights and optional in some HSR services).

- The noise onboard HS trains and commercial aircraft is also an important element of internal comfort. Some measurements show that noise levels are higher in aircraft than in HS train(s) operating at comparable speed(s), i.e. a maximum of 70-82 dBA for aircraft and 62-69 dBA for HS train(s) (UIC 2011).
- The impact on health is much lower in HS trains than in aircraft. At the latter, users/passengers are exposed to inherent stress or to possible injuries due to air turbulence and some effects of modified barometric pressure, etc. Lack of sufficient mobility, which can cause deep vein thrombosis, is not relevant because it is typical of long-haul flights where HSR does not compete with air transport.

2.3.3 Economic Performances of the HSR System

The economic performances of the HSR system include costs and revenues. The former are needed for implementation and operation of the system; the latter cover the former and provide some funds for updating the system and profits for the stakeholders involved.

2.3.3.1 Costs

In general, the total expenditure is on the infrastructure and operating of the HSR system. The infrastructure costs include: (i) the cost of planning the system and acquisition and preparing the land; (ii) the cost of building the lines and stations, including tunnels and bridges; (iii) the cost of supportive facilities and equipment, including the signaling systems, catenaries and electrification mechanisms, communications and safety installations; and (iv) the maintenance costs of the entire infrastructure, the supporting facilities and equipment (UIC 2005). Table 2.5 gives an indication of the costs of already built and planned HSR lines, and these do not include the costs of planning, land acquisition and preparation.

As is seen, the variation in the average infrastructure cost for both already built and under-construction HSR lines is significant in both European and non-European, i.e. two Asian countries. In Europe, the lowest cost has been in France and Spain and much higher in Italy, Germany and Belgium. The average infrastructure cost has been $18*10^6$ €/km. In addition, the cost of building the new HSR lines in Asian countries (Japan, South Korea, except China) has been slightly higher than that in the European countries (De Rus 2009, Pourreza 2011). In addition, the average maintenance cost per unit length of the HSR system infrastructure also varies highly, mainly depending on the length of the line(s). Some estimates indicate that the average maintenance

Table 2.5: Examples of the cost of infrastructure for HSR lines (De Rus and Nombela 2007, Janić 2016, Pourreza 2011)

Country	Cost (10^6 €/km)	
	Built (in-service) lines	Lines under construction
Austria	-	18.5-39.6
Belgium	16.1	15.0
France	4.7-18.8	10.0-23.0
Germany	15.0-28.8	21.0-33.0
Italy	25.0	14.0-65.8
Netherlands	-	43.7
Spain	7.8-20.0	8.9-17.5
Japan	20.0-30.0	25.0-40.0
South Korea	-	34.2

cost in European countries ranges between 13 and 72 $*10^3$ €/yr (Henn et al. 2013, Pourreza 2011).

The operating expenditure generally includes the costs for acquiring, operating and maintaining the rolling stock, the costs of selling services and the costs of administration. The labor, material and energy costs have the largest share in these costs (UIC 2005). Consequently, the average unit cost of operating HSR transport services is mainly influenced by the local pricing of the above-mentioned inputs and the type of HS trains. Some estimates indicate that the average unit cost for 12 types of the HS trains operating in the corresponding European countries is 14.63 €-ct/s-km. The cost of maintenance of rolling stock has shared about 8.5 per cent in this total. Under the assumption that the average load factor was 70 per cent, the total average operating costs of HSR services throughout Europe would be 19.5 €-ct/p-km (De Rus 2009, Pourreza 2011,). However, some other data shows that these direct operating costs may be much lower and depend on the average HS train operating speed (Garcia 2010). In this case, the direct operating costs of an HS train generally pertain to cost of its acquisition (i.e. ownership), rolling stock and cleaning costs, energy, operating personnel and marginal cost of infrastructure use. An example of the dependence of these costs on the train operating speed is as follows (Garcia 2010):

$$C_0(v) = 5.527381 - 0.0192545 * v + 0.0000427 * v^2 \text{ (€-ct/s-km)}$$

where (v) is the average train speed (km/h). Figure 2.26 shows the relationship for the given train operating scenario.

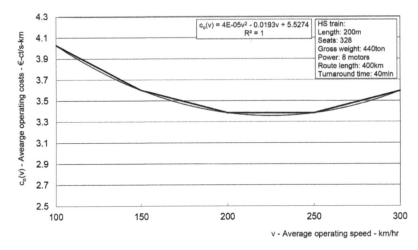

Fig. 2.26: Example of the relationship between the average direct operating cost of a HS train and its speed (Garcia 2010, Janić 2016).

As can be seen, the cost first decreases and then increases more than proportionally with increase in the average speed. In the former case, the cost components on train ownership, maintenance and operating personnel decrease more than the others that increase. In the latter case, the cost of energy and the infrastructure use increase more than the others that decrease.

2.3.3.2 Modelling the Total Costs

A. The total costs of setting up and operating a given HSR line in the (k)-th year of the given period

The net present value of these costs can be estimated as follows:

$$C(k) = \frac{1}{(1+i_1)^k} \left\{ \frac{d}{n*T} [c_c(1+p) + c_m] \right\} \qquad (2.10a)$$

where

i_1 is the discount rate of the cost of infrastructure and other facilities and equipment cost (%);

k is the k-th year starting from the beginning of the period of (n) years of operation of a given HSR line;

$CI(k)$ is the cost of planning, building and regularly maintaining the HSR infrastructure and supportive facilities and equipment in the k-th year of the observed period (€/yr); and

$CRS(k)$ is the cost of acquiring, maintaining and operating the rolling stock/ HSR trains in the k-th year of the observed period (€/yr).

(a) The net present value of the infrastructure cost $CI(k)$ in Equation 2.10a can be estimated as follows (De Rus 2009):

$$CI(k) = \frac{1}{(1+i_1)^k}\left\{\frac{d}{n*T}[c_c(1+p)+c_m]\right\}$$ (2.10b)

where

d is the length of a given HSR line (km);

c_c is the unit construction cost of a given HSR line (€/km) (*see* also Table 2.4);

p is the proportion of the construction costs spent on planning (usually $p = 0.10$);

n is the cycle time, i.e. period of exploitation of a given HSR line (usually $n = 30$-$60T$; $T = 1$ yr); and

c_m is the unit cost of regular maintenance of a given HSR line (€/km).

Equation 2.10b indicates that the infrastructure costs (construction and maintenance) are expressed per unit of length of a given HSR line during one year of its life-cycle, which is about 30 and 60 years for ballast and ballastless tracks, respectively. The maintenance costs of infrastructure referring to its regular periodic inspections and repairs are expressed similarly, i.e. as independent of its wear and tear.

(b) The net present value of the cost of rolling stock $CRS(k)$ in Equation 2.10a evaluated for the k-th year of the observed period can be estimated as follows:

(*i*) Acquisition cost

$$CRS_a(k) = \frac{1}{(1+i_2)^k}[m(k)*s(k)*c_s(k)]$$ (2.10c)

where

i_2 is the discount rate of the cost of rolling stock (%);

$m(k)$ is the number of trains acquired in the k-th year of the observed period (trains);

$s(k)$ is the average seat capacity of a train acquired during k-th year of the observed period (seats/train); and

$c_s(k)$ is the average unit cost of acquiring a train during the k-th year of the observed period (€/seat).

In Equation 2.10c, the number of acquired trains can be determined as follows:

$$m(k) = \max\{0;[f(\Delta t_k)-f(\Delta t_{k-1})]* \tau_{trd}(d,v)\}$$ (2.10d)

where

$f(\Delta t_{k-1}), f(\Delta t_k)$ is the peak-hour transport service frequency scheduled during $(k-1)$ and (k) year of the observed period (dep/hr; $\Delta t_{k-1}, \Delta t_k = 1$ hr); and

$\tau_{trd}(d,v)$ is the average train turn around time on the line of length (d) while operating at an average speed (v) (hr) (d – km; v – km/h).

The peak-hour transport service frequency $f_k(\Delta t_k)$ in Equation 2.10d can be estimated as follows:

$$f(\Delta t_k) = q(\Delta t_k)/[\theta(\Delta t_k)*S(k)] \tag{2.10e}$$

where

$q(\Delta t_k)$ is the passenger demand during the time period Δt_k (passengers/hr); and

$\theta(\Delta t_k)$ is the average load factor of a train operating during the time period Δt_k (≤ 1.0).

(ii) *Operating cost*

$$CRS_0(k) = \frac{1}{(1+i_2)^k}\left[2* f(k)* s(k)* c_o(k)* d\right] \tag{2.10f}$$

where $c_s(k)$ is the average unit cost of operating a train during the k-th year of the observed period (€/seat-km).

(iii) *Maintenance cost*

$$CRS_m(k) = \frac{1}{(1+i_2)^k}\left[m(k)* s(k)* u(k)* C_m(k)\right] \tag{2.10g}$$

where

$u(k)$ is the average utilization of a train in the k-th year of the observed period (km/train); and

$C_m(k)$ is the average unit maintenance cost of a train in the k-th year of the observed period (€/seat-km).

The other symbols are as in the previous equations.

The number of newly acquired trains in Eq. 2.10d depends on the changes in the peak-hour transport service frequency and the train turn around time along the line. The peak-hour transport service frequency in Eq. 2.10(d, e) depends on the volume of passenger demand during the peak-hour, the train seat capacity and its utilization, i.e. average load factor. The train turn around time depends on the length of line, the train operating speed(s) and the time at the beginning and ending of the station/terminus. The train operating costs in Equation 2.10f are proportional to the unit operational cost per seat and the total number of deployed seats during a given period of time (year). The train maintenance costs as seen in Equation 2.10g are proportional to the number of trains and their seat capacity, their utilization during a given period of time (year) and the average unit maintenance cost (€/seat-km).

B. The net present value of the total costs of setting up and operating a given HSR line during its life-cycle

Based on Equation 2.10, these life-cycle costs can be estimated as follows:

$$C(n) = \sum_{k=0}^{n} C(k) = \sum_{k=0}^{n} \left[\left(\frac{1}{(1+i_1)^k} \right) * CI(k) + \left(\frac{1}{(1+i_2)^k} \right) * CRS(k) \right] \quad (2.11a)$$

By dividing the total costs of setting up and operating the given HSR line $C(k)$ in Equation 2.11a by the transport work carried during that time (k-th year), the corresponding average cost per unit of output in terms of pax-km and/or seat-km can be obtained. For the year *(k)* this can be as follows:

$$\overline{C(k)} = (k)/[2 * f(k) * s(k) * \theta(k) * d] \quad (2.11b)$$

where all the symbols are analogous to those in the previous equations.

Equation 2.11b expresses the average costs in terms of €/pax-km. Excluding the average annual load factor, $\theta(k)$ gives the average costs in terms of €/seat-km. As can be seen, the average unit operating cost depends on the net present value of the total costs of setting up and operating the line of a given length and the volume of operations characterized by the HS train's service frequency, their size, i.e. seat capacity and average load factor, all during a given period of time of its life-cycle. The load factor depends on the relationship between the volume of passenger demand and the transport service frequency, and the train size, i.e. seat capacity per frequency. Under conditions of complete absence of the passenger demand, this cost depends exclusively on the costs of capacity components.

C. An example

An application of the above-mentioned models for calculating the net present value of the costs of a given HSR line during its life-cycle is carried out by means of a rather simplified (hypothetical) example.

(a) Inputs

The inputs—characteristics of the line and operations of a single train, traffic scenario and the costs—are given in Table 2.6 (a, b, c).

(b) Output

By using the above-mentioned inputs in Table 2.6 (a, b, c), the non-discounted and the discounted (i.e. the net present) values of the average unit cost of the given HSR line are calculated and shown in Figs. 2.27 and 2.28.

As can be seen, both the non-discounted and the discounted average unit cost decrease more than proportionally over the life-cycle time under the given conditions. In addition, Fig. 2.28 shows the relationship between this average unit cost and the annual volumes of passenger demand assumed to grow over time according to the above-mentioned scenario.

As can be seen, this cost decreases more than proportionally with increase in the annual volumes of passenger demand, thus indicating the existence of economies of scale in the given context.

Table 2.6a: Characteristics of the line and operations of a single train

Length of line (d) (km)	600
Operating speed (v) (km/h)	320
Seat capacity (s) (ETR like train)[1] (seats/ train)	430
Time of providing transport services during the day (τ) (hr/d)	18 (06h- 24h)
Service frequency ($f(\Delta t)$) (trains/hr); depends on demand	-
Train turn around time on the line ($\tau_{trd}(d,v)$) (hr/train)	2*600/320 + (20+20)/ 60 = 4.4
Average load factor $\theta(\Delta t)$ (%)	80
Discount rate for infrastructure and other facilities and equipment (i_1) (%/yr)	1.5
Discount rate for the rolling stock (i_2) (%/yr)	3.0
Life-cycle time (n) (yr)	35

[1]See Table 2.3

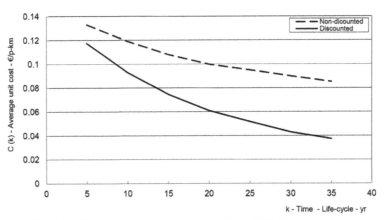

Fig. 2.27: The average unit cost of the given HSR line operating according to the given scenario during its life-cycle.

2.3.3.3 Revenues

HSR systems obtain revenues from different sources, such as mainly transportation-based charging users/passengers, merchandise and others (JR 2014). In particular, the prices/fares for users/passengers are set up to cover the total operating cost if subsidies are not provided as an element for enabling

Table 2.6b: Characteristics of the traffic scenario

Time	Period	Passenger demand[1] growth	Passenger demand[2] volume	Passenger demand volume	Service frequency	Required fleet	New fleet[5]
(yr)	(yr)	(%/yr)	(10^6/yr)	(10^3/d)[3]	(dep/d); (dep/hr)[4]	(trains)	(trains)
0	0 - 5	5	10	27.397	80 (2/4)	9	9
5	5-10	5	12.8	35.068	102 (3/6)	13	4
10	10-15	5	16.3	44.658	130 (4/8)	18	5
15	15-20	5	20.8	56.986	166 (5/10)	22	4
20	20-25	3	24.1	66.027	192 (6/12)	26	4
25	25-30	3	27.9	76.438	233 (6/12)	26	0
30	30-35	3	32.4	88.767	258 (7/14)	31	5

[1]During a 5-year period; [2]At the end of the 5-year period; [3]Per day at the end of the 5-year period; [4]Single direction/both directions; [5]Acquired during the 5-year period.

Table 2.6c: Characteristics of the costs

Infrastructure	Cost value
Construction and building (c_c) (10^6 €/km)	26.6
Maintenance (c_m) (10^3 €/km)	35.3
Rolling stock/HS trains	
Acquisition (c_s) (10^3 €/seat)	45-50
Operations (c_o) (average) (€-ct/s-km)	13.3
Maintenance (C_m) (€/s-km)	0.0124
Annual utilization (u) (10^5 km/seat)	5

stronger competition with other transport modes, such as conventional rail and particularly air transport, both on the above-mentioned competitive lines/ routes. Figure 2.29 shows some examples of the relationship between fares of the HSR and APT in Europe, China and Japan depending on the travel distance.

As can be seen, the fares of HSR services are most dispersed in Europe and much less in China and Japan. In all the three regions, they generally increase with increase in the travel distance at a decreasing rate. In addition, the prices in Japan are highest and those in China lowest. The fares of competitive APT services are also very dispersed in all the three regions. They generally decrease at a decreasing rate with increase in the travel distance. In the given example, they are highest in China and lowest in Europe.

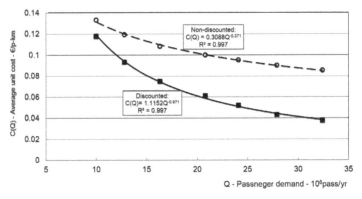

Fig. 2.28: Relationship between the average unit cost and the volumes of passenger demand of the given HSR line operating according to the given scenario.

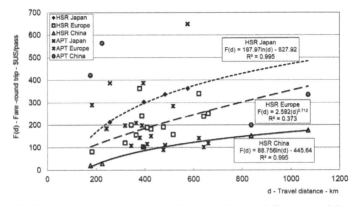

Fig. 2.29: Examples of the relationship between the travel distance and the round trip fare charged one week in advance by the HSR and APT (Air Passenger Transport) (Feigenbaum 2013).

With numerous exceptions, the fares of HSR services are generally lower on the shorter and higher at the longer travel distances than those of the APT, thus confirming the above-mentioned competing potential of HSR in these markets. In addition, the average unit fares of the HSR services have generally decreased more than proportionally with increase in the volume of passenger demand as shown in Fig. 2.29. If these fares are supposed to cover the costs, they also indicate the existence of economies of scale as explicitly suggested in Equation 2.11b and shown in Fig. 2.30.

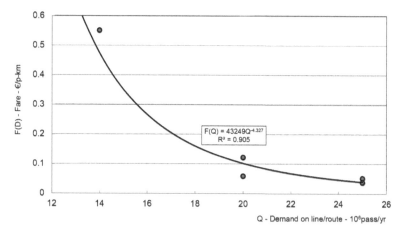

Fig. 2.30: Relationship between the unit price of HSR service and the volume of passenger demand on five Chinese lines/routes (Wu 2013).

2.3.3.4 Balancing Revenues and Costs

As the other systems, HSR systems intend to operate profitably, i.e. by generating sufficient total revenues to cover their total costs. Figure 2.31 shows an example of the profitability of a company operating both HSR and conventional rail services.

As can be seen, despite relatively high variations, the profitability has generally increased with increase in the volume of the company's output during the given period of time. This case could be used as an example of how an HSR system can prove profitable in the medium- to long-term period of time—by carefully balancing the revenues and costs while increasing the scale of operations.

2.3.4 Social Performances of the HSR System

The social performances of the HSR system include impacts and effects. The impacts embrace noise, congestion and safety, i.e. traffic incidents and accidents and their social costs-externalities. The effects generally refer to the

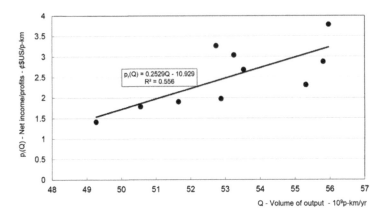

Fig. 2.31: Relationship between the annual volumes of output and the unit profits—Central Japan Company (Period 2004-2013) (Janić 2016, JR 2014).

system's contribution to the overall social-economic welfare, i.e. regional and national GDP (Gross Domestic Product), including local direct and indirect employment and mitigating the overall social and environmental impacts of the transport sector.

2.3.4.1 *Impacts*

The HSR system generally impacts noise, congestion and safety, i.e. traffic incidents and accidents.

(a) Noise

HS trains operating at high speeds generate noise. This consists of rolling, aerodynamic, equipment and propulsion sound. An HS train's noise can generally impact three categories of land use activities: quiet land with intended outdoor use, land with residence buildings and land with daytime activities (businesses, schools, libraries, etc.). The noise mainly depends on the level generated by the source, i.e. the moving HS train and its distance from an exposed observer(s). Figure 2.32 shows a scheme of changing distance and time on exposure of an observer to noise by an HS train.

The shaded polygon represents the HS train of length (L) passing by an observer (small triangle at the bottom) at the speed (v). He/she starts to consider an approaching train when it is at a distance (β) from the point along the line, which is at the closest right-angle distance (γ) from him/her. The consideration stops after the train moves behind the above-mentioned closest point again for the distance (β). Under such circumstances, the distance between the observer and passing HS train changes over time as follows:

$$\rho^2(t) = (L/2 + \beta - v * t)^2 + \gamma^2 \text{ for } 0 < t <= (L + 2* \beta)/v \qquad (2.12a)$$

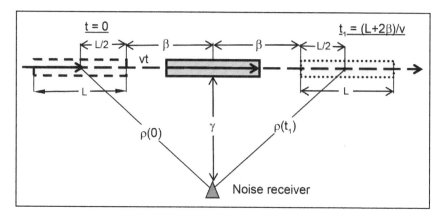

Fig. 2.32: Scheme for determining the noise exposure of a receiver by a passing HS train (Janić and Vleugel 2012).

where the last term represents the duration of the noise event, i.e. the time for the train to pass by the observer. The length of the HS train is given in Table 2.3. If the level of noise received from the train passing by the observer at the speed (v) at the shortest distance (γ) is $L_{eq}(\gamma, v)$, the level of noise during time (t) can be estimated as follows:

$$L_e\,[\rho(t),\, v] = L_e(\gamma,\, v) - 8.6562\ln[\rho(t)/\gamma] \tag{2.12b}$$

The second term in Equation 2.12b represents the noise attenuation with distance over an area free of barriers. The total noise exposure of an observer from ($\lambda(\tau)$) successive trains passing by during the time (τ) can be estimated as follows:

$$L_{eq}[\lambda(t)] = 10\log\sum_{k=1}^{\lambda(\tau)} 10^{\frac{L_e[k,\rho(t),v]}{10}} \tag{2.12c}$$

As a standard approach, the noise from HS trains is measured at the right-angle distance of $\gamma = 25$ m from the track(s). Figure 2.33 shows some results of such measurements across Europe, depending on the HS train operating speed.

As can be seen, at the given speed, the noise from different categories of HS trains and corresponding lines varies in a range of about 4 dBA. In addition, in all cases the noise increases linearly with increase in the maximum operating speed under the given conditions. It needs to be mentioned that the time of exposure to this noise decreases with increase in the speed. For example, if the HS train is just facing an observer, i.e. $\beta = 0$ m, the train's length $L = 200$ m and the maximum speed $v = 250$ km/h, the time of exposure to the maximum

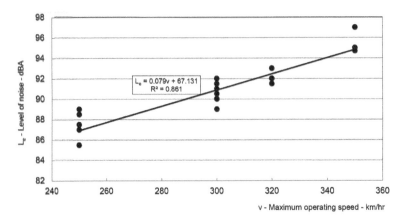

Fig. 2.33: Relationship between the passing-by noise at the right-angle distance of 25 m and the maximum operating speed of HS trains (Belgium, France, Germany, Spain, Italy) (Gautier and Letourneaux 2010).

noise will be about $t_1 = 3$ s. For a speed of 350 km/h, this time will be about $t_1 = 2$ s. In considering the actual exposure to noise to the population located close to railway line(s) and passing HS trains, it is necessary to take into account the noise-mitigating barriers. These are usually set up to protect the above-mentioned land use activities by absorbing the maximum noise levels by about 20 dB(A) (single barrier) and 25 dB(A) (double barrier). When frequent HSR services are carried out along particular lines/routes, their noise becomes persistent over time and can be estimated from Equation 2.12c.

(b) Congestion

On account of applying the separation rules and designing time-table(s) on particular lines/routes and the entire HSR network accordingly, the HSR system is free of congestion and consequent delays due to the direct mutual influence of vehicles on each other while 'competing' to use the same segment of the line/route at the same time. However, substantial delays due to some other reasons can propagate (if it is impossible to absorb and neutralize them) through the affected HS train(s) itinerary as well as along the dense line/routes also affecting otherwise non-affected services. Under such conditions, the severely affected service is usually cancelled in order to prevent further increase and spread of delay. On the one hand, this contributes to maintaining punctuality but on the other, it compromises the reliability of the overall services. Nevertheless, the already mentioned figures indicate that both reliability and punctuality of the HSR system services worldwide have been very high and in some cases, extremely high (the Japanese HSR system is an example of the latter).

(c) Safety, i.e. Traffic Incidents/Accidents

Experience so far indicate that HSR and APT have been the safest transport modes in which traffic incidents/accidents rarely occurred, usually due to unknown reasons. This means that the number of traffic incidents/accidents and related injuries, deaths and scale and cost of damage to property both of the systems and third parties, for example, per billion s-km and/or p-km carried out over a given period of time, have been extremely low. In particular, the high safety of HSR has been provided also *a priori* by designing completely grade-separated lines and other supportive built-in safety features at both infrastructure and rolling stock. This implies that the safety is achieved on account of increased investment and maintenance costs. In addition, the HSR operators and infrastructure managers have continuously practiced a risk management training approach aimed at maintaining a high level of safety, particularly with increase in the maximum speed(s). Nevertheless, the HSR systems in different countries are not completely free of traffic incidents/ accidents. For example, some relevant statistics for the TGV system in France indicate that there have not been accidents with fatalities (deaths) and severe injuries of users-passengers, staff and/or third parties since the HSR services started in the year 1981, despite the trains carrying about ten million passenger-km annually. However, some incidents have occurred on the HSR lines/routes, such as broken windows, opening of the passenger doors during operating at cruising speed, a couple of fires onboard, collision with animals and concrete blocks on the tracks, and a terrorist attempt to bomb the tracks. The incidents and accidents of TGV trains operated on conventional tracks have been more frequent with fatalities, injuries and damage to property, albeit all at a relatively low scale. In these cases, the HS trains have been exposed to external risks as their conventional counterparts (http://www.railfaneurope. net/tgv/wrecks.html). Similarly, since started in 1960s, Japan's Tokaido Shinkansen HS services[2] have also been free of accidents, causing user/ passenger and staff fatalities and injuries due to derailments and collisions of trains. This has been achieved despite the services being exposed to the permanent threat of relatively frequent earthquakes. However, level-crossing incidents/accidents have occurred on the other five lines of the Shinkansen network but at a permanently decreasing rate over time as shown in Fig. 2.34 (JR 2012).

---

[2] The Tokaido Shinkansen line/route of length of 552.6 km connects Tokyo and Shin Osaka station free of level crossings. The trains operate at the maximum speed of 270 km/h, covering the line/route in 2 hours and 25 min. The route/line capacity is 13 trains/hr per direction. The number of passengers carried is about 386 thousand/d and 141 million/ yr (in 2011) (JR 2012).

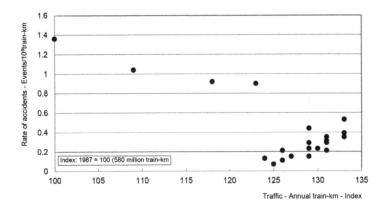

Fig. 2.34: Relationship between the annual traffic growth and the rate of accidents at level crossings—Japanese passenger trains (Period: 1987-2009) (JR 2012).

As can be seen, this rate has generally been very low per million passenger kilometers in each year of the observed period. In addition, it has been relatively constant with increase in the volume of train operations by about 20 per cent as compared to the year 1987 (Index = 100). With a further increase in the annual volume of train operations by an additional 15 per cent, i.e. from index 120 to index 135, this rate has been much lower than previously but with an increasing tendency, although at a very low scale within the given range of annual volumes of train operations (about 0.7-0.4). One of the most important factors for such improvement of the safety records has been the introduction of safety education and training programs primarily aimed at the staff dealing with operations and maintenance of both infrastructure and rolling stock (JR 2012).

Nevertheless, fatal accidents with deaths and injuries of users-passengers and staff have occurred at the HSR systems in Germany, Spain and China (one incident in each country). Table 2.7 gives the main characteristics of these three accidents.

Table 2.7: The main characteristics of the fatal HSR accidents
(NDTnet 2000, Qiao 2012, Puente 2014)

Country/system/ number of trains	Date	Cause	Passengers on board	Fatalities	Injuries
Germany/ICE/1	3/06/1998	Wheel disintegration	287	101	88
China/2	23/07/2011	Railway signal failure	1630	40	>210
Spain/Alvia/1	34/07/2013	Excessive speed on bend	222	>79	139

The accident on the German ICE HS train is known as the 'Eschede train disaster', named after the place where it occurred—near the village of Eschede in Celle district of Lower Saxony, Germany. The HS train derailed and crashed into a road bridge, which collapsed latter on with the impact mentioned in Table 2.7. This was the worst rail accident in the railway history of the Federal Republic of Germany and the worst HSR accident in the world. The wheel disintegration as the direct cause was due to a single crack in the wheel. When it failed, the train derailed at a switch.

The accident in China involved the collision of two HS trains operating on the Yongtaiwen line on a viaduct in the suburbs of Wenzhou, Zhejiang province, China with the impact given in Table 2.7. The collision caused derailment of both the trains, of which four cars fell off the viaduct. The main cause of the collision was the failure of the signaling system.

The third accident, known as the 'Santiago de Compostela rail disaster', occurred in Spain. It occurred when an HS train operating on the Madrid-Ferrol route derailed due to entering a bend at an excessively high speed (double the allowed speed of 80 km/h). The derailment took place about 4 km outside the railway station at Santiago de Compostela (in the northwest of Spain) with the impacts given in Table 2.7. As in Germany, this accident was the worst on Spanish railways in the past 40 years.

(d) Cost of the Social Impacts—Externalities

Quantifying the above-mentioned social impacts of the HSR system and other transport modes in monetary terms as externalities has usually represented an ambiguous and often politically-lead task. Nevertheless, some estimates of these externalities for the HSR system and other transport modes in Europe were carried out. These estimates indicate that the total externalities of a HSR system are 22.9€/1,000 p-km. In this total, noise and traffic incidents/ accidents or externalities shared about 22 per cent and 2 per cent, respectively. Since the HSR system was free of congestion, the corresponding externality was not considered. On the other hand, the total externalities of APT were estimated to be 52.5€/1,000 p-km, of which noise and traffic incident/accident externalities shared about 4 per cent and 3 per cent, respectively (INFRAS/ WWW 2000, UIC 2010b).

2.3.4.3 Effects

In general, the effects of the HSR system on the socio-economic development and welfare have been direct and indirect, both at global-country or continent and local-regional scale. For example, on the global scale, the direct effects include the contribution of the investments in the HSR system to the GDP (Gross Domestic Product), which in the case of Europe is estimated to be about 0.25 per cent. At the same time, at the regional scale, the contribution is estimated to be much higher—about 3 per cent of regional GDP (De Rus and Nombela 2007, Preston 2013). In addition, both direct and indirect effects of the HSR system primarily at the local-regional scale include direct and indirect employment. The former employment includes employment for building and maintaining the HSR system infrastructure and manufacturing, operating and maintaining the rolling stock and supported facilities and equipment, i.e. the main system's components. The latter employment relates to supplying the system with different kinds of daily consumables and energy on the one hand and that generated thanks to the system's existence, on the other. These are non-rail related economic and business activities around and at the HSR stations, such as business services (banking, insurance and advertising), information and retail services, research and development, higher education, tourism and political institutions (UIC 2011). On the larger scale, these businesses have created urban (both business and housing) agglomerations around the HSR stations, which themselves have induced additional demand for HSR services. This development has been taking place mainly at the HSR stations already located in the larger and wider urban agglomerates connected by the HSR lines/routes and also within them. For example, inclusion of the city of Lille (France) in the HSR line/route Paris-Brussels has brought enormous economic development to the city itself and its region in terms of increase in business and tourism activities and related employment. In the UK, substantial economic activities have occurred in cities two hours from the London area, thanks to HSR (Baron 2009). The benefits from these economic activities have been much higher in cities with a service-oriented economy than in those with primarily a manufacturing-oriented one (Albalate 2010). In addition, the German regions with cities of Montabaur and Limburg, with populations of 12,500 and 34,000 respectively, have recorded growth of GDP of about 2.7 per cent due to the increase in market accessibility to the larger cities of Frankfurt and Cologne, thanks to HSR services (Boqué 2012). In Japan, HSR has generated growth of population in cities by about 1.6 per cent compared to those bypassed where this growth has been about 1 per cent. This growth has taken place primarily in cities with an IT industry and higher education. In addition, the (indirect) effects of the HSR system include contributions to general improvement in mutual accessibility of the particular regions-cities it connects and mitigating of the overall social and environmental impacts of the transport sector. The latter is achieved by taking over, through both competition and cooperation, the newly generated and partially existing user-passenger demand from the less socially and environmentally efficient modes, mainly road and APT, on the short- and medium-haul lines/routes.

2.3.5 Environmental Performances of the HSR System

The environmental performances of the HSR system generally include energy consumption and related emissions of GHG, land used for the system's infrastructure and their social costs externalities.

2.3.5.1 Impacts

(a) Energy Consumption and Emissions of GHG

Only energy consumption and related emissions of GHG by operating HSR services are considered. This implies that those from building the infrastructure (lines) and manufacturing supporting facilities and equipment and rolling stock (trains) are not (UIC 2010c).

In general, HS trains consume electricity primarily for accelerating up to the operating/cruising speed and for overcoming rolling/mechanical and aerodynamic resistance to motion at that speed. This also includes the energy for overcoming the resistance of grades and curvatures of tracks along the given line/route. In addition, energy is consumed for power equipment on board the train. In particular, during the acceleration phase of a trip, electric energy is converted into kinetic energy at an amount proportional to the product of the train's mass and the square of its speed. A part of this energy is recovered during the deceleration phase before the train's stop by means of regenerative breaking. During the cruising phase of a trip, energy is mainly consumed to overcome the rolling/mechanical and aerodynamic resistance, which for a given type of HS train can be expressed as follows (Raghunathan et al. 2002):

$$R = R_M + R_A = (a + b* v) * W + C * v^2 \qquad (2.13)$$

where

R_M, R_A are the rolling/mechanical and aerodynamic resistances respectively (N) (N – Newton);

W is the weight of the train (ton);

v is the operating/cruising speed of the train (km/h); and

a, b, c are the coefficients to be estimated experimentally.

The equation essentially reflects a Davis equation with the corresponding coefficients. It indicates that the aerodynamic resistance generally increases with the square of operating/cruising speed while the rolling mechanical resistance increases linearly with increase in this speed but also with the weight of the HS train. Some experiments carried out for Shinkansen Series 100 HS of trains gave an estimation of the total resistance depending on the cruising/operating speed as follows:

$R(v) = 12.484 + 0.04915v + 0.001654v^2$ [$R(v)$ in N and v in km/h] (Raghunathan et al. 2002).

In addition, some other research suggests that the first term of the resistance function is not dependent on the speed but only on the weight of the train since the influence of speed is already contained in the corresponding coefficient (UIC 2010b). The above-mentioned relationship emphasizes the importance of reducing both the weight of the train and its aerodynamic resistance in order to achieve savings in the energy consumption during the longest phase of a given trip—cruising at high speed.

Estimates of the average energy consumption by different types of HS trains including the acceleration/deceleration/cruising phase of a trip have differed and changed over time, thanks to the above-mentioned permanent endeavors for improving both characteristics (aerodynamic and weight) and operations of HS trains. One such figure for Eurostar trains indicates that the average energy consumption has been about 0.041 kWh/s-km (s-km - seat-kilometer). In addition, Table 2.8 gives some recent estimates of the energy efficiency of different types of HS trains.

Table 2.8: Average specific energy consumption of different types of HS trains (ATOC 2009)

Train type	Operating speed (km/h)	Seating capacity (seats)	Energy consumption (kWh/s-km)[1]
Shinkansen Series 7000	300	1323	0.029
AVG	300	650	0.033
TGV Reseau	300	377	0.031
TGV Duplex	300	545	0.032
Pendolino Class 300	300	439	0.033
Eurostar Class 323	300	750	0.041

[1]s-km – seat-kilometer

As can be seen, the Japanese Shinkansen is the most and Eurostar HS train(s) the least energy efficient. One of the reasons is the relative large difference in the seating capacity.

Emissions of GHGs by HS trains in terms of CO_{2e} [carbon dioxide equivalents—CO_2 (carbon dioxide), CH_4 (methane), and NO_2 (nitrogen dioxide)] mainly depend on the above-mentioned energy efficiency and the composition and emission rates of the primary sources for producing electric energy in the countries they operate. Table 2.9 gives an indication for particular HS trains operating in different countries based on Table 2.8.

Table 2.9: Average emissions of GHGs by different types of HS trains
(Defra/DECC 2011)

Train type	Country	EC – Energy consumption[1] (kWh/s-km)	ER – Country's emission rate[2] (gCO$_{2e}$/kWh)	EMS = EC*ER Emissions of GHGs (gCO$_{2e}$/s-km)
Shinkansen Series 7000	Japan	0.029	443	12.8
AVG	Spain	0.033	369	12.2
TGV Reseau	France	0.031	71	2.2
TGV Duplex	France	0.032	71	2.3
Pendolino Class 300	Italy	0.033	411	13.6
Eurostar Class 323	UK France	0.041	280	11.5

[1]Energy consumption from Table 2.8; [2]Emission rates from the consumed electric energy.

As can be seen, the greatest average emissions of GHGs are in Japan and the lowest in France. In particular, in Europe, the desire is to decrease these emissions to an average of 5.9 by the year 2025, to 1.5 by the year 2040 and to 0.9 gCO$_{2e}$/s-km by the year 2055. This is expected to be achieved through further improvements in the energy efficiency of the HS trains and their operations on the one hand, and by changing the type and composition of the primary sources for producing the electric energy, on the other. In the latter case, the aim is to produce as much electric energy as possible from renewable decarbonized primary sources. For comparison, the fuel efficiency and related emissions of the GHGs by the APT competing with the HSR on the short- to medium-haul lines/routes are expected to improve over the forthcoming decades. For example, the emission rate of CO$_2$ is expected to decrease from today's average of 97 gCO$_2$/s-km to 62 gCO$_2$/s-km by the year 2025 and to 47 and 41 gCO$_2$/s-km by the years 2040 and 2055, respectively (the emission conversion factor is: 1 g of Jet A fuel = 3.18 gCO$_2$; the aircraft types considered are similar to today's A319 and B737-800 models). These improvements are expected to be achieved by improving the aircraft airframe and engine efficiency. Beyond the year 2050, further improvements may be expected by means of introducing alternative fuels, such as liquid hydrogen (ATOC 2009, Janić 2014b). In addition, the emission rate of an average passenger car as an additional competitor to HSR on the short- to medium-haul lines/routes is around 140 gCO$_2$/km. This is likely to decrease to about 130 gCO$_2$/km by the year 2020. However, the new cars to be launched in the meantime are expected to have an emission rate of about 120 gCO_2/km, which is according to the EU proposals. In addition, this could be reduced to about 80 gCO_2/km thanks to the more massive introduction of hybrid cars by the year 2030 and to about 57 gCO_2/km during the period between the years 2040 and 2055 when electric or fuel-cell cars are supposed to really contribute to the more significant reduction in the above-mentioned emission rates. As in HS trains, this will be carried out in parallel to changing of the structure of the primary sources for producing electric energy (ATOC 2009, Janić 2014a). Nevertheless, the above-mentioned figures indicate that the HSR system will remain superior in terms of energy efficiency and related emissions of GHG (CO_{2e}) as compared to its competitors—passenger cars and short- and medium-haul commercial aircraft.

(b) Land Use

The HSR infrastructure generally occupies much less land than its road-highway counterpart. For example, if the width of an HSR line is (w) and the length (l), the total occupied land can be estimated as follows:

$$A = w * l \qquad (2.14)$$

For example, if $w = 25$ and length $l = 1$ km for an HSR line, the total area of directly taken land will be $A = 2.5$ ha (ha – hectare) (the average gross area of taken land is 3.2 ha). For a highway with three lanes in both directions and width $w = 75$ m and length $l = 1$ km, the directly taken land is $A = 7.5$ ha (the average gross area of taken land is about 9.3 ha, i.e. three times greater than that of an HSR line). In addition, the utilization of taken land for both modes is quite different. The capacity of HSR line/route in both directions is two times 12-14 trains/hr, i.e., 24-28 trains/hr. If each train carries 600 passengers on board, the intensity of land use will be 24-28 × 600/2.5 = 5760-6720 pass/hr/ha. In case of the above-mentioned highway with the capacity of 4500 veh/hr and the occupancy rate of 1.7 pass/car, the intensity of land use will be 1020 pax/hr/ha, which is about six to seven times lower than that of the HSR (UIC 2010b).

2.3.5.2 Cost of the Environmental Impacts—Externalities

Energy consumption and related emissions of GHG and land use as externalities have also been considered in the HSR system. As in the case of social externalities, the HSR system was found to be rather superior when compared to other competing transport modes, such as APT and road cars. Some estimates indicate that the air pollution associated with climate change shares about 26 per cent and land use about 30 per cent in the total HSR system externalities of 22.9 €/1,000 p-km. After including the above-mentioned share of the social externalities, the rest to 100 per cent is the share of urban, up-, and downstream externalities. The corresponding figures for APT are 86 per cent and 2 per cent for air pollution and land use, respectively. After including the share of social externalities, the rest to 100 per cent is the share of urban, up- and downstream externalities in the total of 52.5 €/1,000 p-km (INFRAS/WWW 2000, UIC 2010b).

2.4 Concluding Remarks

In this chapter, the main characteristics of transport systems are addressed by using the case of an HSR system. The main reasons for such a choice are as follows: (i) innovativeness in terms of high speed on tracks compared to its conventional counterpart, which require completely new infrastructure, supporting facilities and equipment, rolling stock and traffic control and management system including operational rules and procedures; (ii) relatively fast penetration throughout different continents; and (iii) inherent complexity, which could provide a sufficiently generic framework to be applied to the similar elaboration of the mass transport systems operated by other transport modes, of course with necessary modifications.

The framework for elaborating the HSR system includes its main components and performances. The main components include the infrastructure, rolling stock, supporting facilities and equipment, traffic control and management system and operating staff. The main performances considered are operational, economic, social and environmental. The operational performances include demand, capacity and their dynamic relationship reflected through the quality of transport services provided to users-passengers. The economic performances include the cost and revenues of setting up and operating the HSR system and revenues earned by charging users-passengers. The social performances include the impacts and effects of the system on society. The former include noise, congestion and traffic incidents/accidents, i.e. safety, while the latter include global and local direct and indirect contributions (benefits) to the economy in the widest sense. The environmental performances include energy consumption and associated emissions of GHG and land use with both direct and indirect impacts on the environment.

The particular performances are elaborated in a descriptive and analytical manner independent of the most influential factors. In the latter case, some analytical models of particular performances are presented. In addition, where considered appropriate, comparison of the performances of the HSR system with those of other (competing) transport systems operated by other modes has been carried out.

Finally, the HSR has shown to be a mass interurban transport system serving user-passenger demand generally through competition or cooperation with its conventional rail counterpart, bus and air transport system efficiently, effectively and safely.

References

Albalate D., Bel G. 2010. *High Speed Rail: Lesson for Policy Makers from Experiences Abroad*. Research Institute of Applied Economics, University of Barcelona, Barcelona, Spain.

ABB. 2014. *Powering the World's High Speed Rail Networks*. ABB ISI Rail, Geneva, Switzerland.

Anderson T., Lindvert D. 2013. Station Design on High Speed Railway in Scandinavia: A Study of How Track and Platform Technical Design Aspects are Affected by High Speed Railway Concepts Planned for the Oslo Göteborg Line. MSc Thesis. Chalmers University of Technology, Göteborg, Sweden.

ATOC. 2009. *Energy Consumption and CO_2 Impacts of High Speed Rail*. Association of Train Operating Companies Ltd, London, UK.

Baron T. 2009. *High Speed Rail Contribution to Sustainable Mobility*. HAL Id., http://dumas.ccsd.cnrs.fr/dumas-00793156

Boqué R.J. 2012. *High-Speed Rail: Economic Evaluation, Decision-making and Financing*. Master-Arbeit. Technische Universität Dresden, Fakultät Verkehrswissenschaften 'Friedrich List', Institut für Gestaltung von Bahnanlagen, Dresden, Germany.

Carol D.C. 2011. *High-speed Rail is not about Trains*. NC. 1-NETWORK—High Speed Rail 73: 6-7.

Cascetta E., Coppola P. 2011. *High Speed Rail Demand: Empirical and Modelling Evidence from Italy*. European Transport Conference, Glasgow, UK.

Connor P. 2011. *Rules for High Speed Line Capacity or How to Get Realistic Capacity Figure for a High Speed Rail Line*. Info Paper No. 3. Rail Technical Webpages. PRC Rail Consulting Ltd, Sutton Bonington, Loughborough, UK.

Crozet I. 2013. *High Speed Rail Performance in France: From Appraisal Methodologies to Ex-Post Evaluation*. Discussion Paper No. 2013-26. The Roundtable on Economics of Investments in High Speed Rails. International Transport Forum, 18-19 December 2013, New Delhi, India.

CSP. 2014. *China Statistical Yearbook 2014*. China Statistical Press. National Bureau of Statistics of China, Beijing, China.

Defra/DECC. 2011. *Guidelines to Defra/DECC's GHG Conversion Factors for Company Reporting*. http://www.defra.gov.uk/environment/economy/business-efficiency/reporting/

De Rus G., Nombela G. 2007. Is Investing in High-Speed Rail Socially Profitable? *Journal of Transport Economics and Policy* 41 Part 1: 3-23.

De Rus G. 2008. *The Economic Effects of High-speed Rail Investments*. OECD-ITF Transport Research Centre, Discussion Paper, 2008-16.

De Rus G. (Ed.). 2009. Economic Analysis of High Speed Rail in Europe. *Informes 2009*. Economía y Sociedad, Fundación BBVA, Bilbao, Spain.

EC. 1996. *Interoperability of the Trans-European High Speed Rail System*. Directive 96/48/EC. European Commission, Brussels, Belgium.

EC. 2014. *EU Transport in Figures. Statistical Pocketbook 2014.* European Commission, Publications Office of the European Union, Luxembourg.

Feigenbaum B. 2013. *High-Speed Rail in Europe and Asia: Lessons for the United States.* Policy Study 418. Reason Foundation, Los Angeles, California, USA.

Fu J., Nie L., Meng L., Sperry R.B., He Z. 2015. A Hierarchical Line Planning Approach for a Large-Scale High Speed Rail Network: The China Case. *Transportation Research* A 75: 61-83.

Garcia A. 2010. *Relationship between Rail Service Operating Direct Costs and Speed.* Study and Research Group for Economics and Transport. FFE Fundación Ferrocarriles Españoles Operation, UIC (International Union of Railways), Madrid, Spain.

Gautier P.E., Letourneaux F.P. 2010. *High Speed Trains External Noise: A Review of Measurements and Source Models for the TGV Case up to 360 km/h.* http://pdf-ebooks.org/ebooks/high-speed-trains-pdf.html

Henn L., Sloan K., Douglas N. 2013. *European Case Study on the Financing of High Speed Rail.* Proceedings of Australasian Transport Research Forum. 2-4 October 2013, Brisbane, Australia.

Hunyadi B. 2011. *Capacity Evaluation for ERTMS (European Rail Traffic Management System) Level 2 Operation on HS2.* Bombardier Transportation Rail Control Solutions, Bombardier Inc., Montréal, Canada.

INFRAS/WWW. 2000. *External Costs of Transport: Accident, Environmental and Congestion Costs in Western Europe.* INFRAS Consulting Group for Policy Analysis and Implementation, Zürich, Switzerland, WWW University of Karlsruhe, Karlsruhe, Germany.

IR. 2012. *The Benefits of High-speed Rail in Comparative Perspective.* Invensys Rail, www.invensysrail.com/whitepapers/hsh-research-report.pdf

Janić M. 1988. A Practical Capacity Model of a Single Track Line. *Transportation Planning and Technology* 12: 301-318.

Janić M., Vleugel J. 2012. Estimating Potential Reductions in Externalities from Rail-Road Substitution in Trans-European Freight Transport Corridors. *Transportation Research* Part D 17: 154-160.

Janić M. 2014a. *Advanced Transport Systems: Analysis, Modelling and Evaluation of Performances.* Springer-Verlag, London, UK.

Janić M. 2014b. Estimating the Long-Term Effects of Different Passenger Car Technologies on Energy/Fuel Consumption and Emissions of Greenhouse Gases in Europe. *Transportation Planning and Technology* 37: 409–429.

Janić M. 2016. A Multidimensional Examination of the Performances of HSR (High Speed Rail) Systems. *Journal of Modern Transportation*, Springer, DOI 10.1007/s40534-015-0094-y

JR. 2012. *Data Book 2011.* Central Japan Railway Company, Nagoya, Japan.

JR. 2014. *Visitors Guide 2014.* Central Japan Railway Company, Nagoya, Japan.

JR. 2015. *Annual Report 2015.* Central Japan Railway Company. Nagoya, Japan.

Kido M.L. 2005. Aesthetic Aspects of Railway Stations in Japan and Europe, as a Part of Context Sensitive Design for Railways. *Journal of the Eastern Asia Society for Transportation Studies* 6: 4381-4396.

Kovacevic S. 1988. *Eksploatacija Zeleznica I and II.* Zavod za Novinsko-Izdavacku i Propagandnu Delatnost, Belgrade (R Serbia), Yugoslavia.

NDTnet. 2000. ICE Train Accident in Eschede – Recent News Summary. *The e-Journal of Nondestructive Testing & Ultrasonics*: 5 (2), http://www.ndt.net/news/2000/eschedec.htm/

Nishiyama T. 2010. *High-Speed Rail Operations in Japan*. International Practicum on Implementing High Speed Rail in the United States. APTA, American Public Transport Association, New York, USA.

Ollivier G., Sondhi J., Zhou N. 2014. *High-Speed Railways in China: A Look at Construction Costs*. China Transport Topics No. 9. World Bank Office, Beijing. China: 1-8.

Pourreza S. 2011. *Economic Analysis of High Speed Rail*. NTNU (Norwegian University of Science and Technology), Trondheim, Norway.

Preston J. 2013. *The Economic of Investment in High Speed Rail: Summary and Conclusions*. Discussion Paper No. 2013-30. International Transport Forum (ITF/OECD), Paris, France.

Profillidis V. 2006. *Railway Management and Engineering*. Ashgate Publishing Limited, Aldershot, UK.

Profillidis A.V., Botzoris N.G. 2013. High Speed Railways: Present Situation and Future Prospects. *Journal of Transportation Technologies* 3: 30-36.

Puente F. 2014. Driver Error 'Only Cause' of Santiago Accident, Says Report. *IRJ – International Journal of Railways*. http://www.railjournal.com/index.php/high-speed/wenzhou-crash-report-blames-design-flaws-and-poor-management.html/ Falmouth, Cornwall, UK.

Qiao H. 2012. Wenzhou Crash Report Blames Design Flaws and Poor Management. *IRJ – International Journal of Railways*. http://www.railjournal.com/index.php/high-speed/wenzhou-crash-report-blames-design-flaws-and-poor-management.html/, Falmouth, Cornwall, UK.

Raghunathan S.R., Kim H.D., Setoguchic T. 2002. Aerodynamics of High-Speed Railway Train. *Progress in Aerospace Sciences* 38: 469-514.

Siemens. 2014. Velaro C.N. *High Speed Trains for China Railways*. Mobility Division, Siemens AG, Berlin, Germany.

Takagi K. 2011. Expansion of High Speed Rail Services: Development of High-Speed Railways in China. *Japan Railway & Transport Review* 57: 36-41.

Takai H. 2013. *40 Years Experiences of the Slab Track on Japanese High Speed Lines*. Presentation. Railway Technical Research Institute, Tokyo, Japan.

Tomii, N. 2010. *How the Punctuality of the Shinkansen has been Achieved*. WIT Transactions on the Built Environment 114: 111-120.

UIC. 2002. *Feasibility of 'Ballastless' Track. UIC Report*. Infrastructure Commission – Civil Engineering Support Group, International Union of Railways, Paris, France.

UIC. 2005. High Speed Rail's Leading Asset for Customers and Society. *UIC Publications*. International Union of Railways, Paris, France.

UIC. 2010a. *Necessities for Future High Speed Rolling Stock*. Report January 2010. International Union of Railways, Paris, France.

UIC. 2010b. *High Speed Rail: Fast Track to Sustainable Mobility*. International Union of Railways, Paris, France.

UIC. 2010c. High Speed, Energy Consumption and Emissions. *UIC Publications*. International Union of Railways, Paris, France.

UIC. 2011. High Speed Rail and Sustainability. *UIC Publications*. International Union of Railways, Paris, France.

UIC. 2014. *High Speed Lines in the World*. Updated 1st September 2014. UIC High Speed Department, International Union of Railways, Paris, France.

UNIFE. 2014a. *The ERTMS Memorandum of Understanding: A Cross-Sector Agreement to Ensure ERTMS' Success*. The European Railway Industry, Brussels, Belgium.

UNIFE. 2014b. *The ERTMS Levels: Different Levels to Match Customer's Needs*. The European Railway Industry, Brussels, Belgium.

UNIFE. 2014c. *The ERTMS from the Drivers' Point of View: How ERTMS Facilitates Train Operations for Drivers*. The European Railway Industry, Brussels, Belgium.

UNIFE. 2014d. *The ERTMS Deployment Statistics – Overview*. The European Railway Industry, Brussels, Belgium.

USDOT. 1999. *Assessment of Potential Aerodynamic Effects on Personnel and Equipment in Proximity to High-Speed Train Operations*. U.S. Department of Transportation, Federal Railroad Administration, DOT/FRA/ORD-99/11, DOT-VNTSC-FRA-98-3. Washington DC, USA.

USDOT. 2009. *Vision for High-Speed Rail in America*. U.S. Department of Transportation, Federal Railroad Administration, Washington DC, USA.

Wendell C., Vranich, J. 2008. *The California High Speed Rail Proposal: A Due Diligence Report*. Reason Foundation (with Howard Jarvis Taxpayers Association and Citizens Against Government Waste). Los Angeles, California, USA.

Wu J. 2013. *The Financial and Economic Assessment of China's High Speed Rail Investments: A Preliminary Analysis*. Discussion Paper No. 2013/28 prepared for the Roundtable on The Economics of Investment in High Speed Rail. 18-19 December 2013. New Delhi, India, International Transport Forum, Paris, France.

Yanase N. 2010. *High Speed Rolling Stock in Japan: International Practicum on Implementing High Speed Rail in the United States*. APTA (American Public Transportation Association), Washington DC, USA.

http://www.railfaneurope.net/tgv/wrecks.html
http://en.wikipedia.org/wiki/Eurostar
http://en.wikipedia.org/wiki/High-speed_rail_in_Germany
http://en.wikipedia.org/wiki/ICE_3
http://en.wikipedia.org/wiki/Siemens_Velaro
http://www.trainweb.org/tgvpages/tgvindex.html
http://en.wikipedia.org/wiki/New_Pendolino
http://en.wikipedia.org/wiki/List_of_high-speed_trains
http://demo.oxalis.be/unife/ertms/?page_id=42
http://en.wikipedia.org/wiki/European_Rail_Traffic_Management_System)
http://www.johomaps.com/eu/europehighspeed.html
https://en.wikipedia.org/wiki/High-speed_rail_in_China/

MODELLING TRANSPORT SYSTEMS—I
Operational, Economic, Environmental and Social Performances

3.1 Introduction

This chapter deals with modelling the operational, economic, environmental and social performances of select transport systems. This entails analyzing and modelling the utilization of capacity of the given airport runway system, the full internal (operational) and external (social and environmental) costs, i.e. externalities of the intermodal rail/road and road freight transport networks and the effects of substitution of road by equivalent intermodal rail/road freight transport services in the freight transport corridor(s). At the airport runway system, the operational performances represented by the 'ultimate'[1] capacity, its utilization and related quality of service due to the average aircraft delays are analyzed and modelled for the given conditions. These are specified by the current level of aircraft/flight demand and pattern of use of the runway system at a large hub airport.

At the road and intermodal rail/road freight transport networks, the economic, environmental and social performances that are influenced by operational performances, are modelled. The operational performances are represented by the characteristics of road trucks and intermodal trains and

---

[1] The 'ultimate' capacity means the maximum number of aircraft/flights accommodated on the given runway system during a given period of time (usually one hour) under conditions of constant demand for service (Janić 2001).

their operations in door-to-door delivering of goods/freight shipments under given conditions. The economic performances are represented by the operating costs of both the modes, while the environmental performances by the impacts and their costs, i.e. externalities, such as energy/fuel consumption and related emissions of GHG (Green House Gases). Finally, the social performances include noise, congestion and safety, i.e. traffic incidents/accidents and their corresponding costs or the externalities. The full costs, including the operational costs and externalities of both road and intermodal rail/road freight transport networks, are compared in order to identify the break-even distance reflecting the preference of each mode. Specifically, the prospective savings in the environmental and social externalities by substituting road with intermodal rail/road freight transport services in the given freight transport corridor are analyzed and modelled through the corresponding performances of both the modes—primarily operational, environmental and social performances, including the related costs of the latter two under given conditions. Therefore, in addition to this introductory section, this chapter consists of three other sections. Section 3.2 deals with analyzing and modelling operational performances, i.e. utilization of capacity of the runway system at a large airport. Section 3.3 discusses modelling the full costs of the intermodal (rail/road) freight transport network and its road counterpart. Section 3.4 concentrates on modelling the environmental and social effects/ impacts from substitution of road by intermodal (rail/road) freight transport services in the given freight transport corridor. The chapter ends with a section devoted to concluding remarks.

3.2 Runway System Capacity at a Large Airport

3.2.1 Background

At present, many large European and US airports operate under different operational, economic and environmental constraints (FAA 2002). Depending on the level of dominance, these constraints determine the operational, economic and/or environmental capacity at these airports. Under such conditions, ATFM (Air Traffic Flow Management) measures are expected to match one (or all) these capacities to the expected demand at the tactical, operational and strategic level. This implies that, on the one hand, the ATFM needs to provide reasonable utilization, i.e. preventing underutilization of the available airport airside (primarily runway system) capacity. On the other, it aims at preventing or mitigating over-burdening of this capacity to prevent aircraft/flight congestion and delays. Tactical measures include slot allocation, GH (Ground Holding) and charging congestion. Operational measures include aircraft sequencing during landings and take-offs and rerouting of potentially affected aircraft flight. Finally, strategic measures include physical increase in the airside capacity by building new airport airside infrastructure—usually new runway(s) and apron/gate complex (Janić, 2001, 2012). The most recently built fourth runway at Frankfurt main airport and the long and increasing pressure for building a third (parallel) runway at London Heathrow (LHR) airport are typical examples (details in Chapter 7). In addition, before building new runway(s) in the medium- to long-term future, particularly the landing capacity of existing runway system(s) could be increased by adopting innovative technologies and operational procedures as per the scope of the European SESAR (Single European Sky ATM Research) and US NextGen (Next Generation) programs (Janić 2012) (details in Chapter 4).

However, when a new runway is built, the question of the level of utilization of its capacity is kept in mind. This section describes the analysis and modelling of utilization of the runway system capacity at a large European hub, the Amsterdam Schiphol (AMS) airport (The Netherlands), where a new runway has been built recently. In particular, utilization of this newest runway is under focus. In addition, the aircraft/flight delays are estimated on the basis of the present level of demand and the cases of operations of the runway system (Janić 2014).

3.2.2 The System and Problem—Capacity Utilization at a Large Airport

3.2.2.1 General

Each airport consists of a landside and an airside area. The landside area embraces the surface transport access systems connecting the airport to its catchment area and the passenger (and freight) terminal system. The airport airside area comprises the airspace around the airport, called the 'airport zone' or 'terminal airspace', and the ground infrastructure including runways, taxiways and apron/gate complex (Ashford et al. 1997, Janić 2001). In this section, the Amsterdam Schiphol airport is taken into consideration. The airport operates as the secondary hub of Air France-KLM airline and the related SkyTeam alliance (in addition to the primary hub, the Paris Charles de Gaulle airport). The Air France-KLM hub-and-spoke network hosted by this airport consists of several clusters ('waves') of inbound and outbound flights scheduled during the day, which create corresponding demand peaks in the airside area, i.e. the runway system and apron/gate complex and the landside area, i.e. passenger terminal and airport ground access systems.

3.2.2.2 Runway System Configuration

The runway system of Schiphol airport consists of six runways, of which five (long) ones can accommodate all types of commercial aircraft including the largest Airbus A380. The sixth (shortest) runway mainly handles general aviation aircraft. Figure 3.1 shows the simplified scheme of the airport airside area.

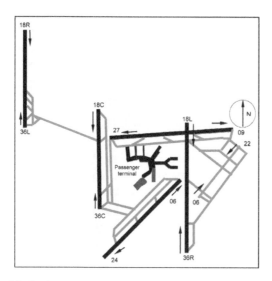

Fig. 3.1: Simplified scheme of the runway system at Amsterdam Schiphol (AMS) airport (ASA 2012, Janić 2014, http://www.schiphol.nl/index_shg.html).

As can be seen, three of the five long runways run parallel with a north-south orientation: Polder RWY18R/36L, ZwanenburgRWY18C/36C and AalsmeerRWY18L/36R. The Polder runway was opened recently in 2003. The main geometrical characteristics of the above-mentioned runways are given in Table 3.1 (ASA 2012).

Table 3.1: Some geometrical characteristics of the runway system at Schiphol airport (ASA 2012)

Runway	Orientation	Dimension (Length/Width) (m)
Polder	RWY18R/36L	3800/60
Aalsmeer	RWY18L/36R	3400/45
Zwanenburg	RWY18C/36C	3300/45
Buitenveldert	09-27	3453/45
Kaag	06-24	3500/45
Schiphol-Oost	04-22	2014/45

The above-mentioned parallel runways are separated by 2.1 kms (7000 ft) and 2.8 kms (9200 ft), respectively, thus enabling their independent operations for both landings and take-offs (ASA 2012, De Neufville and Odoni 2003, http://www.schiphol.nl/index_shg.html).

3.2.2.3 Runway System Operation

In general, the runway system at Schiphol airport operates under the influence of the prevailing weather conditions, the pattern of arrival and departure demand requesting service during the peak and off-peak periods of time and the environmental-social constraints.

(a) Prevailing Weather Conditions and Air Traffic Demand

The prevailing weather conditions at Schiphol airport influencing operation of the runway system include wind (direction, speed) and visibility. North and south winds prevail. Precipitation in the form of rain or snow generally affects visibility, leading to temporary closure of some runways, thus impacting the performances on the contaminated runways, for example, heavy rain or snow can call for temporary closure of the affected runway(s). In particular, visibility is categorized as 'good', 'marginal' or 'low'. Consequently, respecting visibility and the characteristics of inbound and outbound air traffic demands, the above-mentioned runway system of 5+1 runways can theoretically operate as follows: (a) under good to marginal visibility conditions: (i) in 13 and 18 different combinations while serving the inbound and outbound peak demand, respectively; (ii) in seven different configurations while serving the inbound and outbound off-peak demand, respectively; (b) under low visibility conditions: (i) in 13 and four different combinations while serving the inbound and outbound peak demand, respectively; and in 11 and four different combinations while serving the inbound and outbound off-peak demand, respectively. Such a number of combinations of operating the six-runway system has become possible after opening the sixth Polder runway (18R/36L) in 2003 (ASA 2012).

The air traffic demand at Schiphol airport in terms of the annual number of operations (landings and take-offs) has generally fluctuated during the past ten years as shown in Fig. 3.2.

As is seen, after a decrease in the year 2003, the highest annual values have been recorded during the 2006-8 period, followed by a sharp decrease during the 2009-10 period. The values then recovered in the years 2011-12. In addition, the average number of landings and take-offs per hour of the day also fluctuates as shown in Fig. 3.3.

As can be seen, the hourly landing and take-off demand obtained from the airport/ATC (Air Traffic Control) statistics, available only in a rather aggregate form, has been mainly scheduled into five inbound and outbound clusters ('waves') of flights constituting the corresponding peaks. These clusters are typically spread over one to two hours with the longest ones in the

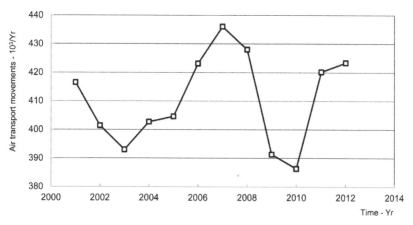

Fig. 3.2: Development of the annual number of operations at Schiphol airport (Janić 2014, Van den Hoven 2012, ASA 2012).

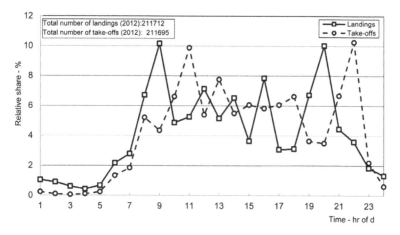

Fig. 3.3: An example of distribution of the average number of operations per hour of the day at Schiphol airport (ASA 2012, Janić 2014, Van den Hoven 2012).

morning and in the evening, respectively. In addition, the landing peaks have taken place earlier than their take-off counterparts by about one hour. The first (morning) and the last (evening) peak for both landing and take-off demand have been the highest. In general, both landing and take-off peaks have shared about 6-10 per cent of the total daily traffic in the corresponding hours.

In each of the above-mentioned runway system configurations serving the above-mentioned peak demand, the following is assumed: before opening the Polder runway, two runways were used, each exclusively for one type of operations; after opening the Polder runway, two runways were exclusively used for one type and one runway for the other type of operations. In the configurations serving the off-peak demand, one runway is exclusively used for each type of operations similarly as before opening Polder (EC 2001). In the given cases, the BGAS (Baangebruiks Advies System) has been used by the airport's approach and tower ATC (Air Traffic Control) as an advisory tool to determine the best runway configuration, respecting the traffic peak and off-peak demand, prevailing weather conditions, and availability of runways. The experience so far indicates that the airport's runway system is frequently operated in two modes (1 and 2), each with two inbound and outbound configurations. Specifically, mode 1 is applied under prevailing north- and mode 2 under prevailing south-wind conditions. These three-runway and the two-runway configurations (without Polder), both operating in the above-mentioned modes, are given in Table 3.2.

Table 3.2: Typical operating modes and configurations of the runway system at Schiphol airport (ASA 2012, BAS 2012, EC 2001, Janić 2014, Van den Hoven 2012)

Mode/Wind direction	Number of runways	Configuration	Type of peak demand	Type of operations/ runway(s) in use
1/North	2	'1+1'	Inbound	Landings: 06 or 36R Take-offs: 36C
2/South	2	'1+1'	Outbound	Landings: 18C Take-offs: 24 or 18L
1/North	3	'2+1'	Inbound	Landings: 06+36R Take-offs: 36L or 36C
	3	'1+2'	Outbound	Landings: 06 or 36R Take-offs: 36L+36C
2/South	3	'2+1'	Inbound	Landings: 18R+18C Take-offs: 24 or 18L
	3	'1+2'	Outbound	Landings: 18R or 18C Take-offs: 24+18L

However, strong westerly or easterly crosswinds with speeds greater than about 20 kts (knots), depending on the aircraft size, can prevent use of the above-mentioned operating modes and configurations, thus requiring use of the other rather numerous possible runway system configurations.

(b) Environmental and Social Constraints

Environmental-social constraints, such as noise, can also influence the selection of the best runway system operational mode and configuration. Independent of the weather conditions, the operational mode and configuration of the runway system is greatly influenced by the noise constraints aimed at preventing affection of the nearby densely inhabited area. This implies that just because of these constraints, particular runways are not used in both directions for landings and take-offs during the day. For example, the Polder runway accommodates only take-offs to and landings from the north (i.e. 18R and 36L, respectively) because of the noise burden on the city of Hoofdorp. The Aalsmeer runway is used only for landings from and take-offs to the south (i.e. 18L and 36R) in order to mitigate the noise burden on cities like Badhoevedorp and Amsterdam. Consequently, the runway system operating mode and configuration are chosen as a result of a compromise between the required efficiency and effectiveness for serving demand and the need for rather uniformly distributing the total noise burden over the potentially affected (urban) areas. In addition, apart from the Polder runway (RWY18R/36L) from the north, the Zwanenburg runway (RWY18C/36C) from the south and the Kaag runway (RWY06/24), operation of all other runways is restricted during the night hours, i.e. from 11 p.m. to 6 a.m. (ASA 2012). This indicates that the new Polder runway is not free from operational constraints.

3.2.3 Modelling Utilization of the Runway System Capacity

3.2.3.1 Some Related Research

The related research explicitly dealing with utilization of the airport airside capacity consisting of runway, taxiway and apron/gate complex system has been relatively scarce. This capacity utilization is considered implicitly in the scope for analyzing and modelling of the airport airside—runway, taxiway and apron/gate system—capacity and related aircraft/flight delays of the already congested airports (Janić 2009a). The light- to moderate-congested airports are not under focus simply due to low to moderate demand and consequently the non-existence of the capacity problem. The same applies to airports that have recently obtained new airside infrastructure, particularly additional runway(s), aimed at mitigating substantial airside congestion and related aircraft/flight delays. The main reason behind such an approach was 'not dealing with things and issues considered as no problem in the short-term' (De Neufville and Odoni 2003, Janić 2009a, b). This paper aims at changing the above-mentioned practice by emphasizing the need for analyzing and modelling the airport airside (runway system) capacity, its utilization and related aircraft/flight delays, permanently, and independently of their current severity. This could also be useful in heavily congested airports struggling since long for expansion by building new runway(s), such as, for example, London's Heathrow airport.

3.2.3.2 Objectives and Assumptions

The main objectives are as follows:

- Investigating utilization of the runway system capacity at a given (large) airport by developing a sufficiently generous methodology, which could, with necessary modifications, be applied to other airports;
- Investigating utilization of the most recently built new runway under given conditions: prevailing weather, the time pattern of daily demand and the environmental and social constraints; and
- Estimating the average delay per atm (air transport movement)—landing or take-off under given conditions.

For such a purpose, the existing analytical models for estimating the ultimate runway system capacity and related aircraft/flight delays are appropriately modified based on the following assumptions:

- The demand for service on the given runway system configuration is constant and uniformly distributed during specified periods of time in the day (quarter of an hour, hour, etc.). This is common at airports where the dominant airline/alliances carry out hub-and-spoke operations consisting of waves of incoming and outgoing flights almost uniformly spread during the duration of the waves;
- The main factors influencing the capacity are considered as parameters rather than as variables;
- Each runway in the given mode and configuration operates independently of the others, i.e. in the segregated mode, serving exclusively one type of operations—landings or take-offs;
- The available runway system capacity in the given mode and configuration is matched with the expected demand in terms of the time and intensity/volume. Consequently, such an allocation prevents uncontrolled escalation in landing and take-off delays;
- The average delay of a landing or take-off aircraft/flight is considered exclusively as an outcome of the relationship between the corresponding demand and capacity of the runway system. This implies that the other causes influencing arrival and departure delays in the terminal airspace around the airport and on the ground (apron/gate complex and the network of taxiways) are not taken into account; and
- The information/data on the daily pattern of demand reflects a significant proportion of the total demand and as such, can be considered relevant for the purpose.

3.2.3.3 Structure of the Models

(a) General

The proposed methodology for investigating utilization of the given runway system capacity includes the model of 'ultimate' arrival (landing) and departure (take-off) capacity of a single runway, the model of 'ultimate' capacity of the runway system in use and the model of corresponding average aircraft/flight delays.

In general, the 'ultimate' capacity of a single or a system of a few simultaneously operating runways is usually expressed by the maximum number of air traffic movements (landings and/or take-offs) accommodated during a given period of time (usually one hour) under conditions of constant demand for service. In general, given the runway system configuration in use, this capacity is mainly influenced by the characteristics of the aircraft fleet and ATC (Air Traffic Control) separation rules. The former implies that landing and take-off speeds depend on the aircraft maximum take-off weights, their relative proportions in the landing/take-off fleet mix and the runway landing/take-off occupancy time(s). The aircraft weights determine the aircraft wake-vortex categories. The latter are specified for landing and take-off sequences of particular aircraft wake-vortex categories as the ICAO (International Civil Aviation Organization), IFR (Instrument Flight Rules) and the US FAA (Federal Aviation Administration) VFR (Visual Flight Rules). In general, they can be applied according to the weather, for instance, IFR is 'more conservative' during marginal to low and VFR during good visibility conditions (BAS 2012).

(b) Capacity of the Runway System

The model for estimating capacity of a single runway is given as follows:
'Ultimate' landing capacity of a single runway: The ultimate landing capacity of a single runway is calculated as follows (De Neufville and Odoni 2003, Janić 2009a):

$$\mu_a = T / \overline{t_a} \tag{3.1a}$$

where T is the time period for which the landing capacity is calculated (usually one hour) and $\overline{t_a}$ is the minimum average inter-event time on the runway threshold for all combinations of the aircraft landing sequence(s).

The time $\overline{t_a}$ in Equation 3.1a can be estimated as follows:

$$\overline{t_a} = \sum_{ij} p_{ia} t_{ij} p_j \tag{3.1b}$$

where

$_a t_{ij}$ is the inter-arrival time between leading aircraft (i) and trailing aircraft (j) in the landing sequence (ij) (s); and

p_i, p_j is proportion of the aircraft types (i) and (j) in the landing traffic mix, respectively.

The time $_a t_{ij}$ in Equation 3.1b can be estimated as follows:

$$
_d t_{ij} = \begin{bmatrix} \max(t_{ai}; \dfrac{\delta_{ij}}{v_j}); v_i \leq v_j \\[3mm] \dfrac{\delta_{ij}}{v_j} + \gamma(\dfrac{1}{v_j} - \dfrac{1}{v_i}); v_i > v_j \end{bmatrix} \tag{3.1c}
$$

where

t_{ai} is the runway occupancy time by the landing aircraft (i) (s);

δ_{ij} is the ATC minimum wake-vortex distance-based separation rules (nm – nautical mile);

γ is the length of the common final approach path (nm); and

v_i, v_j is the speed of aircraft (i) and (j), respectively, along the path γ (kt – knot).

'Ultimate' take-off capacity for a single runway: The ultimate take-off capacity of a single runway can be calculated as follows (De Neufville and Odoni 2003, Janić, 2009a):

$$
\mu_d = T / \bar{t}_d \tag{3.2a}
$$

where

T is the time period for which the take-off capacity is calculated (usually one hour); and

$\bar{t}_d$ is the average minimum inter-event time at the runway threshold for all combinations of the aircraft take-off sequence(s).

The time $\bar{t}_d$ in Equation 3.2a can be estimated as follows:

$$
\bar{t}_d = \sum_{ij} p_{id} \, _d t_{ij} \, p_j \tag{3.2b}
$$

where

$_d t_{ij}$ is the inter-event time between the leading aircraft (i) and the trailing aircraft (j) in the take-off sequence (ij) at the required location after take-off (s); and

p_i, p_j is the proportion of the aircraft types (i) and (j) in the take-off traffic mix, respectively.

The time $_d t_{ij}$ in Equation 3.2b can be estimated as follows:

$$
_d t_{ij} = \max \left[_d t_{ij\min}; _d t_{ij0} - (t_{jd} - t_{id}) - \gamma_d (\dfrac{1}{v_{jd}} - \dfrac{1}{v_{id}}) \right] \tag{3.2c}
$$

where

$_d t_{ij\min}$ is the ATC minimum time-based separation rule between the leading aircraft (i) and the trailing aircraft (j) in the take-off sequence (ij) at the runway take-off threshold(s);

$_dt_{ij0}$ is the required ATC separation between the take-off leading aircraft (i) and trailing aircraft (j) at the end of the common departure path γ_d in the terminal airspace (s);

t_{id}, t_{jd} is the runway occupancy time of the departing leading aircraft (i) and trailing aircraft (j), respectively;

γ_d is the length of the common departure path (nm); and

v_{id}, v_{jd} is the average take-off speed of the aircraft leading aircraft (i) and trailing aircraft (j), respectively (kt).

Based on Equations 3.1 and 3.2, let the runway system at the given airport consist of (N) runways each operating independently of the other, i.e. in the segregated mode, accommodating only one type of atm—landings or take-offs. If, during the given period of time (for example an hour), n of these runways are used for landings and m for take-offs ($N = n+m$), then from Equations 3.1 and 3.2, the total capacity of runway system can be estimated as follows:

$$\mu = n\mu_a + m\mu_d \qquad (3.3)$$

where

μ_a is the 'ultimate' landing capacity of a single runway (atm/hr); and

μ_d is the ultimate take-off capacity of a single runway (atm/hr).

(c) Average Delay

Under conditions of the demand not exceeding the capacity of the runway system and/or of particular runways during the specified period of time, the average delay per an operation-landing or take-off can be estimated as follows (Janić 2009a):

$$W = \frac{\lambda(\sigma^2 + 1/\mu^2)}{2(1 - \lambda/\mu)} \qquad (3.4)$$

where

λ is the average intensity of demand, i.e. the arrival or departure rate (atm/hr);

μ is the average aircraft/flight service rate, i.e. capacity, as the reciprocal of the mean service time for arrivals ($\mu_a = 1/\bar{t}_a$, where $\bar{t}_a$ is the minimum average service time per arrival [Equation 3.1b)] or departures ($\mu_d = 1/\bar{t}_d$, where $\bar{t}_d$ is the minimum average service time per departure [Equation 3.2b)] (atm/hr); and

σ is the standard deviation of service time of an arrival or of a departure(s).

The ratio: $U = \lambda/\mu$ in Equation 3.4 is supposed to generally be less or at most equal to one most of the time. As such, it is commonly considered as the rate of utilization of the 'ultimate' capacity of a single runway (μ_a, μ_d)

or of the runway system (μ). Otherwise, the ratio: $U = \lambda/\mu > 1$ indicates the oversaturation of a single runway system operating at the full utilization of their 'ultimate' capacities (De Neufville and Odoni 2003). Thus, an alternative applicable to the conditions when the demand exceeds the ultimate capacity for a relatively long period of time, i.e. several hours during the day called congestion period, is the deterministic queuing model based on diffusion approximation (Newell 1982). According to this model, the queue of either landing or taking-off aircraft/flights at a certain time (t) of the over-saturated period (τ) can be approximated as follows:

$$Q(t) = A(t) - D(t) \text{ for } t \in \tau \qquad (3.5a)$$

where
$A(t)$ is the cumulative count of demand by time (t) (atm); and
$D(t)$ is the cumulative count of served demand by time (t) (atm).

The average delay per an aircraft/flight waiting in the queue during the congested period (τ) can be approximated based on Equation 3.5a as follows (Newell 1982):

$$\overline{d(\tau)} = [1/A(\tau)] * \int_0^\tau Q(t)dt = [1/A(\tau)] * \int_0^\tau [A(t) - D(t)]dt \qquad (3.5b)$$

where
$A(\tau)$ is the demand of aircraft/flights during the congestion period (τ) (atm).

3.2.4 Application of the Models

3.2.4.1 Inputs

(a) Estimating the Capacity of a Runway System

Approach and landing procedures: An aircraft's final approach and landing on any of the five Schiphol's runways is carried out along STARs (Standard Terminal Arrival Route(s)) defining the aircraft's three-dimensional path from the terminal airspace entry gate to the final approach gate from where the final approach and landing starts towards the assigned runway. The approximate length of STARs in the given case is about 25-30 nm (nautical miles) (1 nm = 1.852 km). The average aircraft speed and altitude at the terminal entry gate is usually about 220-250 kt and 7000-8000 ft, respectively; the speed decreases to about 160 kts and the altitude to 2000 ft until the final approach gate located at an approximate distance of about 6 nm from the runway landing threshold. At that moment, the aircraft intercept the LZZ (Localizer) of ILS (Instrument Landing System). From there, they further adjust the final approach and landing speeds, varying between 100/110 and 140/150 kts and depending on their maximum take-off weight and consequently the wake-vortex categories. During the final approach and landing, the aircraft are separated by the application of the ATC distance-based separation rules, depending on the type of landing sequence, either at the final approach gate or at the landing threshold.

Take-off procedures: After taking-off from either runway, the departing aircraft climb along the three-dimensional SIDs (Standard Instrument Departure) routes towards the terminal exit gates in order to reach the low cruising altitude. The SIDs of different aircraft departing from the same runway can immediately diverge. However, in some cases, they have a common segment requiring maintaining the minimum time separation between successive departing aircraft up to the point of their divergence. The taking-off aircraft are separated by the ATC time-based separation rules applied at the runway take-off threshold (EC 2001).

Landing capacity: The average fleet mix at Schiphol airport consists of small (Fokker 70/100), large (predominantly A318/319/320 and B737) and heavy (B747/767/777 and A310/330/340) aircraft. Their proportion, the average final approach speed, the runway landing occupancy time and the length of final approach path are given in Table 3.3.

Table 3.3: Characteristics of the aircraft landing fleet in the given example (ASA 2012)

A/C(i)	p_i	v_i (kt)	t_{ai} (s)	γ (nm)
Small	0.2	110	55	6
Large	0.6	130	60	6
Heavy	0.2	150	70	6

kt – knot; s – second; nm – nautical mile

In addition, it is assumed that independent of the weather conditions and the runway system configuration in use, the ATC applies exclusively the IFR separation rules given in Table 3.4.

Table 3.4: The ATC separation rules for landing aircraft in the given example (nm) (EC 2001)

(i/j)	Small	Large	Heavy
Small	3	3	3
Large	4	3	3
Heavy	6	5	4

nm – nautical mile

In addition, the time 'buffer' of five seconds is added to each landing sequence to include a rather conservative approach by the ATC controllers while dealing with the inbound peak demand. Consequently, the landing capacity of a single runway operating independently of the others is estimated from Equations 3.1-3.3 as: $\mu_a = 31$ atm/hr. This means that with two runways serving arrivals simultaneously and independently, the landing capacity is doubled, i.e. $\mu_a = 62$ atm/hr.

Take-off capacity: The take-off capacity is estimated like the landing capacity, respecting the characteristics of the take-off aircraft fleet mix in Table 3.5 and the ATC minimum separation rules given in Table 3.6.

Table 3.5: Characteristics of the aircraft taking-off fleet in the given example (ASA 2012, EC 2001)

A/C(i)	$_dp_i$	v_i (kt)	t_{di} (s)	γ_d (nm)
Small	0.2	150	55	3
Large	0.6	170	60	3
Heavy	0.2	190	70	3

kt – knot; s – second; nm – nautical mile

Table 3.6: The ATC separation rules for taking-off aircraft in the given example (s) (ASA 2012, EC 2001)

(i/j)	Small	Large	Heavy
Small	60	60	60
Large	90	60	60
Heavy	120	120	90

s – second

A time 'buffer' of five seconds is again added to each aircraft in the take-off sequence. Then, by Equations 3.1-3.3, the take-off capacity of a single runway operating in the segregated mode is estimated to be: $\mu_d = 41$ atm/hr. The take-off capacity is doubled to $\mu_d = 82$ atm/hr for a pair of the independent runways operating in the segregated mode. Consequently, the capacity envelope of the runway system operating in the configuration '1+1' while serving either the inbound or outbound demand peaks ('2+1' while serving the inbound, and '1+2' while serving the outbound demand peaks) is shown in Fig. 3.4.

The '1+1' configuration implies operation of the runway system at the airport without the new Polder runway (as it was not built), when one of the two existing runways would be used exclusively for landings and the other exclusively for take-offs. The '2+1' configuration implies that two of the three

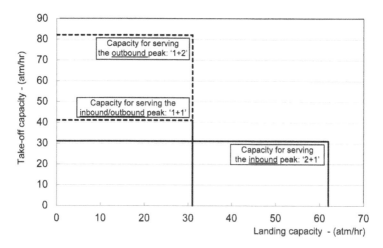

Fig. 3.4: Capacity envelope for the runway system of two and three runways operating in the '1+1', '2+1' and '1+2' configuration, respectively, at Schiphol airport (Janić 2014).

runways are used exclusively for landing and one exclusively for take-off. The '1+2' configuration implies use of one of the three runways exclusively for landing and the other two exclusively for take-offs (Table 3.2).

Total capacity: Based on the number of runways in operation (in this case two and three) and the landing/take-off capacity of each of them, the total capacity of the runway system is estimated by Equation 3.3 as follows:

Configuration '1+1': $\mu = \mu_a + \mu_d = 31 + 41 = 72$ atm/hr;

Configuration '2+1': $\mu = 2\mu_a + \mu_d = 2 \times 31 + 1 \times 41 = 103$ atm/hr; and

Configuration '1+2': $\mu = \mu_a + 2\mu_d = 31 + 2 \times 41 = 113$ atm/hr.

(b) Deriving Demand

The data on the aircraft/flight demand and pattern of its serving using the runway system operating in the above-mentioned modes and configurations at Schiphol airport has been collected primarily from the Casper Aircraft Tracking System (KNMI 2012).

3.2.4.2 Results

(a) Runway System

Demand-capacity relationship: The above-mentioned data on the aircraft/flight demand collected for each hour of a typical day of July 2011 were interrelated with the corresponding estimated capacity of the runway system operating in the specified configurations as shown in Fig. 3.5 (a, b).

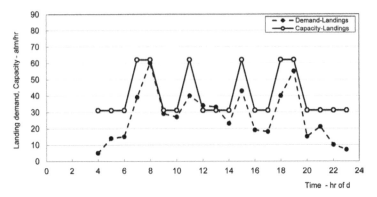

(a) Inbound demand – runway system configuration: '2+1'- segregated mode

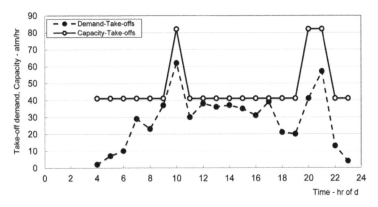

(b) Outbound demand - runway configuration: '1+ 2' – segregated mode

Fig. 3.5: Demand/capacity relationships at Schiphol airport during an average day of July 2011 (Janić 2014).

Figure 3.5a shows that the inbound demand at the airport is scheduled in five peaks. The first (morning) and one before the last (late afternoon/evening) peak demand are the highest. During each period, the peak demand is always lower than the runway system capacity, even during the periods of the highest intensity of demand. This is achieved by applying the runway system configuration '2+1', and consequently maintaining the average inbound aircraft/flight delays under the reasonable (acceptable) limits. Figure 3.5b shows that the outbound demand at the airport is also scheduled into five but not all clearly recognizable peaks (*see* Fig. 3.2). Again, the first (late morning) and the last (evening) peak demand are the highest. The runway system operating in the '1+2' configuration always ensures that the capacity remains greater than the intensity of demand, thus maintaining the average outbound aircraft/flight delays within acceptable limits. It should be mentioned that the capacity of the runway system operating in '1+1' configuration (i.e. when Polder runway was not opened) would be 31 landings and 41 take-offs per hour during the entire day (i.e. a total of 72 atm/hr). Figure 3.5 (a, b) also show that the current, both arrival and departure peak, demand would be higher than the corresponding capacities during a substantial period of the day, thus suggesting that congestion and the aircraft/flight delays could escalate much above the acceptable limits of 15 min.

Average delays: The relationship between the average delay of an inbound and an outbound aircraft/flight and utilization of the runway system capacity during each hour of a typical day of July 2011 is shown in Fig. 3.6. The utilization of the runway system capacity as the demand/capacity ratio has been extracted from Fig. 3.5 (a, b). In addition, the standard deviation of the landing and take-off service time is adopted to be $\sigma \equiv \sigma_a = \sigma_d = 1.5$ min. Then, the average delay per an inbound and outbound aircraft/flight is estimated by Equations 3.4 and 3.5 (b).

As can be seen, in general, as intuitively expected under such conditions, the average delay of an inbound and an outbound aircraft/flight increases more than proportionally with increasing utilization of the capacity of the runway system operating in the given configuration ('2+1' and '1+2'). This confirms the well-known relationship from under saturated queuing systems when mainly stochastic delays of customers requesting service prevail. In addition, the average delay of an inbound aircraft/flight is a bit higher than that of an outbound aircraft/flight, thus indicating a higher utilization of the runway system capacity in the corresponding pattern of operation. Furthermore, when utilization of the runway system approaches about 90 per cent, the average delays increased to about 14-15 min., which is still within the limits of not counting them as real-actual delays (the latter are greater than 15 min.). However, as Fig. 3.6 shows, under the given level of demand, if the runway system operated in the ('1+1') configuration (i.e. if Polder runway was not built), the average delay of both landing and taking-off aircraft/flights would almost explode with an average of 15-45 min. for landing and 15-35 min. for taking-off aircraft/flight. In addition, the regression equations suggest that the take-off average delay, despite being lower in the given example, would increase at a higher rate than its landing counterpart.

(b) The most recently Opened Polder Runway—RWY18R/39L

The above-mentioned figures indicate that the Polder runway opened in the year 2003 prevented uncontrolled escalation of congestion and delays due to the growth in demand (i.e. during the 2003-2011 period). As such it has been preferable and one of the 'basic' runways within the above-mentioned runway system configurations '2+1' and '1+2' (*see* Fig. 3.1). This means that it was preferred under the given conditions to be used during the whole day(s).

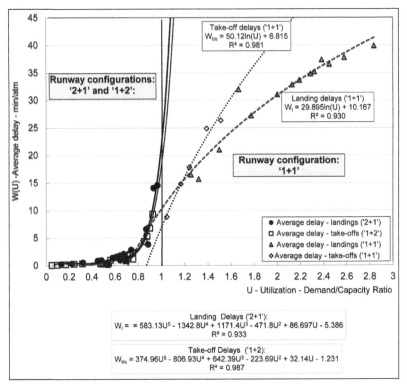

Fig. 3.6: Relationship between the average aircraft/flight delays and utilization of the runway system capacity at Schiphol airport (Janić 2014).

The most important factors for using this runway were the weather conditions (prevailing north/south wind) and endeavors to mitigate the noise menace by spreading it over a much wider area around the airport. Hence some benefits have accrued to Schiphol airport in terms of fewer noise complaints. However, the main disadvantages have been relatively long taxiing-in and taxiing-out periods [and corresponding increase in fuel consumption and emission of GHG (Green House Gases)], calling for additional ATC controllers at the secondary airport Control Tower.

Demand-capacity relationship: The demand-capacity relationship for the Polderban RWY18R/36L runway at Schiphol airport as shown in Fig. 3.1 illustrates the relevant data on the demand during one of the busiest days (21 July 2011). On the given day, this runway was extensively used in both configurations ('2+1' and '1+2') mainly for accommodating the outbound flights, i.e. departures. Therefore, only the capacity of the runway for departures is estimated in Equation 3.2 (a, b, c). Both demand and capacity per hour of the given day are shown in Fig. 3.7.

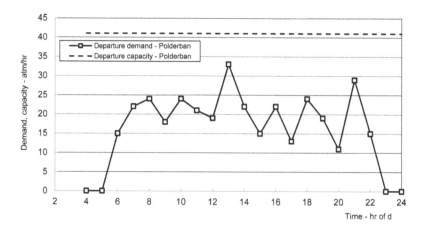

Fig. 3.7: Demand/capacity relationships for Polder RWY18R/36L
at Schiphol airport on 21 July 2011 (Janić 2014).

As is seen, the total demand at the airport varies during the day, reflecting the integrated inbound and outbound peaks. The demand for using the RWY18R/36L runway varies similarly. The total number of operations (atm) at the airport is 914 out of which nearly a half represent the outbound traffic. Out of this, about 70 per cent are accommodated at Polderban. Thus the runway accommodated about 30 per cent of the total airport traffic during the given day. Some additional figures for the 18-24 July 2011 period indicated that the runway accommodated about 30-35 per cent of the total airport traffic and primarily on its outbound flights.

Average delays: The average delay for an outbound aircraft/flight accommodated at the Polderban operating within either configuration of the runway system ('2+1' or '1+2') on 21 July 2011 is estimated by Equation 3.4. The runway departure capacity during the entire day is assumed to be μ = 41 atm/hr, and the standard deviation of the departure-service time is σ = 1.5 min. The demand during each hour of the day is extracted from Fig. 3.7. The results in terms of the relationship between the average departure delay and utilization of the runway departure capacity (i.e. demand/capacity ratio) is shown in Fig. 3.8.

As can be seen, utilization of the take-off capacity on the given runway varies between 30 per cent and 80 per cent, which in turn resulted in the average stochastic delay of an outbound aircraft/flight from about $W = 0.6$ to about $W = 6$ min. This indicates that, under given conditions, the runway was utilized at a low to moderate level, despite accommodating about 70 per cent of the total airport's outbound traffic during the given day. This indicates that there will be sufficient space for accommodating the future growing traffic demand with reasonable (less than 15 min.) average delays.

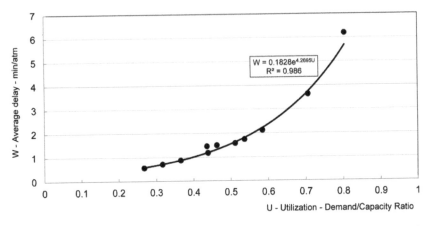

Fig. 3.8: Relationship between the average delay of an outbound aircraft/flight and utilization of the take-off capacity of Polder RWY 18R/36L at Schiphol airport on 21 July 2011 (Janić 2014).

3.2.5 Interim Summary

This section analyzed and modelled the utilization of the runway system capacity of the given large airport as an indicator of its operational performance. For such a purpose, convenient analytical models for estimating the landing and take-off capacity of a single runway and the system of a few simultaneously operating runways in the given mode and configuration, and related aircraft/flight delays were developed. They were applied to the large European hub—Amsterdam Schiphol airport.

The results show that three (of six) runways operating in the segregated mode provide sufficient landing and take-off capacity for accommodating five daily inbound and outbound peaks of corresponding demand with acceptable delays (up to 15 min. at an average capacity utilization rate of about 90-95 per cent). The newly opened Polder runway as the most frequently used runway system operating modes and configurations accommodates about 30-35 per cent of the total mostly outbound (take-off) airport demand. Under such conditions, its utilization sometimes reaches 80-85 per cent, thus generating average aircraft/flight delays of about 6 min., which is much less than the 15 min. considered as an acceptable limit for counting delays. Consequently, it is noticed that Schiphol airport possesses sufficient capacity on the existing runway system to accommodate future growing demand without any substantial increase in the average aircraft flight/delays. This will be the case even if stronger noise (national and international EU or European Union) constraints and consequent use of particular runways are imposed in the near future. As a result, the considered operational performance of the airport appears to be satisfactory at present and will be so in the future. As mentioned above, development in the runway system capacity in the long-term in order to cope with growing demand are elaborated in more detail in Chapter 7 of this book (the case of a new runway for the London airport system).

3.3 Intermodal Rail/Road and Road Freight Transport Networks

3.3.1 Background

Intermodal freight transport provides transportation of consolidated load units, such as containers, swap-bodies and semi-trailers by combining at least two different transport modes (EC 2002a). The vehicles meet at the intermodal terminals and exchange load units according to a given procedure using the transshipment facilities and equipment. In Europe, intermodal freight transport is often perceived as a potentially strong competitor and an environmentally friendlier alternative to road freight transport particularly in the medium- to long-distance corridors-markets. However, the developments to date have not confirmed such expectations. For example, during the 1990-1999 period, the European intermodal freight transport grew steadily from an annual volume of about 119 to about 250 billion t-km[2] and consequently increased its market share in the total freight transport[3] volume from about 5-9 per cent. This happened mainly after enhancement of operations on the Trans-European corridors-markets of length of about 900-1000 kms, which contained about 10 per cent of the goods volumes (tons). During the same period, in the markets-corridors of length of about 200-600 kms containing about 90 per cent of the total goods volumes, the share of intermodal transport was only about 2 per cent and 2-3 per cent in terms of the volumes of t-km and the quantity of goods (tons), respectively. After the year 1999, the above-mentioned market shares have become increasingly flattering. The main reasons are a rather low containerization rate of goods of about 5-6 per cent, deterioration of the quality of services of the intermodal transport (main mode(s)-railways) and

---

[2] In the given case, about 91 per cent of this total was international and 9 per cent domestic traffic. Rail carried out about 20 per cent, inland waterways 2 per cent, and short-sea shipping 78 per cent of the international traffic. About 97 per cent of the domestic traffic was carried out by rail and 3 per cent by inland waterways (EC 1999).

[3] Overall freight transport in Europe grew at an average annual rate of 2 per cent during the 1970-2001 period and reached the volume of about 3,000 billion t-km (ton-kilometers) in 2001, of which about 44 per cent was carried by road, 41 per cent by sea (intra-EU), 8 per cent by rail, and 4 per cent by inland waterways (EC 1999).

further improvement in the efficiency and quality of road transport services (EC 1999, 2000, 2001a; Janić 2007, UIRR 2000).

3.3.2 The System and Problem—Intermodal Rail/Road and Road Freight Transport Network

3.3.2.1 Physical and Operational Characteristics

Analyzing and modelling the full costs of a given intermodal and equivalent road transport network is based on understanding the network size, intensity of operations and technology in use, and the internal and external costs of the individual components.

Both networks are of equivalent size in terms of spatial coverage, the number of nodes and the volumes of demand they serve. Fig. 3.9 shows a simplified scheme. The network nodes are the origins and destinations of goods, i.e. the goods shippers and receivers, respectively. These are usually larger or smaller 'clustered' manufacturing plants, warehouses, logistics centers and/or freight terminals located in larger 'shipper' and 'receiver' areas. Regarding the relative spatial concentration of particular goods shippers and receivers, any large area can be divided into smaller parts called 'zones'.

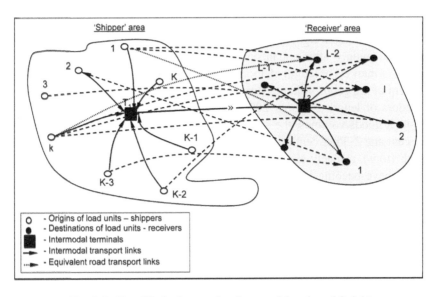

Fig. 3.9: Simplified scheme of an intermodal and road freight transport network (Janić 2007).

In the intermodal transport network, the intermodal terminals are also nodes but only for short-term storage and/or direct transferring of goods.

Goods flow in both networks is consolidated into standardized load units—containers, swap-bodies and semi-trailers.

The transport infrastructure and services of the different (intermodal) and the same (road) transport mode connect particular shipper and receiver 'zones'. The size and type of transport means and the transshipment devices at the intermodal terminals depend on the volume of demand and efficiency and effectiveness of services.

In the intermodal transport network, delivery of load units from particular shippers to receivers is carried out in five steps: (i) collection in the origin 'zone' and transportation to the 'origin' intermodal terminal located in the "shipper" area by road truck(s); (ii) transshipment at the 'origin' intermodal terminal from the road trucks to the vehicles of the main non-road transport mode (rail, inland waterways, air); (iii) the line-haul transportation between the 'origin' and 'destination' intermodal terminal by the main transport mode; (iv) transshipment at the 'destination' intermodal terminal in the 'receiver' area from the mainmode to the road trucks; and (v) distribution from the 'destination' intermodal terminal to the destination 'zone' by road truck(s) (EC 2000).

In the road transport network, delivery of load units between particular shippers and receivers is carried out directly by the same truck in three steps: (i) collection in the origin 'zone'(s); (ii) line-hauling from the border of the origin to the border of the destination 'zone'; and (iii) distribution within the destination 'zone'(s) (Daganzo 1999).

3.3.2.2 Concept of the Full Costs

The full costs of a given intermodal and equivalent road transport network include internal and external costs. The internal costs are imposed on the operators moving load units between the shippers and receivers. The external costs are the costs, which the operations of both networks cause to others—the society and the environment. Both categories of costs can be specified for a particular operational step of both networks. They are generally dependent on the network size characterized by the location, distances and number of nodes, the intensity of activities in the networks conditioned by the volume of demand-load units, efficiency and effectiveness of services and the consumption rates and prices of labor (staff), material (means) and energy. The additional factors particularly relevant for the external costs are the emission rates of pollutants (air, noise), the number of traffic accidents and their actual and prospective harmful effects on society and the environment. Specifically, the network services (vehicles) can interfere and impose delays on other traffic during congestion, causing costs that are treated as external costs.

The internal and external costs are constant in the short-term for a given intensity of the network operations. In addition to factors such as the inter- and intra-modal competition, consumers' preferences and institutional constraints, the full costs might be used as a basis for setting up the prices currently based on the internal costs. Such prices are expected to be higher, which might generally affect the price-sensitive demand and affect modal split. Two options might eventually prevent such developments: (i) improvement of the network's internal efficiency by the organizational measures and consequently reducing the internal costs; and (ii) deploying new technologies and organization, which could reduce the internal and external costs through more efficient and effective utilization of resources with lower rates of energy consumption, emission and noise pollution.

(a) Internal Costs

The collection, distribution, line-hauling and transshipment of load units using the transport means and transshipment devices in the intermodal terminals represent the internal costs of an intermodal transport network. The cost of each component embraces the cost of ownership (depreciation), insurance, repair and maintenance, labor (drivers'-operators' salary packages), energy, taxes, tolls and fees paid for using the transport and intermodal terminal infrastructure. The network infrastructure and mobile means are assumed to be already in place to serve the given volume of demand-load units given the efficiency and effectiveness of services. Thus, the costs of investment in additional infrastructure and/or rolling stock are not taken into account. The internal costs of the equivalent road transport network are analogous to those of the road aspect of the intermodal transport network but always counted for trucks performing the entire door-to-door delivery of load units. In addition, the cost of time of load units while in any of the networks is considered as being dependent on the value and capital depreciation rate of goods and the door-to-door delivery time.

The internal costs of load units, such as depreciation, maintenance, repair and insurance costs are not included since they are assumed to be on the side of their owners—mostly shippers and receivers in this case (EC 2001 a, b; Levison et al. 1996).

(b) External Costs, i.e. Externalities

Activities in each step of the door-to-door delivery of load units in either network generate emissions/burdens on the society and the environment. If intensive and persistent, these burdens can cause damage whose repairing, preventing and/or mitigating costs expressed in monetary terms are treated as external costs. Since not internalized, the external costs are usually estimated indirectly using the concept of 'willingness to pay' for avoiding, mitigating or controlling particular impacts (EC 2001a, Levison et al. 1996). The external costs of both networks embrace the costs of damages by burdens, such as the local and global air pollution, congestion, noise and traffic accidents. In many cases, it appears difficult to precisely extract these from the burdens of other traffic and non-traffic activities. As a result, they are always considered as marginal burdens.

Intermodal Network

Emissions of GHG (Green House Gases): The trucks carrying out collection and distribution of load units usually burn diesel fuel and cause air pollution, whose particular components (air pollutants, i.e. GHG or Green House Gases can cause damage to the surrounding buildings, green areas and people's health locally, but also globally if transported and deposited by weather to remote locations. This implies that these emissions of GHG may have local and global impacts over the area served by a given intermodal transport network. Emissions of GHG by the main transport mode during the line-hauling of load units between intermodal terminals depend on the type of energy used by the main mode. If aircraft, rail diesel engine(s), and/or diesel-powered ships (barges) are used, the air pollution is direct. If electric energy is exclusively used (mainly in railways), emissions of GHG are indirect depending on the composition of sources from which the electric energy is obtained. These are usually remote power plants as the point sources of local emissions. The operation of the main mode vehicles along the long links creates both local and global emissions of GHG and consequently air pollution. Emissions of GHG generated by operations of the intermodal terminals are mainly indirect because the electric energy produced at the remote plants is used for the electric cranes transshipping the load units.

Congestion: The trucks engaged in collection and distribution of load units usually move in the densely urbanized and/or industrialized 'zones'. They may experience congestion and consequent private delays. However, they may also impose delays on other vehicles whose costs are counted as externals. The main transport mode is assumed to be free of congestion, i.e. the advance scheduling of departures excludes vehicle interference. In addition, introducing new services (departures) does not cause shifting or rescheduling of existing ones. That means the services-departures do not interfere and impose delays counted as externals on each other. The load units are also assumed not to impose costs of delays on each other while being handled at the intermodal terminals.

Noise: Trucks involved in the collection and distribution of load units generate noise, which generally annoys the nearby population if it exceeds the tolerable limit(s). If persistent, the noise can affect work productivity or lead to some disease in the long-term. Noise generated by the line-hauling of load units between two intermodal terminals can be considered similarly. Noise from the intermodal terminals is not considered since it is assumed to be a part of the ambient (urban) noise.

Traffic Incidents/Accidents, i.e. Safety: Traffic incidents and accidents cause damage and loss of property of the network operators and of the third parties in addition to the loss of life and injuries of the affected people. They are considered separately for each step and the transport mode of the intermodal transport network due to the different frequency, character of occurrence and consequences. Accidents at the intermodal terminals are not considered since spilling-out events are very rare (EC 2002a, Levison et al. 1996).

Road Network

The same categories of external costs and methods of their consideration are used for the three operational steps of the road transport network. Specifically, particular burdens, damages and associated costs are considered during the use of diesel-powered trucks along the entire door-to-door distance(s).

3.3.3 Modelling the Full Costs

3.3.3.1 Some Related Research

The research and policy efforts related to promoting intermodal freight transport particularly in Europe have been increasing over the past two decades (Bontekoning et al. 2004, EC 1999, ECMT 1998). The most recent overview classifies seven research topics as follows: collection and distribution of load units, rail line-haul transportation, technologies/transshipment at the intermodal terminals, unification and standardization of load units, management and control, economic policy and planning (Bontekoning et al. 2004). In particular, substantial research is directed to optimizing the costs of collection and distribution of load units with the share of about 30-40 per cent of the total internal costs of intermodal transport in Europe (Daganzo 1999, EC 2001b). In addition to research on the line-haul rail transportation aimed at optimizing labor, energy consumption and the capital stock (rolling stock, load units and trucks), the feasibility of real-life cases of intermodal rail-truck initiatives in the European Union have been evaluated (UIRR 2000). Research on the technologies/transshipment at intermodal terminals and standardization of load units has evaluated the effects of prospective improvements in the efficiency and effectiveness of the intermodal chain(s) (Ballis and Golias 2002, EC 2001b). In addition to the organization, management and control, the internal and external costs and conditions of competitiveness of intermodal transport have been investigated (EC 2001a, b; Morlok et al. 1995). Research on policy has mainly considered enhancement of the market position of intermodal transport in Europe by internalizing externals (EC 1999, 2001a, Morlok et al. 1995, Forkenbrock 2001).

In a general sense, this section based on the work of Janić (2007) continues the work of EC (2001a), UIRR (2000), Daganzo (1999), and Hall (1993).

3.3.3.2 Objectives and Assumptions

The main objectives of this research are as follows:

- Developing the analytical models for estimating the full (internal-operational and external) costs of an intermodal rail/road freight transport network and its pure road counterpart; and
- Applying these models to assess the real potential of the intermodal rail/ road freight transport network to compete with currently dominant road networks under given conditions. These are specified by the goods/freight shipments 'door-to-door' delivery distance(s) and size of the vehicles (trains and trucks deployed).

The models developed are based on the following assumptions:

(a) Intermodal Network

Collection and distribution

- The vehicles of the same capacity and load factor collect and/or distribute load units in a given zone(s) (Daganzo 1999);
- Each vehicle makes approximately a tour of the same length at a constant average speed;
- The collection step starts from the vehicle's initial position, which can be anywhere within the 'shipper' area and ends at the origin intermodal terminal. The distribution step starts from the destination intermodal terminal where the vehicles may be stored in a pool and ends in the 'receiver area' at the last receiver (Daganzo 1999, Morlok et al. 1995);
- The headways between the arrivals and departures of the successive vehicles (and load units) at the origin and from the destination intermodal terminal, respectively, are approximately constant and independent of each other (Daganzo 1999).

Line-haul between two terminals

- The headways between the successive departures of the main mode's vehicles between two intermodal terminals is constant, reflecting the practice of many non-road transport operators in Europe to schedule regular weekday services (Daganzo 1999, EC 2001a);
- The capacity of the main mode's vehicles is constant. It may consist of the separate capacity modules or 'units', each with an approximately equal

capacity. In the rail case, this implies 'shuttle' or direct trains composed of the same (flat) cars (EC 2000); and

- The average speed and anticipated arrival/departure delays of all departures of the main mode are constant and approximately equal.

(b) Road network

- The road vehicles-trucks of similar capacity and load factor transport load units between particular origin and destination 'zones' (Daganzo 1999);
- Load units are loaded on to each truck for exclusively one (given) pair of 'zones'. The area, layout and distance between particular shippers and receivers in particular 'zones' crucially influence the length of the vehicle tour-distance(s). The vehicle speed is constant (Daganzo 1999); and
- The trucks move between the borders of particular pair(s) of the origin and destination 'zones' along the same routes at a constant line-hauling speed (Daganzo 1999).

(c) Caveats on the internal and external costs

As mentioned above, modelling the full costs of intermodal and equivalent road transport network include developing the models, collection of data and the models' application. Developing the models includes identification of the relevant variables and their relationships. The variables reflect the type and format of data needed for the model application. Data collection contains caveats on the methodology of obtaining, the required format and values. In the given case, it has appeared to be not particularly difficult to collect the relevant data on the internal costs mainly due to the availability of relatively reliable statistics and empirical analysis techniques (questioning the stakeholders involved). However, the data on external costs have always been used as estimates obtained from a four-stage process starting from the quantification of emissions/burdens and estimation of their spatial concentration, proceeding with estimation of the prospective damages and ending with assigning monetary values to the damages in both the short- and long-term. Quantification of particular emissions/burdens has been carried out regarding the transport and transshipment technology and the intensity of activities, which in both networks depends on the volume of demand-load units. The concentration of emissions/burdens is estimated keeping in mind the spatial character, size and positioning of both networks in relation to the populated areas and sensitive landscape, flora and fauna. Damages from particular emissions/burdens are usually estimated by using the specific models developed for more general purposes. The costs of damages are usually evaluated indirectly by applying the 'willingness to pay' method due to the fact that the market still does not completely recognize these costs (EC 2001a, Levison et al. 1996). In both networks, the data on the internal and external costs refer to their parts (segments, stakeholders), which differ and operate under different technical/technological, economic-market regulatory and environmental-spatial conditions. Under such circumstances, the aggregation of outcomes from the partial calculations is needed to make the data convenient for using in the proposed analytical model. This aggregation can be carried out: (i) per cost category for an individual activity—the vehicle tour, line-haul service, or single transshipment in the intermodal terminal(s); (ii) per cost category for a given volume of similar activities; (iii) per cost category for the volume(s) of activities in given segment(s) or step(s) of the network(s) operations; (iv) per cost category for a given volume of all activities in the network(s); and (v) for all cost categories and activities in the network(s) under the given circumstances.

Dividing these aggregate values of costs by the volume of demand-load units in the network gives the aggregate average cost values per activity and/ or unit of output—load units, distance or ton-kilometers (t-km).

3.3.3.3 Structure of the Models

Based on the above-mentioned assumptions, the generic structure of the model for calculating particular cost categories (internal, external) and cost type (transport, time, handling, type of externality) for particular steps of operation of both networks is developed as follows (Janić 2007):

Internal cost

- Transport cost = (Frequency) × (Cost/Frequency) (3.6a)
- Time cost = (Demand) × (Time) × (Cost/Unit time/Unit of demand) (3.6b)
- Handling cost = (Demand) × (Cost/Unit of demand) (3.6c)

External cost

- External cost = (Frequency) × (External cost/Frequency) (3.7)

The scheme of both networks are shown on Fig. 3.10 (a, b, c).

The variables in Equation 3.6 (a, b, c) are specific for particular steps of the intermodal transport network as follows: In the collection and distribution step, the 'Frequency' variable relates to the number of vehicle runs needed to collect and/or distribute the given volume of demand-load units. In zone (k), 'Frequency' (f_k) is proportional to the volume of demand-load units (Q_k) and inversely proportional to the product of the vehicle-truck capacity (M_k) and load factor (λ_k). The 'Cost/Frequency' variable relates to the cost of individual vehicle-truck type and is usually expressed in dependence on the distance (i.e. length of tour) as $[c_{ok}(d_k)]$. The distance (d_k) includes the segments between the vehicle's initial position and the first stop (x_k), the average distances between the successive stops (δ_k) and the distance between the last stop and intermodal terminal (r_k) (Fig. 3.10a). A similar reasoning for the trip frequencies and distances can be applied to the distribution step (Fig. 3.10c). In the line-hauling step, the 'Frequency' (f) variable is proportional to the total volume of demand-load units in the network (Q), and inversely proportional to the product of the modular capacity of the main-mode vehicle and load factor (Fig. 3.10b). This 'Frequency' can be determined in a way to minimize the internal and external cost of the transport operator and the time cost of users-load units while in the network (Daganzo 1999). The internal and external cost per departure, [$c\,(w, s)$] and [$c_e(w, s)$], respectively, depends on the vehicle size (weight = w) and the line-hauling distance(s). The unit cost of time of load units at intermodal terminals and line-hauling step (α_{b1}), (α_{b2}) and (α_{b1},), respectively, depends on the value of goods and the capital-discounting rate. The time in the line haul step is proportional to the distance (s) and anticipated delays (D) and inversely proportional to the vehicle speed (v_s).

In the road transport network, the variables in Equation 3.6a have an analogous meaning respecting the fact that trucks operate along the entire door-to-door distance between 'zones' (k) and (l). The variables in Equation 3.6b mean that in the intermodal transport network the time cost in the collection step in zone (k) is proportional to the quantity of load units (Q_k), the unit value of goods time (α_k) and the time of the vehicle tour (t_k), which is proportional to the length of tour (d_k) and the average vehicle speed (v_k); in the line-hauling step the time cost is proportional to the waiting and line haul time and their unit costs; it has also been determined after optimizing the total costs of the line-hauling step with respect to the departure frequency of the main transport mode.

In the road transport network, the time cost (Equation 3.6b) refers to transportation of load units between zones (k) and (l). It is proportional to the quantity of load units (Q_{kl}), the unit value of goods time ($\alpha_{b/k}$) and the time between zones (t_{kl}). This time depends on the distance (s_{kl}), the average vehicle speed (v_{kl}), the anticipated delay (d_{kl}) and the time of stopping to pick-up/deliver the load units in each 'zone' ($t_{s/kl}$).

The variables in Equation 3.6c mean that in the intermodal transport network, the handling cost in the collection step in 'zone' (k) is proportional to the quantity of load units (Q_k), unit handling time and cost (t_{hk}) and (c_{hk}), respectively. This cost is analogous to the distribution step in zone (l). In the line-hauling step, the handling cost is proportional to the total quantity of load units in the network (q) and the unit handling cost at both intermodal terminals (c_{h1}) and (c_{h2}), respectively.

In the road transport network, the handling cost (Equation 3.6c) refers to the 'zones' (k) and (l) which are analogous to those in the collection and distribution step of the intermodal transport network.

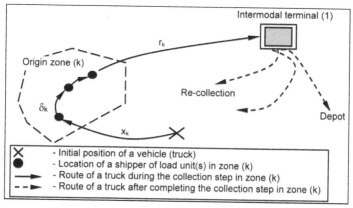

(a) Collection in the 'shipper' area

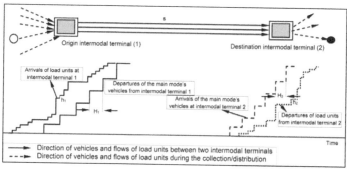

(b) Line haul between two intermodal terminals

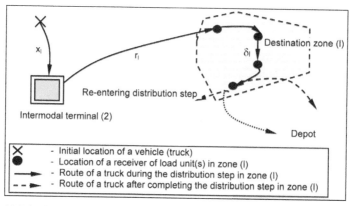

(c) Distribution in the 'receiver' area

Fig. 3.10: Simplified scheme of delivering load units in given intermodal transport network (Janić 2007).

The variables in Equation 3.7 mean that in the intermodal transport network, the external cost in the collection step in 'zone' (k) is proportional to the frequency of trips (f_k) dependent on the quantity of load units (Q_k), the vehicle capacity and load factor (m_k) and (λ_k), respectively and the aggregate external cost per trip $(c_{ek}(d_k))$. For a given vehicle type, this cost depends on the distance (d_k) and costs of the individual burdens and the external cost (Equation 3.7) is analogous to the distribution step in 'zone' (l). In the line-hauling step, the external cost is proportional to the total quantity of load units (Q), the unit aggregate external cost of each intermodal terminal (c_{e1}) and (c_{e2}) and the unit aggregate external cost of each departure-service $[c_e(w, s)]$. In the road transport network, the variables in Equation 3.7 are analogous to those in the collection and distribution step of the intermodal transport network, but again applied to the door-to-door distance between 'zones' (k) and (l). The full cost of both the networks can be obtained by summing up the internal and external costs for each of the above-mentioned steps.

Consequently, the analytical expressions for particular cost components of both networks are given in Table 3.7 (Daganzo 1999; Janić et al. 1999). The analytical procedure of optimizing frequency of the main transport mode between two intermodal terminals in the intermodal transport network, which minimizes the full costs, can be found in the reference literature (Daganzo 1999). Dividing the total costs—Equations 3.12, 3.13 and 3.14—in Table 3.7 by the volume of demand and distance provides the average internal, external and full costs per unit of the network output—t-km, which enables their comparison.

3.3.4 Application of the Models

3.3.4.1 Inputs

The proposed models are applied to the simplified European intermodal rail-truck and equivalent road freight transport network (Ballis and Golias 2002, EC 2000, 2001a, b; Janić 2007).

(a) Load Units, Time Cost and Operating Time of the Networks

Both networks deliver load units of 20 foot or about 6 m (TEU-Twenty Foot Equivalent Unit) as the most common in Europe. Each load unit has an average gross weight of 14.3 metric tons (12 tons of goods plus 2.3 tons of t are) (EC 2001a, Janić 2007). The unit cost of the time of load units in each operating step is adopted as $\alpha_b = 0.028$ €/h-ton[4]. The network operational time is $T = 120$ hr, i.e. five weekdays.

---

[4] The average value of ten chapters of goods groups including the load units transported by the road and rail-truck intermodal transport between particular EU member states is estimated to be 2.08 €/kg. The total discount rate is adopted as 12 per cent, which gives the time cost equal to: $\alpha_b = (2.08€/kg \cdot 14300kg \cdot 0.12)/(8760hr \cdot 14.3 \text{ ton}) = 0.028$ €/hr-ton (EC 2002a).

Table 3.7: Components of the full costs of given intermodal and equivalent road freight transport network (Janić 2007)

	Intermodal network		Road network
	Collection/Distribution[1]	Line-haul[2]	Collection/Distribution/Line-haul
1. Transport (internal) cost	$C_{1/k} = (Q_k/\lambda_k M_k)c_{ok}(d_k)$ (3.8a)	$C_{1/t/min} = \sqrt{2}/2\,c(w,s)\left[\dfrac{QT(\alpha_{b1}+\alpha_{b2})}{c(w,s)+c_e(w,s)}\right]^{0.5}$ (3.8b)	$C_{1/kl} = (Q_{kl}/\lambda_{kl}M_{kl})c_{o/kl}(d_{kl})$ (3.8c)
2. Time cost	$C_{2/k} = Q_k\alpha_{bk}t_k$ (3.9a)	$C_{2/t/min}+C_{2/t}=\sqrt{2}/2\left[\dfrac{QT[c(w,s)+c_e(w,s)]}{(\alpha_{b1}+\alpha_{b2})}\right]^{0.5} + Q\alpha_b(s/v_s+D)$ (3.9b)	$C_{2/kl} = Q_{kl}\alpha_{b/kl}(d_{kl}/v_{kl}+D_{kl}+2t_{s/kl})$ (3.9c)
3. Handling cost	$C_{3/k} = Q_k t_{kh}c_{hk}$ (3.10a)	$C_{3/t} = Q(c_{h1}+c_{h2})$ (3.10b)	$C_{3/kl} = 2Q_{kl}t_{h/kl}c_{h/kl}$ (3.10c)
4. Transport (external) costs	$C_{4/k} = (Q_k/\lambda_k M_k)c_{ek}(d_k)$ (3.11a)	$C_{4/t}+C_{5/t/min}=Q(c_{e1}+c_{e2}) + \sqrt{2}/2\,c_e(w,s)\left[\dfrac{QT(\alpha_{b1}+\alpha_{b2})}{c(w,s)+c_e(w,s)}\right]^{0.5}$ (3.11b)	$C_{4/kl} = (Q_{kl}/\lambda_{kl}M_{kl})\,c_{e/kl}(d_{kl})$ (3.11c)
Sub-total	$C_c = \sum\limits_{i=1}^{4}\sum\limits_{k=1}^{K}C_{i/k}$ (3.12a)	$C_{T/min} = \sum\limits_{i=1}^{6}C_{i/t}$ (3.12b)	$C_{T/kl} = \sum\limits_{i=1}^{4}C_{i/kl}$ (3.12c)
Total	$C_{I/FULL} = C_c + C_{T/min} + C_d$ (3.13)		$C_{R/Full} = \sum\limits_{i=1}^{4}\sum\limits_{k,l=1}^{K,L}C_{i/kl}$ (3.14)

[1] The analogous expressions are used for calculating the sub-total cost C_d for the distribution step; [2] *See also* Daganzo (1999).

(b) Road Collection, Distribution and Line-hauling

In each zone of both networks, the average length of tour and speed of each vehicle assumed to make only one stop during the collection and distribution step is adopted as $d = 50$ km and $u = 35$ km/hr, respectively. In the road network, the average vehicle speed during the line-hauling step, i.e. between the borders of particular origin and destination 'zones, is adopted as $v = 60$ km/hr (EC 2001a, b; Janić 2007).

The vehicle operating cost based on the full vehicle load equivalent of two 20-foot load units is determined by applying the regression technique to the empirical data as follows:

$c_0(d) = 5.4563d^{-0.2773}$ €/vehicle-km (N = 26; $R^2 = 0.7808$; $25 \leq d \leq 1600$ km (EC 2001a, b). The average load factor is $\lambda = 0.85$. The same equation is used for calculating the vehicle operational cost during the collection and distribution step of the intermodal transport network. The average load factor is $\lambda = 0.60$ (EC 2001a, b; Janić 2007). In both networks, the vehicle cost already includes the handling costs of load units.

From the same sources of data, the externalities comprising the local and global air pollution, congestion, noise pollution and traffic accidents are determined in the following regression form: $c_e(d) = 9.884d^{-0.6235}$ €/vehicle-km ($N = 36$; $R^2 = 0.6968$; $25 \leq d \leq 1600$ km) (EC 2001a, b, Janić 2007).

The headways between the arrivals and departures of load units at/from both intermodal terminals during the collection and distribution step, h_1 and h_2, respectively, are assumed to be zero.

(c) Rail Line-hauling

Trains operating between two intermodal terminals consist of $m = 26$ flat cars. Each car weighs about 24 tons, which together with the weight of the engine of about 100 ton gives the weight of an empty train as: $W = 26 * 24 + 100 = 724$ ton. The capacity of each car is equivalent to three TEU, i.e. $M = 3 * 14.3 = 42.9$ ton. Regarding the average load factor per train of $\lambda = 0.75$, the load per train is equal to: $Q = 26 * 42.9 * 0.75 \approx 837$ ton. The gross weight of the train is equal to $w = W + Q = 724 + 837 \approx 1560$ ton (EC 2000, 2001a). The average train speed and the average anticipated delay is: $v_s = 40$ km/hr and $D = 0.5$ hr, respectively (UIRR 2000). The train internal-operating cost is estimated by applying the regression technique to the empirical data related to the European shuttle and direct train services as follows: $c(w,s) = 0.58338$ $(w*s)^{0.7413}$ €/train ($N = 42$; $R^2 = 0.812$; $100 \leq s \leq 1300$ km; $600 \leq w \leq 2000$ ton) (EC 2001a, Janić 2007).

The train externalities caused by local and global air pollution, noise and traffic accidents are estimated by the regression technique applied to the above-mentioned empirical data as follows: $c_e(w,s) = 0.5670 (w*s)^{0.6894}$ €/train ($N = 24$; $R^2 = 0.862$; $100 \leq s \leq 1300$ km; $600 \leq w \leq 2000$ ton) (EC 2001a, Janić 2007).

(d) Intermodal Terminals

The service time of a load unit t in each intermodal terminal is always assumed to be shorter than the interval between its arrivals and departures. The handling cost includes only the transshipment cost of 40 €/load unit at both terminals, which gives the unit handling cost of $c_{h1} = c_{h2} = 40$(€/load unit)/14.3(ton/load unit) $\approx$ 2.8 €/ton (Ballis and Golias 2002, EC 2001b, c, Janić 2007).The external cost of the intermodal terminals includes only the cost of local and global air pollution imposed by the production of electricity for moving cranes used for transshipment of load units as follows: $c_{e1} = c_{e2} = 0.0549$ €/ton (EC 2001a, Janić 2007).

3.3.4.2 Results

The results from the model calculations with the above-mentioned input are shown in Figs. 3.11 and 3.12, and Tables 3.8 and 3.9 (Janić 2007). For the purpose of the sensitivity analysis, the length of hauling distance (i.e. the length of door-to-door distance) and the volume of demand-load units are varied in both networks as parameters. Specifically, the demand is varied using the increments equivalent to the single loaded train (837 ton).

Figure 3.11 shows the dependence of the average internal and full cost of both networks on the length of door-to-door distance. The volume of demand-load units corresponds to the loads of 5 trains/week, i.e. 1 train/day, as the benchmarking case. Such train frequency is the most common in many trans-European intermodal markets-corridors (EC 2000, 2001 a, b, UIRR 2000).

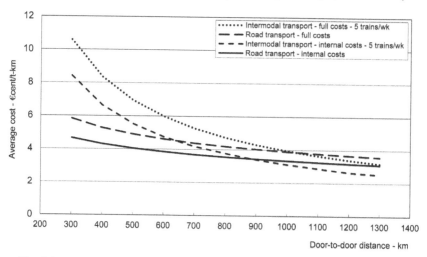

Fig. 3.11: Dependence of the average external, internal and full costs of a given intermodal and road transport network on the door-to-door distance (Janić 2007).

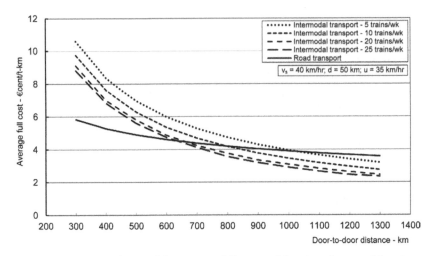

Fig. 3.12: Dependence of the average full costs of the given intermodal and road transport network on the volume of demand-load units and door-to-door distance (Janić 2007).

As can be seen, the average internal and full costs decrease more than proportionally with increase of the door-to-door distance in both networks, indicating the existence of economies of distance. Both costs of the intermodal transport network decrease at a higher rate, equalize with the costs of its road network counterpart at the distance of about 900 km and become increasingly lower afterwards. This indicates that intermodal transport is currently a competitive alternative to long-haul road transport beyond the given 'break-even' distance in some trans-European corridors-markets.

The relationship between the average internal costs of both networks might partially explain the current split between the two modes in Europe. Since the operational cost of road is lower than the operational cost of intermodal transport network over the range of short, medium and even some long-distance markets, the lower road prices based on such lower cost with other market and regulatory factors might seemingly attract more of the voluminous and price-sensitive demand on these distances (about 90 per cent up to 600 km).

The full cost as the sum of the internal and external costs in both networks also decreases more than proportionally with increase in the door-to-door distance. The rate of decrease is again higher in the intermodal transport network, thus enabling equalizing of its costs with the costs of its road counterpart at the 'break-even' distance of about 1050 km. This is longer than in the case of operational costs. Since the volume of demand around these distances is generally low, basing the prices on the higher costs may generally affect this already low and price-sensitive demand and thus make conditions for the intermodal transport to gain higher market shares even more complex. This again raises the question of consistency of EU policies expecting strengthening of the market position of intermodal transport, in addition to investments and subsidies in the non-road transport modes, also through internalizing the externalities (EC 2001a, Janić 2007).

Table 3.8 gives the structure of the full costs of the given intermodal transport network. As can be seen, the shares of the rail/terminal-related external cost increase and the shares of the road-related external cost decrease with increase of the door-to-door distance. In such a context, the share of the road external cost is about twice higher than that of the rail-terminal related external cost. Consequently, the road operational steps at both ends of the intermodal network considerably contribute to its total external cost (about 40-50 per cent).

Table 3.8: The structure of the full costs of given intermodal transport network[1] (Janić 2007)

The cost component	Door-to-door distance (km)					
	300	500	700	900	1100	1300
	Share (per cent)					
• Rail + terminal externalities	6	7	7	7	8	8
• Rail internal-operational	12	13	14	15	16	17
• Terminal internal-operational	17	16	15	14	13	13
• Rail line-hauling + terminal time	17	20	22	24	25	25
• Sub-total	52	56	58	60	62	63
• Road externalities	15	14	13	13	12	12
• Road internal-operational	33	30	29	27	26	25
• Road time	0	0	0	0	0	0
• Sub-total	48	44	42	40	38	37
Total	100	100	100	100	100	100

[1] The volume of demand – load units corresponds to the load of 5 trains/week as the benchmark case.

In absolute numbers, the relatively constant shares of the rail and terminal internal costs are generally comparable to the shares of road internal cost decreasing with increase of the door-to-door distance. The shares of the time cost increase in the rail-terminal and appear negligible in both road operational steps. Consequently, with increase in the door-to-door distance by about 1000 km (i.e. from 300-1300 km), the shares of the main mode generally increase from about 52 per cent to 67 per cent on account of the decreasing shares of the road mode from about 48 per cent to 37 per cent. Table 3.9 gives the relative structure of the full costs of the road transport network.

Table 3.9: The structure of the full costs of given road transport network (Janić 2007)

The cost component	Door-to-door distance (km)					
	300	500	700	900	1100	1300
	Share (per cent)					
Road internal-operational	79	82	83	84	85	86
Road time	1	1	1	1	1	1
Road externalities	20	17	16	15	14	13
Total	100	100	100	100	100	100

As can be seen, with increase in the door-to-door distance, the shares of internal cost increase from about 80 per cent to 86 per cent, the shares of external cost decrease from about 20 per cent to 13 per cent, and the share of time cost remains almost negligible (about 1 per cent).

Figure 3.12 shows the influence of change in the volume of demand-load units and door-to-door distance in both networks on the average full costs.

As can be seen, the average full costs of the road transport network are constant and that of the intermodal transport network decrease with increase in the volume of demand-load units over the range of door-to-door distances. Such diminishing of the full costs shortens the 'break-even' distance for the intermodal transport network. For example, if the demand increases from 5 to 10 trains/week (from one to two trains/day), i.e. by 100 per cent, the 'break-even' distance will shorten from about 1050 km to 800 km (about 30 per cent). If the demand increase is from 10 to 20 trains/week (i.e. by an additional 100 per cent), the 'break-even' distance will additionally shorten from about 800 km to 650 km (i.e. by 23 per cent). This indicates diminishing effects of increasing demand on the shortening of the 'break-even' distance. Consequently, the competitiveness of intermodal transport can eventually be enhanced by increasing the service frequencies of the main mode on shorter distances. This may sound realistic since there might be sufficient demand in these markets to justify such a capacity increase.

3.3.5 Interim Summary

This section has dealt with modelling of the full, internal and external, costs of a given intermodal rail/road freight transport network and its road counterpart. The analytical models for particular cost components were developed and applied to the simplified configurations of both the networks, respecting the operational practice in Europe. As such, these networks represent the transport systems and the full costs of their economic and environmental performances, both dependent on the operational performances.

The results showed that the average internal and full costs of both the networks decrease more than proportionally with increase in the door-to-door goods/freight shipment delivery distance, thus indicating economies of the distance. In the intermodal transport network, the average costs decrease at a decreasing rate with an increase in the volume of the demand, i.e. load units reflecting economies of scale. In the road transport network, this cost is rather constant under the same conditions. In addition, the average full and internal costs decrease with increase in distance at a higher rate in the intermodal than in the road transport network and consequently equalize at the so-called 'break-even' distance(s)—shorter for the internal and longer for the full costs. Since the full costs of intermodal transport decrease and the full costs of road transport remain constant with increase in the volumes of demand-load units, the 'break-even' distance(s) shorten at a decreasing rate.

Despite being based on the caveats on the estimation of all, particularly external costs used in the model, the results can be used to assess some of the implications of the EC (European Commission) policies intending to internalize transport externalities, stimulate modal shift and consequently enhance the market position of the intermodal transport. If the full costs are to be used as the main factor for pricing, the 'break-even' distance will increase and thus push the intermodal transport to compete in longer-distance markets with increasingly diminishing volumes of demand as shown in Fig. 3.13.

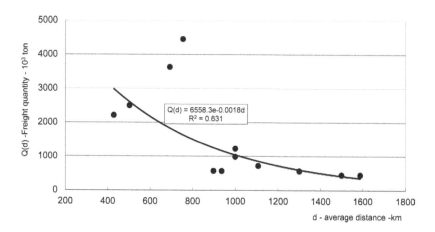

Fig. 3.13: Dependence of quantity of inland goods/freight shipments on distance - the main European intermodal freight transport corridors (EC 2002b).

Nevertheless, in the medium-distance markets (around 600-900 km), intermodal transport might increase the service frequencies in order to meet a more voluminous demand and consequently neutralize the effects of raising the prices after internalizing the external costs. While observing caution due to existing caveats on the particular cost inputs, the results suggest that the EC policy of internalizing externalities might stimulate intermodal transport to improve efficiency and effectiveness under given conditions in the shorter but more promising rather than in the longer-distance markets with insufficient volumes of demand for setting up overall competitive and feasible transport services.

3.4 Rail/Road Mode Substitution in Freight Transport Corridor(s)

3.4.1 Background

The previous section indicated how intermodal rail/road and road transport can compete in terms of the operational and total cost depending on the door-to-door goods/freight delivery distance and size/volume of shipment(s). Such competition however does not mean that any mode can take over all demand flows on the distances where its total costs are the lowest. It simply means that on these distances, such a mode can gain a dominant market share(s). In many cases, the above-mentioned freight transport networks have a corridor-like layout.

As will be seen in Chapter 7, a transport corridor is defined as a relatively long linear strip of land where infrastructure and related transport services of at least one transport mode (e.g. road, rail or inland waterways) are set up. The origins and destinations of relatively constant and voluminous freight transport demand flows and related competitive/complementary services of particular transport modes/operators are located along such a corridor.

In Europe, transport corridors are considered by many national and EU (European Union) policy makers as one of the solutions for ensuring the more sustainable development of freight and consequently the entire transport sector. This implies simultaneous expansion of infrastructure and services on the one side, and the mitigation of their overall impacts on the environment and society, on the other. Consequently, the second Pan-European Transport Conference in Crete (March 1994) defined, in addition to the corridors constituting the European Transport Networks spreading throughout EU countries, nine additional Pan-European transport corridors (passenger and freight transport routes) in Central and Eastern Europe. Some additions were made at the third conference at Helsinki, in 1997 and as a result, these corridors are sometimes referred to as 'Crete corridors' or 'Helsinki corridors'. After the civil war in the former Yugoslavia, a tenth corridor emerged. At the same time, different initiatives supported by EU-funded research started aiming at combining and integrating the two systems (European Transport Networks and Pan-European Corridors) after most of the countries concerned joined the EU.

Many of these initiatives focus on investigating the conditions for increasing the market share of rail freight services operating through both (integrated) systems. A strong argument was that rail freight services were generally viewed as more competitive and more environment and socially-friendly than road truck services in medium- to long-distance markets (in this case, the corridors). However, at least in terms of the competitiveness of rail freight services, the figures do not support the above-mentioned arguments. For example, as Fig. 3.13 shows, at transport distances of about 900-1000 km in the Trans-European Transport Networks, the market share of rail services was only about 10 per cent of the total freight volumes (tons). At distances of up to about 800 km, which account for as much as 90 per cent of the total freight volumes, the market share of rail services is even lower at around 2-3 per cent (tons) and 2 per cent (t-km) (EC 2001a, 2009a). With a view to gradually improving these figures, many railway freight operators have launched innovative services primarily aimed at catching newly emerging transport demand by offering competitive, efficient and effective services. At the same time, they also highlight confirmed figures indicating their higher environmental and social friendliness.

This section deals with estimating savings in some environmental and social externalities substituting road truck with rail freight transport services through competition in a given trans-European corridor (Janić and Vleugel 2012).

3.4.2 The System and Problems—Environmental and Social Externalities

Rail and road freight transport operations impact the environment and society in terms of energy consumption and related emission of GHG (Green House Gases), noise, congestion and safety i.e. traffic incidents/accidents. In cases involving building up of new infrastructure and facilities, land use can also be included. When expressed in monetary terms, these impacts become externalities.

3.4.2.1 Energy Consumption and Emissions of GHG

The freight trains operating in trans-European corridors mainly use electricity for propulsion. The quantity of electricity consumption per given train service depends on the train's weight, operating speed and distance travelled (UIC 2010a). Such trains indirectly spew of greenhouse gases, depending on the sources from which the electricity in question is produced (EC 1997, 2001).

Trucks usually consume diesel fuel at a rate depending on their payload and operating speed, thus spewing greenhouse gases (DB SCHENKER 2010, EC 2005, Janić 2007,UIC 2010b). In both cases, with the exception of electricity generated from renewable sources, non-renewable sources of energy deplete on the one hand and greenhouse gases contributing to global warming are emitted on the other.

3.4.2.2 Noise

In general, the level of noise generated by freight trains depends on their length, weight and speed, as well as on their distance from the observer, or, more specifically, the affected population. Longer, heavier and faster trains passing closer to the affected population generate higher noise levels. However, due to higher speeds, the duration of exposure to the given level of noise is shorter. Barriers along the tracks also help reduce the noise levels (UIC 2010a).

Trucks generate noise which is measured at similar reference locations. Noise generally increases in line with the truck's speed, although exposure is again shorter due to the higher speed (EC 2001a, Hamet and Steiner 2001, Janić 2007, Profilidis 2006).

3.4.2.3 Congestion

Freight trains operating in trans-European corridors are generally assigned time slots, thus preventing their interference with other trains and consequent delays under regular operating conditions. In exceptional cases, particularly long and/or slow freight trains may delay faster freight and/or passenger trains (EC 2008, Janić 2007).

Road trucks usually operate on motorways along dedicated (far right) lanes at limited speeds. Nevertheless, they can cause congestion and consequent time losses for other lighter vehicles in cases where two (particularly dense) flows meet (EC 2001a).

3.4.2.4 Safety, i.e. Traffic Incidents/Accidents

The safety of rail and road freight transport operations, similarly as in other transport modes and operations, depends on traffic incidents/accidents, which happen due to known reasons. They usually cause loss of life, inflict injuries, damage and lead to loss of property for all parties involved as well as for the affected third parties.

The overall rate of traffic accidents in both rail freight and road truck operations has decreased over the past decade in the EU27 Member States

(EC 2009b, ERSO 2007). The rail and road freight services carried out along the trans-European corridors have significantly contributed to this trend (EC 2009b, ERSO 2007).

3.4.3 Modelling and Savings in Externalities by Rail/Road Substitution in a Given Corridor

3.4.3.1 Some Related Research

Related research on the characteristics of operations of rail freight trains and trucks and their competition in trans-European corridors was partially elaborated in the previous section. It was mentioned that it can be broadly divided into studies published in scientific journals and as reports published as part of EC-funded projects. The studies published in the scientific journals indicate that rail/road intermodal transport has emerged as an innovative research field over the past two decades (Janić and Reggiani 2001, Janić 2006) and focus on assessing the full (internal) and external (costs) of both systems in Europe. This latter research indicates the advantages of rail intermodal transport over road-truck transport on medium- to long-haul routes (Janić 2007, 2008; Janić and Vleugel 2012).

The above-mentioned EC-sponsored research was carried out in the scope of actions, such as topical networks, concerted actions and integrated projects. Some of them include COST Transport Actions (European Co-operation in the Field of Scientific and Technical Research), Framework Programs and the Marco Polo Program (EC 2008). The research mainly focused on investigating the potential of rail-truck intermodal transport as an economically and environmentally/socially more feasible alternative to road-truck transport in the medium- to long-haul markets.

3.4.3.2 Objectives and Assumptions

The objectives of the research described in this section are to develop models for estimating savings in the environmental and social impacts and their costs (externalities), which could be achieved by substituting road truck with equivalent rail freight transport services in a given trans-European freight transport corridor. These models are based on the following assumptions:

- The transport corridor is divided into segments characterized by infrastructure, supporting facilities and equipment, rolling stock and operating conditions of each transport mode;
- Rolling stock of both modes is interoperable, which implies its ability to operate smoothly along the corridor independently of the above specifics of its particular segments;

- The capacity of freight trains providing services is constant;
- Substitution of services takes place according to the 'all and/or nothing' principle; and
- The intensity and scale of impacts and their costs of particular transport services of both modes are uniform and constant along the given corridor.

3.4.3.3 Structure of the Models

(a) Substitutive Capacity of the Rail Freight Transport Mode and Possible Effects

The substitutive capacity of freight transport modes operating in segments of a given corridor is based on the given quantity of freight flows to be transported by either mode. If these transport modes are (i) and (j), their service frequencies can be estimated as follows (Janić and Vleugel 2012):

$$f_i(d) = Q(d)/\lambda_i C_i \text{ and } f_j(d) = Q(d)/\lambda_{ij}C_{ji} \tag{3.15a}$$

where

Q is the quantity of freight flows (goods) to be transported either by mode (i) or by mode (j) (tons);

d is the length of the corridor (and/or its segment) (km);

C_i, C_j is the carrying capacity of a service of transport mode (i) and (j), respectively (tons); and

λ_i, λ_j is the load factor of a service of transport mode (i) and (j), respectively.

If transport mode (j) substitutes transport mode (i), the factor of substitution can be estimated as follows:

$$S_{j/i}(d) = f_j(d)/f_i(d) = \lambda_j C_j/\lambda_i C_i \tag{3.15b}$$

Substitution is feasible in terms of the mitigating impacts on the environment and society, if the following conditions are fulfilled:

$$E_{j/k}(d) \le E_{i/k}(d) \tag{3.15c}$$

and

$$\sum_{k=1}^{K} E_{j/k}(d) \le \sum_{k=a}^{K} E_{i/k}(d) \tag{3.15d}$$

where

$E_{i/k}, E_{j/k}$ is (k)-th type of the environmental and/or social impact of mode (i) and mode (j), respectively; and

K is the number of possible types of impacts of both modes.

The other symbols are as in the previous equations.

(b) Environmental and Social Impacts

Energy consumption and emissions of GHG

Rail freight trains

Energy consumption: A freight train substituting for road trucks consumes electricity depending on the train's weight (locomotive + wagons + payload), movement resistance and route length. Consequently, a train operating along one of the segments of the corridor consumes the following amount of energy (EC_{jm}) (kWh)(Profilidis 2006):

$$EC_{jm} = \frac{2.725M_{jm} + 2.724*10^{-3}R_{jm}}{\eta_{jm}} d_m \qquad (3.16a)$$

where

M_{jm} is the weight of a train (ton);

d_m is the length of the segment of the corridor (km);

R_{jm} is the train's resistance along segment of the corridor (d_m)(kp) (kp - kilopond); and

η_{jm} is efficiency of the electric locomotive (0.85-0.95 for most electrical locomotives operating in Europe).

Emissions of GHG: Emissions of GHG are indirect, namely through the production of electricity used in powering freight trains. The corresponding emissions of GHG in terms of CO_{2e} (carbon dioxide equivalents) from the amount of energy consumed by train (EC_{jm}); (*see* Equation 3.16a) can be estimated as follows:

$$EM_{jm} = EC_{jm} * e_{jm} \qquad (3.16b)$$

where

e_{jm} is the rate of emission of greenhouse gases (kgCO$_2$e/kWh).

Road trucks

Fuel consumption: Road trucks consume diesel fuel. When each truck consumes fuel at the rate of (r_{imf}); (l/100 km), the total consumption of the convoy of (f_{im}) trucks over corridor segment (d_m) can be estimated as follows:

$$EC_{jm} = f_{om} * r_{im} * (d_m 100) \qquad (3.16c)$$

where all symbols are as in the previous equations.

Emissions of GHG: Emissions of GHG from a road-truck convoy in terms of CO_{2e} can be estimated as follows:

$$EM_{im} = EC_{im} * e_{im} = f_{om} * r_{im} * (d_m 100) * e_{im} \qquad (3.16d)$$

where

e_{im} is the emission rate per unit of fuel consumed (CO_{2e}/l of fuel).

The rate (e_{im}) usually relates to on-wheel emissions including emissions from manufacturing fuel and emissions from direct fuel consumption as a result of providing the given transport service(s).

Noise

Rail freight trains

Noise generated by a freight train passing an observer can be estimated according to the following reasoning (EC 2008). Assume a freight train of length (S_{mj}) passing an observer at speed (v_{mj}) while operating along segment (d_m) of the corridor. If the level of noise heard by the observer from the *r*-th train passing at the speed (v_{jmr}) at the shortest distance (γ_{jmr}) is $L_{eq}(r, \gamma_{jmr}, v_{jmr})$, the level of noise during period (t_{jmr}) can be estimated as follows (*see* also Fig 2.32 in Chapter 2):

$$L_{eq}[r, \rho_{jmr}(t), v_{jrm}] = L_{eq}(r, \gamma_{jmr}, v_{jmr}) - 8.6562\ln[\rho_{jmr}(t)/\gamma_{jmr}] \qquad (3.17a)$$

The second term in Equation 3.17a represents noise attenuation over an area free of barriers. The total noise exposure of the observer from (N_{jm}) successive trains passing during the period (T) can be estimated as follows:

$$L_{eq}(N_{jm}) = 10\log \sum_{r=1}^{N_{jm}} 10^{\frac{L_{eq}(r, \rho_{jmr(t)}, v_{jmr})}{10}} \qquad (3.17b)$$

where all symbols are as in the previous equations.

Road trucks

In the case of road trucks, length (S_{ir}) in Equation 3.17a reflects the length of a convoy of trucks substituted by a given train. In such a case, if each truck is of length (s_i) and if the distance between successive trucks in the convoy moving along segment (d_m) is ($\delta_{im}(v_{im})$), the total convoy length is as follows:

$$S_{im} = f_{im} * S_i + (f_{im} - 1) * \delta_{im}(v_{im}) \qquad (3.17c)$$

where the number of trucks in convoy (f_{im}) can be estimated from Equation 3.15 (a, b, c, d). In general, the distance between trucks in the convoy ($\delta_{im}; v_{im}$) is an increasing function of speed (v_{im}). Consequently, the time and intensity of exposure of an observer to noise generated by a passing convoy of trucks can be estimated using Equation 3.17 (a, b).

In considering the actual noise exposure of a population located close to the railway line(s) and highways/motorways from passing train(s) and/or truck convoys, the influence of noise barriers should be taken into account.

Congestion

Rail freight trains

In the given context, newly launched freight trains are scheduled within the available time slots along a given corridor segment. As a result, such trains do not affect or cause delays of existing passenger and/or freight train services. Consequently, when operating under regular conditions, they can be considered free of congestion and delays.

Road trucks

When estimating delays due to congestion caused by a convoy of road trucks, it is assumed that the convoy moves along lane p of the highway at an average speed (v_{imp}) and that the flow of cars intending to overtake it along lane q has an average speed of (v_{impq}) $(v_{imq} > v_{imp}.)$ (EC 2008). Consequently, the average waiting time of the first vehicle in the queue following the convoy at a distance $(x_{im/pq})$ before overtaking can be estimated according to the theory of steady-state queues as follows:

$$\overline{w}_{im/pq} = x_{im/pq} * (1/v_{imq} - 1/v_{imp}) \qquad (3.18a)$$

Total time losses, i.e. savings of all cars in the flow of intensity (Λ_{imp}) queuing behind the convoy and waiting for the first car to overtake, can be estimated as follows:

$$W_{im} = (\Lambda_{im/pq} w_{im/pq} - 1) * (w_{im/pq} + \overline{t}_{im/pq}) \qquad (3.18b)$$

where

$\Lambda_{im/pq}$ is the intensity of flow of cars intending to overtake the convoy (veh/hr); and

$t_{im/pq}$ is the average time a car needs to pass the convoy (min.).

Safety, i.e. traffic incidents/accidents

Rail freight trains

The number of potential fatalities (and/or severe injuries) of a train assumed to substitute the convoy of trucks along a given corridor segment (d_m) can be estimated as follows:

$$IA_{jm} = \lambda_{jm} * C_{jm} * d_m * a_j \qquad (3.19a)$$

where

a_j is the freight rail accident/incident rate (fatalities/injuries/t-km).

Other symbols are analogous to those in Equation 3.15 (a, b, c, d).

Road trucks

For a convoy of road trucks assumed to be substituted by trains, the number of potential fatalities (and/or severe injuries) along corridor segment (d_m) can be estimated based on Equation 3.15 (a, b, c, d) as follows:

$$IA_{im} = f_{im} * \lambda_{im} * C_{im} * d_m * a_i \qquad (3.19b)$$

where

f_{im} is the number of trucks substituted by train (*see* Equation 3.15a).
  Other symbols are analogous to those in Equation 3.17a.

3.4.4 Application of the Models

3.4.4.1 Inputs

(a) The Corridor and Transport Services

Geography/layout of the corridor

The above-developed models were applied to the CREAM Trans-European corridor (Customer-driven Rail-Freight Services on a European Mega-Corridor Based on Advanced Business and Operating Models) (EC 2008) whose simplified layout is shown in Fig. 3.14.

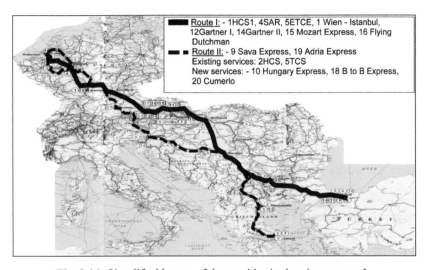

Fig. 3.14: Simplified layout of the corridor in the given example
(EC 2008, Janić and Vleugel 2012).

As can be seen, the corridor begins in the north in the Netherlands and Belgium, spreads through Germany, Austria, Hungary, Romania, Bulgaria, Slovenia, Croatia, Serbia and Macedonia and ends in southern Greece and Turkey. It is about 2,700 kilometers long.

Rail freight train and road truck transport services

In this corridor, new regular (weekly) freight train services were scheduled in 24 markets (routes) as given in Table 3.10.

Table 3.10: New freight train services in the given corridor (EC 2008, Janić and Vleugel 2012)

Route	City/Country	City/Country	Service/Product name	Status
1	Ljubljana/Slovenia	Halkali/Turkey	Bosporus-Europe Express	r
2	Munich/Germany	Ljubljana/Slovenia	Adria Express	r
3	Duisburg/Germany	Ljubljana/Slovenia	Adria Express Network	r
4	Köln/Germany	Ljubljana/Slovenia	Adria Express Network	r
5	Rotterdam/ Netherlands	Duisburg/Germany	/	r
6	Skopje/F.Y.R. of Macedonia	Thessaloniki/ Greece	Intermodal train Skopje – Thessaloniki	r
7	Belgrade/Serbia	Skopje/F.Y.R. of Macedonia	Mixed train Skopje – Belgrade	r
8	Budapest/Hungary	Halkali/Turkey	/	r
9	Genk/Belgium	Oradea/Romania	TRex	r
10	Genk/Belgium	Sopron/Hungary	TRex	r
11	Sopron/Hungary	Oradea/Romania	TRex	r
12	Antwerp/Belgium	Sopron/Hungary	/	r
13	Ludwigshafen/ Mannheim/Germany	Istanbul/Turkey	Multimodal train Turkey – Germany via Trieste	r
14	Ludwigshafen/ Germany	Istanbul/Turkey	Multimodal train Turkey – Germany via Trieste	r
15	Ludwigshafen/ Germany	Wels/Austria	Network "Hungaria Express"	r
16	Neuss/Germany	Wels/Austria	Mozart-Express II	r
17	Duisburg/Germany	Wels/Austria	Mozart-Express II Network	r

(Contd.)

Table 3.10: (*Contd.*)

Route	City/Country	City/Country	Service/Product name	Status
18	Wels/Austria	Budapest/Hungary	"Hungaria Express"	r
19	Pirdop/Bulgaria	Olen/Belgium	Copper anode train Pirdop-Olen	r
20	Köln/Germany	Köseköy/Turkey	Automotive logistics train (via corridor X)	r
21	Köln/Germany	Köseköy/Turkey	Automotive logistics train (via corr. IV)	r
22	Constanza/Romania	Kelheim/Germany	/	r
23	Bucharest/Romania	Halkali/Turkey	ICF Container Train	r
24	Ciumesti/Romania	Valenton/France	Automotive train (Dacia)	r

r – running services

These 24 new services serving 30 O-D (Origin-Destination) markets have been converted into weekly rail and truck services as given in Table 3.11.

Table 3.11: Characteristics of freight train and substituted road truck services in the given corridor (EC 2008, Janić and Vleugel 2012)

Distance cluster (km)	Number of O-D of goods/ services	Frequency (Trains/ Week)	Composition (Wagons/ Train)	Dimensions (train length (m)/weight (t)	Frequency (Trucks/ Week)
< 500 (250)	6	20	20	440/1200/"A"	402
500-1000	12	20	25	500/1200/ "A"	492
1000-1500	5	11	30	600/1400/ "B"	337
1500-2000	2	2	22	440/1200/ "A"	43
> 2000	5	10	29	600/1400/ "B"	285
Total	30	63	-	-	1559

(b) Transport Vehicles

Rail freight trains

Five classes of freight trains of differing weight and length carry out new services in the given corridor as follows: Class 'A' with an average weight of 1,200 ton and a length of 500 m; Class 'B' with an average weight of 1,400 tons and a length of 600 m; Class 'C' with an average weight of 1,600 tons and a length of 700 m; Class 'D' with an average weight of 1,600 tons and a length of 600 m; and Class 'E' with an average weight of 1,200 tons and a length of 700 m. The train load factor is assumed to be 100 per cent. Figure 3.15 shows the differences between train categories 'A' to 'E' and other freight trains operating in Europe.

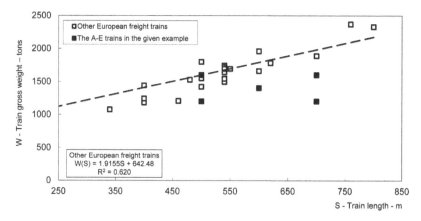

Fig. 3.15: Relationship between the weight and length of European freight trains and the new freight trains in the given corridor (EC 2008, Janić 2007, 2008b).

The maximum operating speed of the freight trains performing new services significantly varies along particular sections of the corridor. For example, on sections between the Netherlands and Belgium and the Austrian-Slovenian border as well as the Hungarian-Romanian border, the maximum speed can be up to 100-120 km/h or even higher. On other sections, the maximum speed is no greater than 40-50 km/h. In any case, the commercial speed, i.e. the speed of delivery of goods, is much lower due to restrictions and regulations.

Both electric and diesel tractions are used for existing and new train services. In case of electric tractions, multi-stream locomotives are used due to differences in tension, despite being more expensive (EC 2008).

Road trucks

In the given case, road trucks are five or more axle vehicles with a maximum gross weight of 40 tons and a maximum length of 18.75 m. Depending on the country, their maximum operating speed is typically limited to 80, 90 or 100 km/hr (EC 2008, 2009a).

(c) Environmental and Social Impacts

Rail freight trains

Energy consumption and emissions of GHG: The energy consumption of a freight train of a gross weight of (M_{jm}) (tons) assumed to operate along the corridor segment (d_m) (km) is estimated as: $EC_{jm} = 0.315 * M_{jm}^{0.6} * d_m$ (kWh) (IFEU 2008). The emission rate of CO_{2e} per unit of produced electricity (kWh) consumed by freight trains varies significantly across certain European countries as it depends mainly on the primary source of electricity production in the countries in question. The average for the countries in which the corridor spreads through is estimated as: $e_{jm} = 0.46$ ($kgCO_{2e}$/kWh) of electricity produced (IEA 2009). Consequently, the emissions of CO_{2e} by a freight train above can be estimated as: $EM_{jm} = 0.315\ M_{jm}^{0.6} * d_m * 0.46 = 0.145\ M_{jm}^{0.6} * d_m$ ($kgCO_{2e}$).

Noise: The noise generated by a freight train, depending on its speed and length, can be estimated as follows (EC 2001a): $L_{eq}(r, 25, v_{jmr}, S_{jmr}) = 51.761 + 0.107v_{jmr} + 0.006S_{jmr}$, dB(A) ($R^2 = 0.968$; $40 < v_{jmr} < 100$ (km/hr)); $350 < S_{jmr} < 700$ (m)). Data on the noise of individual trains is obtained by measuring at a distance of 25 m and at height of 3 m from the source (EC 2001a, http://ec.europe.eu/transport/rail/environment/noise.en.htm).

Congestion: As already pointed out, under the assumption that free slots are always available, new freight train services do not cause congestions and related delays in other freight or passenger trains already operating along particular segments of the corridor.

Safety, i.e. traffic incidents/accidents: The rate of potential fatalities (and severe injuries) arising from traffic incidents/accidents of new freight train services in the corridor is expressed as the average number of fatalities and/or severely injured persons per unit of freight train output (train-km). The data for the EU 27 Member States in 2006 indicates an average accident rate of: $a_j = 3.011 * 10^{-6}$ (fatalities/injuries/train-km) (EC 2009b). Assuming that the average weight of a freight train in the corridor is about 1,200 tons (*see* Fig. 3.15), the average accident/incident rate can be estimated as: $a_j = 2.509 * 10^{-9}$ (fatalities/injuries/t-km).

Road trucks

Energy consumption and emissions of GHG: Road trucks consume diesel fuel at an average rate of: $r_{im} = 0.30\text{-}0.35$ l/km at an average speed between 50 and 90/100 km/hr. The gravity of diesel fuel is 0.82-0.95 kg/l, and its calorific value is 12.777 kWh/kg, which gives an average rate of fuel consumption of about $SEC_{im} = 3.17$ kWh/km, and consequently, the energy consumption over the corridor segment (d_m) of: $EC_{im} = 3.17 * d_m$ (kWh). The rate of emissions of greenhouse gases is $e_{im} = 0.324$ kgCO$_{2e}$/kWh, which gives the total amount of emissions of a truck of: $EM_{im} = 1.03 *d_m$ (kgCO$_{2e}$) (EC 2005, Janić and Vleugel 2012).

Noise: The noise of road trucks is measured at the right-angle distance of 25 meters from the source (i.e. passing truck) and at a height of 3.0m. The typical values, depending on the truck's operating speed, are expressed as follows (EC 2001b, Janić and Vleugel 2012):

$$L_{eq}(r, 25, v_{imr}) = 5.509\ln v_{imr} + 25.36 \text{ [dB(A)]}$$

$$(R^2 = 0.988; \ 10 < v_{imr} < 90 \text{ [km/hr]})$$

Congestion: Congestion and delays caused by a road truck convoy are estimated as follows—a convoy of trucks always consists of: $f_{im} = 1$ truck running at an average speed of: $v_{im/p} = 80$ km/hr. The flow of car(s) with an average intensity of: $\Lambda_{im/pq} = 1$ car/min. queue behind this truck for an average distance of: $x_{im/pq} = 1$ km before starting to overtake at the same speed. The car's normal free speed is taken as: $v_{imq} = 120$ km/hr. The length of the truck is $s_j = 18.75$ m, while the distance between individual trucks in the convoy, if viable, is adopted as: δ_{im} (80) = 80 m. In addition, the distance between the overtaking car(s) and the last and the first truck in the convoy before and after overtaking, respectively, is assumed to be: $\Delta_{mp} = 100$ m. Likewise, the distance between the overtaking car and the first car in the overtaking lane is adopted as $\alpha_{im/pq} = 100$ m. Consequently, the waiting time of a car before starting to overtake the convoy of trucks is estimated as $w_{im/pq} = 0.25$ min., and the duration of overtaking is $t_{im/pq} = 0.5$ min.

Safety, i.e., traffic incidents/accidents: The rate of fatalities and severe injuries in traffic incidents/accidents involving only HGV (Heavy Goods Vehicles) for the EU 27 Member States is considered relevant for the case in question (EC 2009a, ERSO 2007). Thus, this rate is estimated to be $a_i = 2.191*10^{-9}$ (fatalities/t-km; severe injuries are not included due to the lack of reliable and detailed data).

(d) Cost of Particular Impacts—Externalities

In order to assess the effects of the above road-rail substitution in monetary terms, the impacts of both modes are given in their average unit values as externalities in Table 3.12 (CE Delft 2008, Janić 2007).

3.4.4.2 Results

Using the volumes of operations of the new freight train services in Table 3.11 and their equivalents if carried by road truck services on the one hand and the average unit rates of the environmental and social impacts, on the other, the totals of these impacts generated by both modes per week under the given circumstances are calculated and given in Table 3.13 (EC 2008, Janić and Vleugel 2012).

Table 3.12: The average unit cost—externalities of particular rail freight and road truck services in the given corridor (CE Delft 2008)

Externality	Trucks (€cent/v-km)	Freight trains (€cent/v-km)
Noise	0.18	6.75
Emissions of GHG	4.70	34.4
Congestion	0.17	0
Traffic incidents/accidents	0.30	0.19
Total	5.35	41.34

Table 3.13: Saved quantities of impacts by rail/road substitution in the given example (Janić and Vleugel 2012)

Impact	Saved quantity	Unit
Energy consumption		
Road trucks	- 3,186	MWhr/wk
Rail freight trains	- 1,449	MWhr/ wk
Ratio Rail/Road	- 0.455	
Emissions of greenhouse gases		
Road truck	- 1,031	$tonCO_{2e}$/wk
Rail freight trains	- 667	ton CO_{2e}/wk
Ratio Rail/Road	- 0.667	
Cumulative perceived noise level		
Road trucks	- 80.9	dB(A)/wk
Rail freight trains	- 81.0	dB(A)/wk
Ratio Rail/Road	- 1.01	-
Delays due to congestion		
Road trucks	- 291	min/wk
Rail freight trains	- 0.0	min/wk
Ratio Rail/Road	- 0.0	-
Potential fatalities/severe injuries		
Road trucks	- 0.144	events/wk
Rail freight trains	- 0.198	events/wk
Ratio Rail/Road	- 1.375	-

Comparison of the absolute values of impacts of both modes reveals the following (Janic and Vleugel 2012):

- Substitution has a particularly favorable impact on the energy consumption and consequently related emission of GHG;
- Substitution has a neutral effect on noise due to its scale of impact and intensity under comparable conditions;
- Substitution has a particularly favorable effect on congestion; and
- Substitution does not favor any of the modes in terms of safety, i.e. traffic incidents/accidents, despite disparities caused by the calculation, which actually disfavor rail freight services under the given conditions.

Taking the average weight of a road truck of 40 ton and that of a freight train of 1,200 ton, and using the data on externalities in Table 3.12, the average unit externalities are estimated at 0.034€cent/t-km for rail freight trains and 0.134€cent/t-km for road trucks. Combining these with the figures in Table 3.13, the total externalities of both types of services are calculated and given in Table 3.14.

Table 3.14: Externalities of the substituting rail freight and substituted road truck services in the given corridor (Janić and Vleugel 2012)

Distance cluster (the average) km	Trucks (t-km/wk)	Cost of trucks C_{ei} (€/wk)	Freight trains (t-km/wk)	Cost of trains C_{ej} (€/wk)	Ratio C_{ei}/C_{ej}
< 500 (250)	4020000	5387	6050000	2057	0.382
500-1000 (750)	14760000	19770	18000000	6120	0.310
1000-1500 (1250)	16050000	21507	19250000	6545	0.304
1500-2000 (1750)	3010000	4033	4200000	1420	0.352
> 2000 (2350)	26728000	34016	32900000	11196	0.329
Total:	54 638 000	86 521	66 430 000	27386	0.316

As can be seen, in each market segment of the given corridor, externalities of the substituted road truck services appear to be higher than those of the substituting rail freight services. The average difference amounts to approximately 30 per cent in favor of the substituting rail freight services – put differently, their externalities amount to about 70 per cent of those of the substituted road truck services. In addition, providing the given substitution remains in place, these savings will constantly increase in absolute terms over time.

3.4.5 Interim Summary

This section deals with modelling of savings in the external costs, i.e. externalities like energy consumption, emissions of greenhouse gases, noise, congestion and traffic incidents/accidents which could be achieved by substituting road truck with equivalent rail freight services in a given trans-European transport corridor under given conditions. In this context, the corridors have been considered as 'transport systems' and their externalities as the environmental performances based on the operational performances.

The models have been developed and applied using real-life data on operating new freight train services in the trans-European freight transport corridor and secondary data on particular impacts and their costs, i.e. externalities.

The results show that the savings in total externalities arising from such road/rail substitution under given conditions could reach about 30 per cent. In addition, the results confirm the feasibility of the EC (European Commission) transport policy aimed at stimulating more intensive use of rail freight services in markets where they could be fully competitive with their road truck counterparts; also because of their contribution to mitigating the overall environmental and social externalities of the freight transport sector in Europe.

3.5 Concluding Remarks

This chapter deals with modelling the operational, economic, environmental and social performances of transport systems under three distinctive cases. The first case involved analyzing and modelling the utilization of the runway system capacity of a given large airport, where a new runway has recently been built. The models developed relate to estimating the runway system's 'ultimate' capacity and the average delay per an aircraft/flight under given conditions specified by the configuration of runways in use and the level of utilization. The application of the models to the given large hub airport operating a system of several runways shows that the newly built runway has been reasonably utilized together with the existing runways, enabling maintenance of the average of aircraft/flight delays within tolerable accepted limits, both currently and prospectively, the latter implying the traffic growth in the medium- to long-term future period.

The second case relates to modelling the full cost of an intermodal (rail/road) freight transport network and its road counterpart. The full costs consist of the operational and external costs, i.e. externalities. The application of the full cost models of both networks using the inputs from the European freight transport sector shows that the intermodal (rail/road) transport network could be competitive in terms of the full costs on the routes, i.e. door-to-door goods/freight delivery distances of about and over 900-1000 km. However, the volumes of goods/freight flows decrease more than proportionally with increase in this distance, which puts the intermodal (rail/road) networks counting on high volumes for the full trains into a rather complex situation, i.e. being competitive in the markets with insufficient volumes of demand justifying profitable and market-attractive services.

The last case deals with modelling savings in the external costs, i.e. externalities, which could be achieved by substitution of the road by equivalent rail or intermodal rail/road freight transport services in the given European freight transport corridor. This substitution is carried out either through competition or complementarity of the two modes under given conditions. The results show that substantive savings in the overall externalities is achieved if substitution were carried out already on a very limited scale. But this also has raised the question of how a competitive freight transport market could allow or even drive such substitution and what transport policy measures could stimulate this.

The common ground for the above-mentioned cases were twofold: first, they have been elaborated by the modelling approach; second, they explicitly indicate the mutual dependability of the considered operational, economic, environmental and social performances independent of the transport system and scope.

References

ASA.2012. *Traffic Review 2012: Market Development*. Aviation Statistics & Forecasting. Amsterdam Schiphol Airport. Schiphol, The Netherlands.

Ashford N., Stanton M., Moore C. 1997. *Airport Operations*. A Wiley-Interscience Publication, John Wiley & Sons.Inc., New York, USA.

Ballis A., Golias J. 2002. Comparative Evaluation of Existing and Innovative Rail-Road Freight Transport Terminals. *Transportation Research* A36: 593-611.

BAS. 2012. *BewonersAanspreekpunt Schiphol*. http://www.bezoekbas.nl/

Bontekoning Y., Macharis C., Trip J.J. 2004. Is a New Applied Transportation Field Emerging?—A Review of Intermodal Rail-Truck Freight Transport Literature. *Transportation Research* A38: 1-34.

CE Delft. 2008. *Handbook on Estimation of External Costs in the Transport Sector. Produced within the Study Internalization Measures and Policies for All External Cost of Transport (IMPACT)*. CE Delft-Solutions for Environment Economy and Technology, Delft, The Netherlands.

Daganzo C.F. 1999. *Logistics System Analysis*. Third Edition. Springer, Berlin, Germany.

DB SCHENKER. 2010. *The Environmental Performance 2009/10: Report*. DB Schenker Logistics, Eco Program, Essen, Germany.

De Neufville R.D., Odoni A. 2003. *Airport Systems: Planning, Design and Management*. McGraw Hill Book Company, New York, USA.

EC. 1997. Estimating Emissions from Railway Traffic. *Report on the MEET (Methodologies for Estimating Air Pollutant Emissions from Transport)*. EC 4 Framework Program, European Commission, Brussels, Belgium.

EC. 1999. *The Common Transport Policy—Sustainable Mobility: Perspectives for the Future*. Economic and Social Committee and Committee of the Regions, European Commissions, Directorate General DG VII, http://europa.eu.int/en/comm/dg07/tif/, p. 22.

EC. 2000. *The Way to Sustainable Mobility: Cutting the External Cost of Transport*. Brochure of the European Commission. European Commission, Brussels, Belgium.

EC. 2001. *Study of Optimisation of Procedures for Decreasing the Impact of Noise II—SOURDINE II*. SII-WP2_D2-1-Validation Methodology Report (Appendixes)_ v10, Competitive and Sustainable Growth Program, European Commissions, Brussels, Belgium.

EC. 2001a. *Real Cost Reduction of Door-to-Door Intermodal Transport—RECORDIT*. European Commission. Directorate General DG VII, RTD 5 Framework Program, Brussels. Belgium.

EC. 2001b. *Improvement of Pre- and End-Haulage – IMPREND*. European Commission, Directorate General DG VII, RTD 4 Framework Program, Brussels, Belgium.

EC. 2002a. *EU Intermodal Transport: Key Statistical Data 1992-1999*. European Commission, Office for Official Publications of European Communities, Luxembourg.

EC. 2002b. *EU Statistical Pocketbook – Transport*. European Commission. Office for Official Publications of European Communities, Luxembourg.

EC. 2005. *Energy and Fuel Consumption from Heavy Duty Vehicles*. COST 346 Final Report on the Action. European Cooperation in the Field of Scientific and Technical Research, European Commission, Brussels, Belgium.

EC. 2008. *Customer-driven Rail-Freight Services on a European Mega-Corridor Based on Advanced Business and Operating Models (CREAM)*. Different Deliverables. European Commission, Brussels, Belgium.

EC. 2009a. *EU Energy and Transport in Figures, Statistical Pocket Book*. Directorate General for Energy and Transport, Brussels, Belgium.

EC. 2009b. *Rail Transport Accidents Decreasing in 2007: Statistics in Focus*. 52/2009 Transport, European Commission, Brussels, Belgium, p. 3.

ECMT.1998. *Report on the Current State of Combined Transport in Europe*. European Conference of Ministers of Transport, Paris, France.

ERSO. 2007. *Traffic Safety: Basic Figures 2007—Heavy Goods Vehicles and Buses*. European Road Safety Observatory, Safety Net Transport. http://ec.europa.eu/transport/wcm/road_safety/erso/index.html.

FAA. 2002. *Airport Capacity Benchmarking Report 2001*. Federal Aviation Administration, Washington D.C., USA.

Forkenbrock D.J. 2001. *Comparison of External Costs of Rail and Truck Freight Transport*. Transportation Research A35: 321-337.

Hall R.W. 1993. *Design for Local Area Freight Networks.* Transportation Research B27: 70-95.

Hamet, J.F. and Steiner, V. 2001. *Modelling Pass-by Noise of Heavy Trucks by Power Unit Noise and Rolling Noise.* The International Congress and Exhibition on Noise Control Engineering, The Hague, The Netherlands, p. 6.

IEA. 2009. *CO_2 Emissions from Fuel Combustion. Statistical Report.* International Energy Agency, Paris, France.

IFEU. 2008. *Energy Savings by Light Weighting. Final Report.* IFEU, Heidelberg, Germany.

Janić M. 2001. *Air Transport Systems Analysis and Modelling.* Gordon & Breach Science Publishers, Amsterdam, The Netherlands.

Janić M., Reggiani, A. 2001. Integrated Transport Systems in the European Union: An Overview of Some Recent Developments. *Transport Reviews* 21: 469-497.

Janić M. 2006. Sustainable Transport in the European Union: Review of the Past Research and Future Ideas. *Transport Reviews* 26: 81-104.

Janić M. 2007. Modelling the Full Costs of Intermodal and Road Freight Transport Networks. *Transportation Research* D 12: 33-44.

Janić M. 2008. An Assessment of Performances of European Long Intermodal Freight Trains (LIFTs). *Transportation Research* A42: 1326-1339.

Janić M. 2009a. *The Airport Analysis, Planning and Design: Demand, Capacity and Congestion.* Nova Science Publishers, Inc. New York, USA.

Janić M. 2009b. The EU-US 'Open Skies' Agreement: The Long-Term Managing Capacity at London Heathrow Airport (UK). *Journal of Airport Management* 3: 245-265.

Janić M. 2012. Modelling Effects of Different ATC (Air Traffic Control) Operational Procedures, Separation Rules and Service Disciplines on Runway Landing Capacity. *Journal of Advanced Transportation* 48: 556-574.

Janić M., Vleugel J. 2012. Estimating Potential Reductions in Externalities from Rail-Road Substitution in Trans-European Transport Corridors. *Transportation Research* D 17: 154-160.

Janić M. 2014. Investigating Utilization of the Runway System Capacity at a Large Airport. *Journal of Airport Management* 8: 71-88.

KNMI. 2012. *Uurgegevens Meetstation Schiphol*

Levison D., Gillen D., Kanafani A., Mathieu J.M. 1996. The Full Cost of Intercity Transportation—A Comparison of High-Speed Rail, Air and Highway Transportation in California. *Research Report* UCB-ITS-RR-96-3. Institute of Transportation, University of California, Berkeley, USA.

Morlok E.K., Sammon J.P., Spasovic L.N., Nozick L.K. 1995. Improving Productivity in Intermodal Rail-Truck Transportation. pp. 407-434. *In:* Harker, P. (Ed.). *The Service Productivity and Quality Challenge.* Kluwer Academic Publishers, Boston, USA.

Newell G.F. 1982. *Application of Queuing Theory.* Chapman and Hall. London, UK.

Profilidis V.A. 2006. *Railway Management and Engineering.* (3rd ed.). Ashgate Publishing Limited, Aldershot, UK.

Schiphol. 2012. http://www.schiphol.nl/index_shg.html.

UIC. 2010a. *Railway Noise in Europe: A 2010 Report on the State of the Art.* International Union of Railways. Paris, France.

UIC. 2010b. *Introduction to EcoTrans World-Overview Basic.* International Union of Railways. Paris, France.

UIRR. 2000. *Developing a Quality Strategy for Combined Transport.* Final Report. PACT Program. International Union of Combined Rail-Road Transport Companies. Brussels, Belgium.

Van den Hoven R.N. 2012. *Investigating Operating Regimes of Using Runways at Large Airports: Case Schiphol.* BSc Thesis. Faculty of Civil Engineering and Geosciences. Delft University of Technology, Delft, The Netherlands. http://ec.europe.eu/transport/rail/environment/noise.en.htm.

MODELLING TRANSPORT SYSTEMS—II
Influence of New Technologies
on Performances

4.1 Introduction

This chapter deals with modelling the influence, i.e. effects/impacts, of new and innovative technologies and related operations on the performances of transport systems. In this case, the considered transport systems are a supply chain(s)[1] served by mega vehicles and an airport runway system consisting of two closely-spaced parallel runways where innovative operational procedures (for landings) are supported by the new ATC/ATM (Air Traffic Control/Air Traffic Management) technologies.

At the supply chain(s), the mega vehicles are considered to be the largest in terms of size and carrying capacity for goods/freight shipments. These can be mega container ships, large cargo aircraft, long freight trains and mega trucks (Janić 2014a). The infrastructural, technical/technological, operational, economic, social and environmental performances of the given supply chain(s) are analyzed and modelled. The models are applied to one of the global intercontinental supply chains served by mega container ships and their smaller counterparts, the latter just for the purpose of comparison and comparative evaluation of the chain's performances. At the airport runway system, its 'ultimate' capacity is influenced by the operational procedures supported by the new technologies and corresponding ATC/ATM separation rules between landing aircraft. The model(s) is applied to estimating the

---

[1] An alternative term for 'supply chain(s)' can be 'logistics network(s)' as used in Chapter 5 of this book. However, regardless of which one is used, it is always preferred to use the selected term(s) consistently.

capacity gains of the system of two closely-spaced parallel runways of a large congested international airport.

The objectives of dealing with the above-mentioned cases are twofold. The first is to indicate the prospective benefits but also some disadvantages and controversies in the expected performances of supply chain(s) served by mega vehicles. The second objective is to indicate the potential flexibility of actually constrained transport infrastructure to acquire capacity and thus accommodate additional demand, thanks to the innovative technologies and corresponding operational procedures. Consequently, Section 4.2 describes the modelling of performances of the above-mentioned supply chain(s). Section 4.3 deals with modelling the capacity of the system of two closely-spaced parallel runways operating under the above-mentioned conditions. In a certain sense, this complements the modelling of the airport capacity as described in Chapter 3. The last section provides some concluding remarks.

4.2 Supply Chain(s) Served by Mega Vehicles

4.2.1 Background

One of the numerous definitions of a supply chain is as follows: "A supply chain is a network of facilities and distribution options that performs the functions of procurement of materials, transformation of these materials into intermediate and finished products and the distribution of these finished products to customers." In this section, a supply chain is considered as a physical network producing, handling, transporting and consuming goods/freight shipments consolidated into TEUs [Twenty Foot Equivalent Unit(s)]. Generally, these goods/freight shipments need to be delivered from their ultimate suppliers to their ultimate customers efficiently, effectively and safely. The ultimate suppliers and customers, such as large production/consumption plants, distribution centers, sea-ports, airports, large surface modal (rail, road) and intermodal (rail/road/barge) terminals usually generate and attract substantial flows of these (consolidated) goods/freight shipments. As such, they operate as the hub nodes of global (continental and intercontinental) freight transport network(s). In many cases, these substantial goods/freight flows to be transported between particular hub nodes can justify the more frequent, if not even regular, use of larger mega freight transport vehicles (http://www.investopedia.com/terms/s/supplychain.asp, http://en.wikipedia. org/wiki/Supply_chain).

Generally speaking, the size and payload capacity of the freight transport vehicles operated by various transport modes such as road, rail, air, sea and intermodal and serving a variety of supply chains, have increased over time. The main driving force of such increase include: (i) the growing volumes and diversity of freight transport demand in combination with its increased internalization, globalization and consequent rate of consolidation, i.e. containerization, (ii) strengthening competition in the freight transport markets forcing transport operators in almost all modes to permanently improve the efficiency, effectiveness and safety of their services, (iii) increasing importance of the economics of freight transport and related logistics, (iv) raising concerns on the impacts of the freight transport sector and its particular modes on the environment and society, and (v) innovative design, materials and manufacturing processes of the vehicles, supportive facilities and equipment and infrastructure. Figure 4.1 shows an example of the relationships between the demand and capacity of global maritime container transport (UNCTAD 2013).

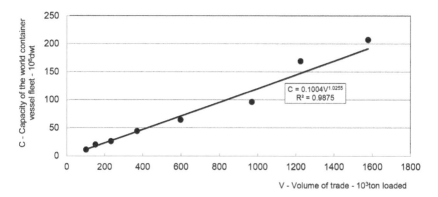

Fig. 4.1: Relationships between the containerized freight volumes of seaborne trade and the capacity of the container ship fleet (1980-2013) (dwt – dead weight ton) (UNCTAD 2013).

As can be seen, the capacity of the global container fleet has increased more than proportionally, driven by the need for satisfying growing goods/freight containerized demand during the observed period (1980-2013). In addition, the average size of container ships ordered has also increased over time as shown in Fig. 4.2.

On the one hand, larger freight transport vehicles with a greater payload capacity usually run fewer services and corresponding vehicle-kilometers while transporting the given quantities of goods/freight shipments under the given conditions. On the other, these vehicles usually have greater empty weight, energy (fuel) consumption, total cost per service in addition to constraints in accessing particular transport (usually loading/unloading) locations and providing sufficient goods/freight shipments for profitable services, i.e. load factor. The latter applies particularly to a specific category of these vehicles referred to as 'mega' freight transport vehicles and considered as the largest in terms of their external dimensions, gross weight and payload capacity, when compared to their closest (smaller) counterpart(s). They are easily recognizable within each transport mode: road—mega trucks, rail/intermodal—long freight trains, air—large cargo aircraft, and sea—large container ships.

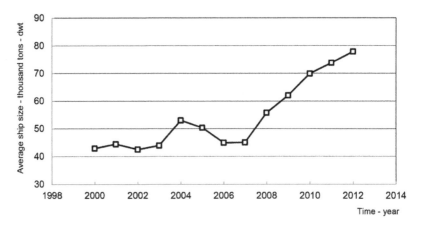

Fig. 4.2: Development of the global maritime container ship fleet – tonnage on order (2000-2013) (dwt–dead weight ton) (UNCTAD 2013).

4.2.2 The System and Problem—Performances of Supply Chain(s)

4.2.2.1 Definition and Categorization

Supply chain performances include their inherent ability to deliver goods/freight shipments from the ultimate suppliers/senders to the ultimate customers/receivers generally efficiently, effectively and safely, i.e. as planned under the given conditions. Consequently, similarly as for other transport systems, the supply chain performances can be classified as infrastructural, technical/technological, operational, economical, environmental and social (Janić 2014a). Regardless of the supply chain type and the characteristics of the freight transport vehicles serving them, these performances are inherently interrelated and interact with each other as shown in Fig. 2.13 (Chapter 2). It is seen that in a 'top-down' consideration, the infrastructural performances can generally influence the technical/technological performances and consequently create a mutual influence between these and all other performances. In a 'bottom-up' consideration, the social and environmental performances can influence the infrastructural and technical/technological performances and consequently create a mutual influence of these and all other performances.

4.2.2.2 *Characterization*

Supply chain performances are generally characterized as follows:

- *Infrastructural performances* relate to the physical/spatial characteristics of the chain's producing, storing and consuming plants of goods/freight shipments and the infrastructure of the various transport modes (road, rail, inland waterways air, sea and intermodal) connecting them;
- *Technical/technological performances* reflect the capacity of production, storage and consumption plants including those of the supportive facilities and equipment for loading/unloading, handling and storing goods/freight shipments before and after their transportation throughout the chain(s). The latter are installed at and around the corresponding plants. Additional performances relate to the dimension (length, width, height, overall configuration), weight (gross, tare, payload), number, size and location of the loading/unloading door(s), engines (power, energy/fuel) and the technical speed of the freight transport vehicles serving the chain(s);
- *Operational performances* relate to the chain's production/consumption cycle. These include the number or quantity of goods/freight shipments to be transported within the chain under the given conditions, the frequency of orders of goods/freight shipments and related transport services, the required vehicle fleet, i.e. the type and number of vehicles deployed to serve the chain(s) under the given conditions and the (technical) productivity of the transport services;
- *Economic performances* generally encompass the total and average costs, which generally include the chain's inventory, handling and transportation costs of the goods/freight shipments;
- *Environmental performances* are considered to be energy (fuel) consumption and related direct and indirect emissions of GHG (Green House Gases) and the area of land/space used/taken by the chain(s); and
- *Social performances* relate to the noise, congestion and safety of the chain(s). Excessive noise generated by producing, storing, transporting and consuming goods/freight shipments at and between the chain(s) hub supplier(s) and the hub customer(s), respectively, can burden the neighboring population. Congestion mainly occurs during transportation of goods/freight shipments, most frequently near the hub supplier(s), the (hub) customers and along the route(s) between them. Safety reflects the risk of incidents/accidents in the chain(s) that can cause damage and/or loss of property and/or goods/freight shipments and human injury and/or loss of life.

4.2.3   Modelling Performances of Supply Chain(s)

4.2.3.1 Some Related Research

Substantial previous research exists either directly or indirectly dealing with particular performances of supply chains. The body of research closely related to that presented in this section can be classified into three categories: (a) the general performances of supply chain(s); (b) the role and influence of transport operations on the overall performances of supply chain(s); and (c) the sustainability (greening) of supply chain(s).

(a) General performances

Research on the general performances of supply chain(s) mainly focuses on understanding the relationship between SCM (Supply Chain Management) practice and SCP (Supply Chain Performances (SCP). Here the performances and their measures focus on the strategic, operational and tactical level (Gunasekaran et al. 2004), reliability, responsiveness, cost and assets (Huang et al. 2005, Lai et al. 2002), the overall chains' goals (Otto and Kotzab 2003), instruments for measuring collaboration between the chain's suppliers and retailers (Simatupang and Shridharan 2005), performances of the suppliers (Giannakis 2007) and integration of the performance management process for delivering services into the customer/supplier yards (Forslund and Jonsson 2007). In addition, this research also includes measuring performances of the supply chain(s) under uncertainty by applying fuzzy logic (Olugu and Wong 2009) and setting up the criteria for developing supply chain performance measurement systems (PMS), including identifying the barriers to their implementation (Fauske et al. 2006).

(b) Role and influence of transport operations on the performances

Research on the role and influence of transport operations on the performances of supply chain(s) mainly focuses on understanding the relationships between the transport and logistics operations and potential improvements through the goods/freight shipment(s) delivery speed, quality of service, operating costs, use of facilities and equipment and energy savings (Tseng et al. 2005), modelling the performances of various spatial and operational configurations of the goods/freight collection/distribution networks (Janić 2005, 2014a) and understanding the potential interactions between the location of the European manufacturing industry, related services and logistics and freight transport (EC 1999).

(c) Sustainability, i.e., 'greening' of supply chain(s)

Research on the sustainability (i.e. 'greening') of supply chain(s) mainly focuses on defining the management of green supply chain(s) by means of integrating environment-thinking into supply chain management, including product design, material sourcing and selection, manufacturing processes, delivery of the final product to the consumer and the end-of-life management of the product after its use (Janić et al. 1999, Srivastara 2007, Stevels 2002). In addition, this body of research also investigates the potential initiatives, driving forces/actions and barriers to implementing 'greening' initiatives by transport and logistics companies in order to reduce the environmental impacts of transport and logistics activities carried out within the given supply chain(s). These could all lead to the achievement of sustainable (green) logistics and supply chain management (Evangelista et al. 2010, WEF 2009).

4.2.3.2 Objectives and Assumptions

The objective is to develop a methodology consisting of particular analytical models of performances of the given supply chain(s) served by various freight transport vehicles, including mega freight vehicles. Consequently, any such model should primarily enable a sensitivity analysis of the chain's performances in light of the characteristics of the various vehicle categories serving it. In the present context, the given supply chain has a generic (spatial) configuration. This implies that it consists of a single hub supplier, a single hub customer and the transport infrastructure connecting them. Goods/ freight shipments consolidated into TEUs are transported between two hubs by various vehicles, including mega freight transport vehicles. The spoke suppliers connect to their hub supplier by smaller vehicles delivering smaller shipments of TEUs. The hub customer connects to the spoke customers by smaller vehicles delivering smaller shipments of TEUs. Therefore, the models of performances of the above-mentioned (generic) supply chain(s) are based on the following assumptions (Daganzo 2005, Hall 1993, Janić et al. 1999, Janić 2005, 2014b):

- The hub supplier of a given supply chain is ultimately the production location, i.e. origin, of the goods/freight shipments; the hub customer is ultimately the consumption location, i.e. destination;
- The chain's production/consumption cycle taking place during the specified period of time satisfies the series of successive orders of goods/ freight shipments to be transported between the hub supplier and the hub customer exclusively by various vehicle fleets, including that of mega vehicles; this implies that, independent of the size of vehicles in the fleet, there is always sufficient demand justifying the operational (service frequency) and economical (load factor) feasibility of their use;
- The size of a goods/freight shipment(s) is always less than or at most equal to the payload capacity of a vehicle serving the given chain(s);
- The fleet serving a given supply chain(s) consists of vehicles of the same size/payload capacity operating with the same load factor;
- The infrastructural and technical/technological performances of the

above-mentioned supply chain(s) are assumed to be given as inputs for the models, thus implying considering only the chain's operational, economic, environmental and social performances; and

- The exclusive use of the given fleet of vehicles to serve the supply chain implies the 'all-or-nothing principle' of serving demand under the given conditions.

4.2.3.3 Structure of the Methodology

(a) Generic supply chain configuration

The generic configuration of a supply chain(s) served by any kind of freight transport vehicles is represented as a H-S (Hub-and-Spoke) transport network whose main nodes are the hub supplier and the hub customer connected by the transport link(s) between them as shown in Fig. 4.3 (a, b).

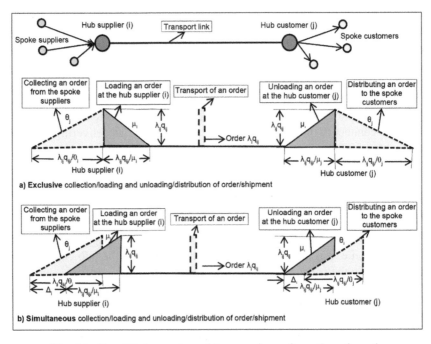

Fig. 4.3: Simplified overview of the generic configuration of supply chain(s) (Janić 2014b).

The spokes 'feeding' the hub supplier and those 'fed' by the hub customer are also shown. As can be seen, the inventories of goods/freight shipments take place at the hub supplier, the hub customer and along the route between them. Figure 4a shows case of the exclusive and Fig. 4b of the simultaneous collecting and loading of goods/freight shipments at the hub supplier, and their exclusive unloading and distributing at the hub customer, respectively. 'Exclusivity' implies that the entire shipment is collected before its loading begins and that the entire shipment is unloaded before its distribution begins. 'Simultaneously' implies that both collecting and loading of goods/freight shipment(s) on the one end and its unloading and distribution on the other end of the chain can be partially or fully carried out at the same time. The inventories of goods/freight shipments and related costs can be effectively managed in such a manner.

(b) Operational performances

The operational performances of the above-mentioned supply chain are considered to be: (i) the transport service frequency exclusively: (a) serving the given demand, and (b) enabling the specified services during the chain's production/consumption cycle; (ii) the size of deployed vehicle fleet; and (iii) (technical) productivity.

- Transport service frequency (dep/TU):
  (a) Serving the given demand:

$$f_{ij}(\tau) = \frac{Q_{ij}(\tau)}{\lambda_{ij} q_{ij}} \qquad (4.1a1)$$

  (b) Enabling the specified services during the chain's cycle:

$$f_{ij}^*(\tau) = \tau/h_{ij}(\tau) \qquad (4.1a2)$$

From Equation 4.1a2, the total quantity of goods/freight shipments, which can be transported within the chain during time (τ), is determined as:

$$Q_{ij}(\tau) = \beta_{ij}(\tau) * \left[\min \left(f_{ij}(\tau); f_{in}^*(\tau) \right) \right] * (\lambda_{ij} q_{ij}) \qquad (4.1a3)$$

- The size of deployed vehicle fleet (vehicles/cycle)

$$N_{ij}(\tau) = \beta_{ij}(\tau) * \left[\min \left(f_{ij}(\tau); f_{in}^*(\tau) \right) \right] * t_{ij}(d_{ij}) \qquad (4.1b)$$

If each vehicle operating in both directions within the chain is full, its average turnaround time $t_{ij}(d_{ij})$ in Equation 4.1b can be estimated as follows:

$$t_{ij}(d_{ij}) = \tau_{ij} + \tau_{ji} = \Delta_{i1} + \frac{\lambda_{ij} q_{ij}}{p_{i1} \mu_{i1}} + \frac{d_{ij}}{s_{ij} * v_{ij}(d_{ij})} + D_{ij} + \Delta_{j1} + \frac{\lambda_{ij} q_{ij}}{p_{j1} \mu_{j1}} +$$

$$\Delta_{j2} + \frac{\lambda_{ij} q_{ji}}{p_{j2} \mu_{j2}} + \frac{d_{ji}}{s_{ji} * v_{ji}(d_{ji})} + D_{ji} + \Delta_{i2} + \frac{\lambda_{ji} q_{ji}}{p_{i2} \mu_{i2}}$$

$$(4.1c)$$

- Technical productivity (TEU, m³, or t-km/TU)

$$TP_{ij}(\tau) = Q_{ij}(\tau) * s_{ij} * v_{ij}(d_{ij}) \qquad (4.1d)$$

where

TU	is the time unit (hr, d);
τ	is the duration of the chain's production/consumption cycle (TU);
$Q_{ij}(\tau)$	is the quantity of goods/freight shipments to be transported from the hub supplier (i) to the hub customer (j) during the chain's production/consumption cycle (τ) (ton, m³, or TEU/TU);
λ_{ij}, q_{ij}	is the average load factor and the payload capacity, respectively, of a vehicle serving the chain (ij) (ton, m³, or TEU/veh);
$h_{ij}(\tau)$	is the average time between the scheduled vehicle departures between the hub supplier (i) and the hub customer (j) during time (τ) (TU);
$\beta_{ij}(\tau)$	is the proportion of transport services realized during the chain's production/consumption cycle of duration (τ);
τ_{ij}, τ_{ji}	is the average time a vehicle spends operating in the direction (ij) and (ji), respectively (TU/veh);
Δ_{i1}, Δ_{j1}	is the time between starting a vehicle's loading at the hub supplier (i) and its unloading at the hub customer (j), respectively (TU);
Δ_{j2}, Δ_{i2}	is the time between starting a vehicle's loading at the hub customer (j) and its unloading at the hub supplier (i), respectively (TU);
d_{ij}, d_{ji}	is the length of the chain's route, i.e. the distance between the hub supplier (i) and the hub customer (j) and vice versa, measured along the transport infrastructure link connecting them, respectively (km);
$v_{ij}(d_{ij}), v_{ji}(d_{ji})$	is the vehicle's average (planned) operating speed on the distances (d_{ij}) and (d_{ji}), respectively (km/TU or kt (knot); 1 kts = 1 nm/hr; nm – nautical mile = 1.852 km);
D_{ij}, D_{ji}	is the average delay per transport service due to the traffic conditions on the route connecting the hub supplier (i) and the hub customer (j) and back, respectively (TU);
μ_{i1}, μ_{j1}	is the loading and unloading rate of a vehicle at the hub supplier (i) and the hub customer (j), respectively (ton, m³, or TEU/TU);
p_{i1}, p_{j1}	is the proportion of the vehicle's loading and unloading rate used at the hub supplier (i) and the hub customer (j), respectively ($p_{i1}, p_{j1} \le 1.0$);

μ_{j2}, μ_{i2}	is the loading and unloading rate of a vehicle at the hub customer *(j)* and the hub supplier *(i)*, respectively (ton, m^3, or TEU/TU);
p_{j2}, p_{i2}	is the proportion of the vehicle loading and unloading rate used at the hub customer *(j)* and the hub supplier *(i)*, respectively $(p_{j2}, p_{i2} \leq 1.0)$; and
s_{ij}, s_{ji}	is the portion of the maintained average vehicle planned operating speed under some kind of irregular operating conditions along the distance (d_{ij}) and (d_{ji}), respectively, caused by disruptive event(s) $(s_{ij} \leq 1.0)$.

Equation 4.1a1 indicates that the transport service frequency is adjusted to serve the demand of goods/freight shipments generated during the chain's production/consumption cycle. Equation 4.1a2 implies that the demand of goods/freight shipments is always available and uniformly distributed over a specified period of time and thus the transport service frequency is adjusted to serve it at regular time intervals. The vehicle's loading and unloading rates μ_{i1}, μ_{j1}, μ_{j2}, and μ_{i2} in Equation 4.1c depend on the number of loading/unloading devices (usually cranes) engaged and the loading/unloading rate of each. In addition, Equation 4.1c indicates that the vehicle turnaround time can be affected during loading at the hub supplier *(i)*, unloading at the hub customer *(j)* and while operating between them in both directions. If any such impact lasts a prolonged period of time, then Equation 4.1b indicates that a larger fleet may be needed to serve the supply chain(s) under the given conditions. Equation 4.1d also indicates that the (technical) productivity of the supply chain can also be affected by the service frequencies on the one hand, and by the speed of the services realized, on the other.

(c) Economic performances

The economic performances of the given supply chain are considered to be the (i) inventory, (ii) handling, and (iii) transport (a) total and (b) average costs of a goods/freight shipment(s) served by the chain. If the size of goods/freight shipment corresponds to the vehicle payload capacity, the costs are determined as follows:

- Inventory costs (€ or $US)

$$C_{ij/INV}(\lambda_{ij}q_{ij}) = IT_i(\lambda_{ij}q_{ij}) * \alpha_i + (\lambda_{ij}q_{ij}) * \frac{d_{ij}}{s_{ij} * v_{ij}(d_{ij})} * \alpha_{ij}$$
$$+ IT_j(\lambda_{ij}q_{ij}) * \alpha_j \qquad (4.2a)$$

The first and third term in Equation 4.2a represent the inventory costs of a goods/freight shipment at the hub supplier *(i)* and at the hub customer *(j)*, respectively. The second term represents the inventory, i.e. the shipment's costs of time while in transportation between the hubs (i) and (j). From Fig. 4.4, the goods/freight shipment inventory time in Equation 4.2a at the hubs (i) and (j), respectively, is determined as follows:

$$
IT_i(\lambda_{ij}q_{ij}) = \begin{cases} \dfrac{1}{2}(\lambda_{ij}q_{ij})^2 \left[\dfrac{1}{r_i\theta_i} + \dfrac{1}{p_i\mu_i} \right] & \text{if} \quad (a) \\[4mm] \max\left\{ 0; (\lambda_{ij}q_{ij})^2 \left(\dfrac{1}{p_i\mu_i} - \dfrac{1}{2r_i\theta_i} \right) + (\lambda_{ij}q_{ij})\Delta_{i1} \right\} & \text{if} \quad (b) \end{cases} \tag{4.2b}
$$

and analogously

$$
IT_i(\lambda_{ij}q_{ij}) = \dfrac{1}{2}(\lambda_{ij}q_{ij})^2 \left[\dfrac{1}{r_j\theta_j} + \dfrac{1}{p_j\mu_j} \right] \quad \text{if} \quad (a) \tag{4.2c}
$$

$$
\max\left\{ 0; (\lambda_{ij}q_{ij})^2 \left(\dfrac{1}{p_j\mu_j} - \dfrac{1}{2r_j\theta_j} \right) + (\lambda_{ij}q_{ij})\Delta_{j1} \right\} \quad \text{if} \quad (b)
$$

- Handling and transport costs (€ or $US)

$$
C_{ij/H-TRA}(\lambda_{ij}q_{ij}) = c_i * (\lambda_{ij}q_{ij}) + c_{ij}(\lambda_{ij}q_{ij}) * (\lambda_{ij}q_{ij}) * d_{ij} + c_j * (\lambda_{ij}q_{ij}) \tag{4.2d}
$$

- Total (inventory + handling + transport) costs (€ or $US)

$$
C_{ij}(\lambda_{ij}q_{ij}) = C_{ij/INV} * (\lambda_{ij} * q_{ij}) + C_{ij/H-TRA} * (\lambda_{ij} * q_{ij}) \tag{4.2e}
$$

- Average total costs (€ or $US/TEU-km or ton-km)

$$
\bar{c}_{ij}(\lambda_{ij}q_{ij}) = C_{ij}(\lambda_{ij}q_{ij}) / [(\lambda_{ij}q_{ij}) * d_{ij}] \tag{4.2f}
$$

where

θ_i, θ_j is the rate of collecting and distributing goods/freight shipments at the hub supplier (i) and the hub customer (j), respectively (tons, m^3 or TEU/TU);

r_i, r_j is the proportion of the rate of collecting and distributing goods/freight shipments used at the hub supplier (i) and the hub customer (j), respectively ($r_i, r_j \le 1.0$);

c_i, c_j is the handling (loading/unloading/transshipment) costs of a goods/freight shipment at the hub supplier (i) and the hub customer (j), respectively (€/(ton, m^3, or TEU)); and

$\alpha_i, \alpha_{ij}, \alpha_j$ is the costs of goods/freight shipment inventory time while at the hub supplier (i), in transportation and at the hub customer (j), respectively (€/(ton or m^3 or TEU/hr or d)).

The other symbols are analogous to those in Equation 4.1 (a, b, c, d). By replacing the size of shipment ($\lambda_{ij}q_{ij}$) with the quantity of goods/freight generated during the chain's production/consumption cycle (Q_{ij}), the corresponding economic performances can be estimated from Equation 4.2 (a, b, c, d). In addition, this equation indicates that the goods/freight shipment inventory time and related costs can be compromised in any handling phase in the chain, i.e. during collecting, loading, transporting, unloading and distributing.

(d) Environmental performances

The environmental performances of a given supply chain(s) are considered to be: (i) the energy (fuel) consumption and related emissions of GHG, and (ii) land/space used/taken.

- Energy (fuel) consumption and emissions of GHG

  The total and average fuel consumption, respectively, from Equation 41a3, are estimated as follows:

$$FC_{ij}(\tau) = \left\{ \beta_{ij}(\tau) * \left[\min\left(f_{ij}(\tau); f_{in}^{*}(\tau)\right) \right] \right\} * FC[q_{ij};$$
$$v_{ij}(d_{ij})] * d_{ij} \ (\text{l, kg, ton, or kWh})$$

$$AFC_{ij}(\tau) = FC_{ij}(\tau)/[Q_{ij}(\tau) * d_{ij}] \ (\text{l, kg, ton or kWh/TEU} - \text{km}$$
$$\text{or ton-km}) \tag{4.3a}$$

The total and average emissions of GHG, respectively, are determined based on Equation 4.3a as follows:

$$EM_{ij}(\tau) = \sum_{k=1}^{K} FC_{ij}(\tau) * e_k \ (\text{kg or ton}) \tag{4.3b}$$

$$AEM_{ij}(\tau) = EM_{ij}(\tau)/[Q_{ij}(\tau) * d_{ij}] (\text{kg /TEU} - \text{km} \ \text{or} \ \text{ton-km})$$

where

$FC[q_{ij}; v_{ij}(d_{ij})]$ is the energy (fuel) consumption of a vehicle of the payload capacity (q_{ij}) serving the supply chain (ij) at the speed $v_{ij}(d_{ij})$ on the distance (d_{ij}) (l, kg, or KWh/km);

e_k is the emission rate of the (k)-th GHG from the consumed energy (fuel) of a vehicle serving the supply chain (ij) (kg of GHG/ l, kg, or KWh); and

K is the number of various GHG emitted from the consumed energy (fuel) by a vehicle serving the supply chain (ij).

The other symbols are analogous to those in the previous equations.

Land used/taken

The land used/taken by a given supply chain is expressed as the area of land or space at the supplier and the hub customer intended to park vehicles during their loading and unloading, respectively. If the frequency of vehicles during the production/consumption cycle of the supply chain (*ij*) is determined from Equation 4.1 (a1, a2), then the number of required parking stands for vehicles at the hub supplier (*i*) and the hub customer (*j*), respectively, per cycle is estimated as follows:

$$n_i(\tau) = \beta_{ij}(\tau) * \left[\min \left(f_{ij}(\tau); f_{in}^*(\tau) \right) \right] * \tau_{i1} = \beta_{ij}(\tau) * \qquad (4.3c1)$$

$$\left[\min \left(f_{ij}(\tau); f_{in}^*(\tau) \right) \right] * \left(t_{i2} + \Delta_{i1} + \frac{\lambda_{ij} q_{ij}}{p_{i1} \mu_{i1}} \right)$$

and

$$n_j(\tau) = \beta_{ij}(\tau) * \left[\min \left(f_{ij}(\tau); f_{in}^*(\tau) \right) \right] * \tau_{j1} = \beta_{ij}(\tau) * \qquad (4.3c2)$$

$$\left[\min \left(f_{ij}(\tau); f_{in}^*(\tau) \right) \right] * \left(\Delta_{j1} + \frac{\lambda_{ij} q_{ij}}{r_{j1} \mu_{j1}} + t_{j2} \right)$$

where

τ_{i1}, τ_{j1} is the average occupancy time of a parking stand during handling vehicle(s) at the hub supplier (*i*) and the hub customer (*j*), respectively, (TU);

t_{i2}, t_{j2} is the time of unloading a vehicle from the previous task at the hub supplier *(i)* and loading it for the forthcoming task at the customer (*j*), respectively (TU).

The other symbols are analogous to those in the previous equations.

Equation 4.3 (c1, c2) assumes that the same parking stand is used for both loading and unloading of the vehicle(s). Otherwise, the terms t_{i2} and t_{j2} can be neglected. In addition, the terms (Δ_{i1}) and (Δ_{j1}) indicate that the vehicle(s) can occupy the parking stand while waiting for loading and unloading operations, respectively, to begin with. Thus, the number of required parking stands for loading and unloading vehicles mainly depends, in addition to the service frequency and size of freight/goods shipment, on the actual loading and unloading rate(s), i.e. the corresponding times. From Equation 4.3 (c1, c2), the net area of land or space taken for parking vehicles at the hub supplier (*i*) and the hub customer (*j*), respectively, not including space for maneuvering, is determined as follows:

$$A_i(\tau) = n_i(\tau) * (L_{ij} * w_{ij}) \text{ and } A_j(\tau) = n_j(\tau) * (L_{ij} * w_{ij}) \qquad (4.3d)$$

where

L_{ij}, w_{ij} is the length and width of the vehicle's footprint relevant for dimensioning the parking stand (m, m).

(e) Social performances

The social performances of a given supply chain are considered to be (i) noise; (ii) congestion; and (iii) safety (i.e. the risk of potential traffic incidents/accidents), which are all primarily related to the chain's transport operations (Janić and Vleugel 2012).

Noise

Noise is generally generated by the transport vehicles (trains, trucks, barges and aircraft) serving the supply chain while passing an exposed observer. The noise mainly depends on the level generated by the source, i.e. moving vehicle and its distance from an exposed receiver. This distance changes over time, during the vehicle's passing, as follows:

$$\rho_{ij}^2(t) = (L_{ij}/2 + \beta_{ij} - v_{ij}t)^2 + \gamma_{ij}^2 \text{ for } 0 < t \le (L_{ij} + 2\beta_{ij}/v_{ij}) \qquad (4.3e)$$

The noise to which the above-mentioned receiver is exposed by the passing vehicle is determined as follows:

$$L_{eq}[\rho_{ij}(t), v_{ij}] = L_{eq}(\gamma_{ij}, v_{ij}) - 8.6562\ln[\rho_{ij}(t)/\gamma_{ij}] \qquad (4.3f)$$

The noise from $f_{ij}(\tau)$ successive passing vehicles over the period (τ), i.e. during the chain's production/consumption cycle, is determined as follows:

$$L_{eq}[f_{ij}(\tau)] = 10\log\sum_{r=1}^{f_{ij}(\tau)} 10^{\frac{L_{eq}(\rho_{ij(t)}, v_{ij})}{10}} \qquad (4.3g)$$

where

$L_{eq}(\gamma_{ij}, v_{ij})$ is the noise of a passing vehicle at the speed (v_{ij}) and distance (γ_{ij}) (dB(A) – decibels);

v_{ij} is the speed of a passing vehicle serving the supply chain (ij) (km/h); and

γ_{ij}, β_{ij} is the shortest (right angle) and slant distance, respectively, between the noise source, i.e. moving vehicle serving the supply chain (ij) and the exposed observer (m).

The other symbols are analogous to those in the previous equations.

The second term in Equation 4.3f represents the noise attenuation over an area free of barriers between the noise source, i.e. the moving vehicles serving the given supply chain and an exposed receiver (*see* also Fig. 2.32 in Chapter 2). Sea ships are excluded from consideration mainly due to the nature of their operations on high seas.

Congestion

Congestion depends on the type of vehicle/transport mode serving the given supply chain (*ij*). In general, freight trains, aircraft and sea ships are given time slots for accessing and using the transport infrastructure around and between the hub supplier(s) and the hub customer(s) (rail/intermodal terminals, airports, sea port terminals), thus substantially diminishing their contribution to the overall congestion.

For example, trucks serving the supply chain (*ij*) cause congestion and consequent time losses of individual vehicles/cars trailing behind since the latter are not able to overtake them along the road(s) connecting the hub supplier (*i*) and the hub customer (*j*). The time a vehicle/car spends before overtaking a truck serving the supply chain (*ij*) can be estimated using the theory of stochastic and deterministic queuing systems. This assumes that the vehicles/cars are waiting to enter the road segment currently occupied by a truck, in which case they represent the arriving customers. The time the truck occupies the road segment represents their service time. Consequently, the average time a vehicle/car waits before starting to overtake the given truck is estimated as follows (Janić and Vleugel 2012, Van Woensel and Vandaele 2007):

$$
W_{q/ij/c} = \begin{bmatrix} L_{ij/t}/(v_{ij/t} - L_{ij/t} * \Lambda_{ij/c}) & \text{if} & \Lambda_{ij/c} < v_{ij/t}/L_{ij/t} \\ (1/2)\Delta t_{ij}(L_{ij/t}\Lambda_{ij/c}/v_{ij/t} - 1) & \text{if} & \Lambda_{ij/c} > v_{ij/t}/L_{ij/t} \end{bmatrix} \quad (4.3h)
$$

where

$L_{ij/t}$ is the length of a truck serving the supply chain (*ij*) including the safe front and rear buffer distance (space) from the other vehicles (m);

$v_{ij/t}$ is the average speed of a truck serving the supply chain (*ij*) (m/s);

$\Lambda_{ij/c}$ is the intensity of flow of vehicles/cars intending to overtake, i.e. to 'occupy the space' currently occupied by the truck serving the supply chain (*ij*) (veh/s); and

Δt_{ij} is the time in which the intensity of the flow of vehicles/cars to overtake the truck serving the supply chain (*ij*) exceeds the truck service time (s).

The total waiting time of vehicles trailing behind all trucks serving the given supply chain can be calculated by multiplying the transport service frequency during the chain's production/consumption cycle and the average waiting time determined through Equation 4.3h.

Safety (i.e. the costs of risk of vehicle loss in an accident)

The costs of risk of vehicle loss (including its load) in an accident occurring during the production/consumption cycle of the given chain is estimated as follows:

$$C_{RAC}(q_{ij}) = a_{ij} * IP(q_{ij})$$ (4.3i)

where

a_{ij} is the probability of an accident causing the loss of a vehicle and its load while serving the supply chain (*ij*) (event/TU); and

$IP(q_{ij})$ is the insurance premium for a vehicle of the payload capacity (q_{ij}) serving the supply chain (*ij*) (€ or $US/veh).

The other symbols are analogous to those in the previous equations.

4.2.4 Application of the Models

4.2.4.1 *Case*

The above-mentioned models of supply chain performances are applied to the case of the supply chain between North Europe and the Far East served by liner container shipping services. The hub supplier is assumed to be the port of Rotterdam—APM Terminals Rotterdam (The Netherlands) and the hub customer is assumed to be the port of Shanghai—Yangshan Deepwater Port Phases 1/2 or 3/4 (People's Republic of China). Currently, this is one of the world's busiest chains (sea trading routes),[2] an overview of which is provided in Fig. 4.4.

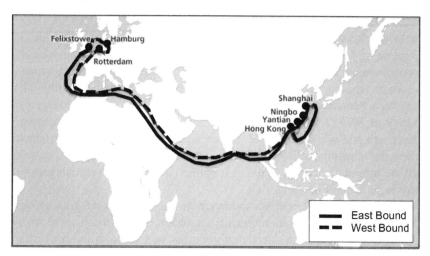

Fig. 4.4: Simplified overview of the given supply chain geography: Rotterdam–Shanghai liner shipping route. (Janić 2014b, http://www.ship.gr/news6/hanjin28.htm)

---

[2] This chain (sea trading route) included in the WCI (World Container Index) together with the remaining 10 most voluminous global container chains (sea trading routes) shares about 35 per cent of their total volumes (TEUs) (http://www.worldcontainerindex.com/).

In addition, Fig. 4.5 shows the development of container cargo flows on the trading routes between Europe and Asia in both directions over time.

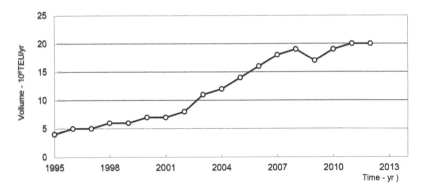

Fig. 4.5: Development of the container cargo volumes on the Europe-Asia-Europe trading route over time (UNCTAD 2013).

The given route accounts for almost 70 per cent of this total. The container terminals at both ports of the given route enable access and operations of large container ships, including the currently largest 'Triple E Maersk'. The collection and distribution of goods/freight shipments (TEUs) at both ports is carried out by rail/intermodal, road, inland waterway (barge) and feeder (including short-sea) vessel transport modes (Zhang et al. 2009). Two scenarios of operating the given chain (route) are considered: the first is the exclusive use of container ships of the capacity of 4000 TEU (or the current 'Panamax'); the other is the exclusive use of container ships of the capacity of 18000 TEU (i.e. 'Neo Panamax' represented by the Triple E class ship introduced by Maersk in 2013) (AECOM/URS 2012, http://www.worldslargestship.com/). The length and width (beam) of the container ships, as well as their above-mentioned capacity, are specified by design (http://en.wikipedia.org/wiki/List_of_largest_container_ships). A simplified overview of both categories of ships is shown in Fig. 4.6.

In addition, only direct transportation of the containerized goods/freight shipments in the single direction of the chain is considered. Due to the specifics of the given case, social performances, such as noise and congestion, as defined in the above-mentioned models, are not considered. However, this does not compromise the quality and generality of the models' application.

4.2.4.2 Inputs

The input data for application of the proposed models to the given supply chain are collected from the case itself and the other various sources given in Table 4.1.

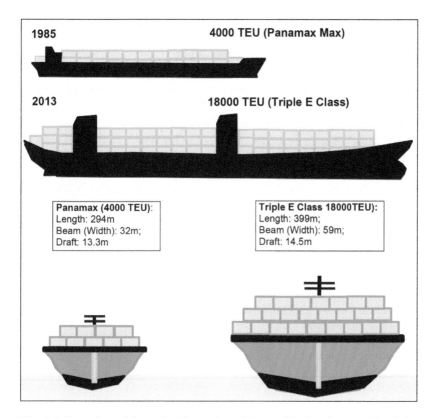

Fig. 4.6: Overview of the scale of container ships used in the given supply chain (Janić 2014b, PR 2011).

The number of containers (TEU) per the chains' production/consumption cycle of one year is determined by assuming the service frequency of Triple E class ships of 1 dep/week, of Panamax class ships of 5 depts/week and the average load factor of both ship classes of 0.80. These give the total annual number of 748800 TEUs to be transported within the chain according to the specified scenarios implying use of exclusively one class of ships under the given conditions. This is, however, only about one-sixth of the total annual number of TEUs transported within the chain (http://www.worldcontainerindex.com/).

The rates of collection and distribution of goods/freight shipments (TEUs) are set up in light of the service schedule of various inland transport modes serving the ports (terminals) at both ends of the chain (route) (Zhang et al. 2009). The container loading and unloading rates are set up based on the empirical evidence provided by both ports/terminals. In general, both 'Panamax' and 'Triple E' class ships are loaded/unloaded by using three or

Table 4.1: Input data for application of the models of performances to the given supply chain – the Rotterdam (The Netherlands) – Shanghai (China) liner shipping route (Janić 2014b)

Input variable	Notation/Unit	Value
• Duration of the chain's production/consumption cycle	τ (year(s))	1
• Number of containers per chain's production/consumption cycle	Q_{ij} (TEU/yr)	748,800
• Container ship capacity	q_{ij} (TEU/ship)	4000; 18,000
• Container ship length	L_{ij}(m)/q_{ij}(TEU/ship)	294 (4,000); 399 (18,000)
• Container ship beam (width)	w_{ij}(m)/q_{ij} (TEU/ship)	32 (4,000); 59 (18,000)
• Container ship load factor	λ_{ij}	0.80 (4,000); 0.80 (18,000)
• Time between the ships' scheduled departures between hubs (days)	h_{ij}/q_{ij}(TEU/ship)	1.5 (4,000); 7(18,000)
• Collection rate of containers at the hub supplier port	θ_i (TEU/d)	1,100
• Proportion of used collection rate of containers at the hub supplier port	r_i	1.0
• Distribution rate of containers at the hub customer port	θ_j (TEU/d)	1,100
• Proportion of used distribution rate of containers at the hub customer port	r_j	1.0
• Loading rate of containers at the hub supplier port	μ_i (TEU/hr)	92 (3-4 cranes) 215 (7-8 cranes)
• Proportion of used loading rate of containers at the hub supplier port	p_{i1}	1.0
• Unloading rate of containers at the hub customer port	μ_j (TEU/hr)	94 (3-4 cranes) 215 (7-8 cranes)
• Proportion of used unloading rate of containers at the hub customer port	p_j	1.0
• Time between collecting and loading containers at the hub supplier port beginning	Δ_i (d)	1
• Time between unloading and distributing containers at the hub consumer port beginning	Δ_j (d)	1

(Contd.)

Table 4.1: (*Contd.*)

Input variable	Notation/Unit	Value
• Average occupancy time of a berth by a ship at the hub supplier port	τ_{i1} (d)/μ_i (TEU/hr)/ q_{ij}(TEU/ship)	1.45/ 92/(4,000) 6.52/92/(18,000) 2.79/215/(18,000)
• Average occupancy time of a berth by a ship at the hub customer port	τ_{j1}(days)/μ_i (TEU/hr)/ q_{ij}(TEU/ship)	1.41/ 94/(4,000) 6.38/94/(18,000) 2.79/215/(18,000)
• Operating distance between the hub ports	d_{ij} (nm)	10525
• Average operating speed of container ship	v_{ij} (kt)	20 (Slow steaming) 15 (Super slow steaming)
• Portion of the maintained average ship's operating speed	s_{ij}	1.0
• Proportion of realized transport services	β_{ij}	1.0
• Average delay per realized transport service	D_{ij} (d)	0.0
• Container inventory costs at the hub ports	α_i, α_j (€/TEU-d)	124; 124
• Container costs of time in transportation	α_{ij} (€/TEU-d)	10.6
• Container handling costs at the hub supplier port	c_i (€/TEU)	185
• Container handling costs at the hub customer port	c_j (€/TEU)	58
• Container ship operating costs	c_{ij} (€cents/TEU-nm)/ v_{ij}(kts)/q_{ij}(TEU/ship)	9.90/20; 5.49/15 (4,000) 2.01/20; 1.13/15 (18,000)
• Average fuel consumption of container ship	$f_{c/ij}$ (ton/d)/ v_{ij} (kt)/q_{ij}(TEU/ship)	221/20; 111/15 (4,000) 249/20; 150/15 (18,000)
• Average emission rate of GHG (Green House Gases) of container ship	e_{ij}(tonCO$_{2e}$/d)/ v_{ij}(kt)/q_{ij}(TEU/ship)	688/20; 346/15 (4,000) 775/20; 467/15 (18,000)
• Risk of accident of container ship	a_{ij} (probability of 1 event/year)	$8.876 * 10^{-4}$

four cranes simultaneously (Mongelluzzo 2013). In addition, it is considered that the 'Triple E' class ships are loaded/unloaded by up to seven or eight cranes simultaneously at both ends of the chain (route) (SCG 2013). All selected crane rates are considered to be fully operational over the period of 24h/day.

The time between docking and loading and unloading of ships starting at the corresponding ports is chosen as an illustration, although the chosen duration could be reasonable in light of the administrative procedures to be carried out after the ship docks at the berth. The ships are assumed to operate along the route at constant (slow or super slow steaming) speed(s) without any substantial variations (http://www.sea-distances.org/). This implies that all transport services are assumed to be perfectly reliable, i.e. without delays along the route and consequently at the destination.

The inventory costs of container(s) during collection and loading at the hub supplier port (Rotterdam) and unloading and distribution at the hub customer port (Shanghai) are estimated based on the average retail value of goods in containers and the typical share of the inventory costs (25 per cent) in that value (REM Associates 2014). The costs of container time during transportation are considered as an average for the goods/freight shipments carried out by sea (VTI, 2013).

The handling costs of containers at both port terminals are based on empirical evidence (EC 2009). The costs of container ship(s) operating on high seas are estimated in light of the effects of cruising/operating speed(s) on the fuel consumption, fuel price (assumed constant) and the share of fuel cost in the total ship's operating costs (AECOM/URS 2012, Cullinane and Khanna 2000, Davidson 2014, Janić 2014b, Sys et al. 2008, Stopford 2003, http://www.scdigest.com/ontarget/13-09-12-1.php?cid=7401).

The fuel consumption of container ship(s) is estimated as the quantity of fuel used per day while operating on high seas at the given operating/cruising speed. In addition, the corresponding emissions of CO_2 (carbon dioxide) as the predominant GHG in the total emissions of GHG are calculated using the emission rate of $e_k = 3.114$ gCO_2/g of fuel [No. 6 Diesel or HFO (Heavy Fuel Oil)]. The fuel consumption and related emissions of CO_2 during the ships' time at berths in the ports are not taken into account (AECOM/URS 2012, Janić 2014b, Rodrigue 2013a, http://www.scdigest.com/ontarget/13-09-12-1.php?cid=7401).

Finally, the risk of incidents/accidents causing a loss of one container ship per period of time (one year) is estimated as the product of two probabilities: (i) the probability of losing a container ship in a freight ship accident; and (ii) the probability of such an accident occurring within the given chain/route (region). The former is estimated as the quotient of the total number of lost container ships (35) and the total number of lost (freight/cargo) ships in accidents (1,547), while the latter probability is estimated as the quotient of the number of ships lost in accidents that occurred along and near the given chain (route) and the total number of ships lost at 10 geographical locations worldwide (0.51). Both probabilities are estimated using the relevant data for the period 2001-2013 (Allianz 2013, UNCTAD 2013).

4.2.4.3 Results

The results of the application of the models of performances to the given supply chain, based on the input data in Table 4.1, are shown in Figs 4.7, 4.8, 4.9 and 4.10.

(a) Infrastructural and technical/technological performances

The infrastructural and technical/technological performances of the given supply chain are specified in the form of inputs for the models of other performances as given in Table 4.1. The former implicitly assume the given demand of goods/freight shipments during the chain's production/consumption cycle, the availability of the berths in both port terminals to accommodate container ships of any size including mega ships and the length of sailing route between the hub ports of the chain. The latter include the container ship characteristics (payload capacity and dimension) and the number and rate of loading/unloading devices (cranes) of container ships, including the reliability of their daily operation.

(b) Operational performances

The operational performances of the given supply chain, such as service frequency, fleet size and technical productivity are shown in Fig. 4.7 (a, b, c), respectively.

Figure 4.7a shows that the transport services provided by smaller ships need to be about five times more frequent than those provided by mega ships in order to transport the required number of containers (TEUs) in the given supply chain under the given conditions. Figure 4.7b shows that such higher service frequency requires about three times greater fleet of smaller ships than that of mega ships. Both fleets need to be further increased (by about 35 per cent and 20 per cent, respectively) if operating at the super slow (15 kts) instead of the slow (20 kts) steaming speed. Figure 4.7c shows that the technical productivity of mega ships is higher than that of their smaller counterparts in proportion to the difference in their size/capacity. However, the productivity of both classes of ships decreases (by about 33 per cent) as the operating/cruising speed falls.

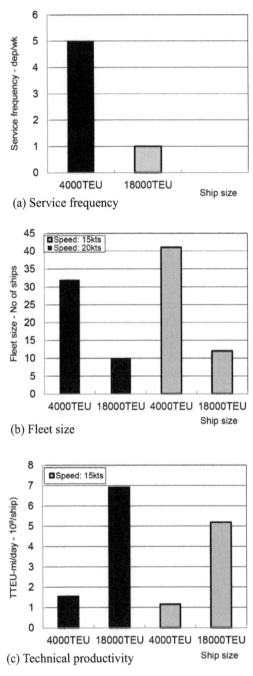

(a) Service frequency

(b) Fleet size

(c) Technical productivity

Fig. 4.7: Operational performances of the given supply chain (Janić 2014b).

(c) Economic performances

The economic performances of the given supply chain such as the average ship (transport) costs, the average costs of the supply chain including the inventory costs during collecting/loading and unloading/distributing of containers (TEUs) and the average chain's costs including only the inventory costs during loading and unloading of containers (TEUs) are shown in Fig. 4.8 (a, b, c), respectively.

Figure 4.8a shows that, in relative terms, if only transport costs are considered, mega ships are about five times more cost efficient than their smaller counterparts while operating on high seas at either steaming speed (20 kts or 15 kts). This unit cost difference appears to be in line with the differences in the ships' size/capacity, thus confirming the existence of substantial economies of scale of the mega ships under the given conditions. Figure 4.8b shows the total chain average costs consisting of the inventory and handling costs of collecting/loading and unloading/distributing containers (TEUs) at hub ports, their time costs in transportation and transport costs. In such a case, a fleet of smaller ships serving the chain will be more cost efficient (by about 52 per cent and 79 per cent) than a fleet of mega ships at either the slow (20 kts) and super slow (15 kts) steaming speed, respectively. Speeding up the loading and unloading of the fleet of mega ships at the hub ports decreases this difference in favor of the fleet of smaller ships to about 30 per cent (at slow) and 52 per cent (at super slow) steaming speeds. In addition, reducing the steaming speed decreases the chain average costs much more when served by a fleet of smaller than by a fleet of mega ships by about 24 per cent and 1-1.5 per cent, respectively. Figure 4.8c shows that the chain total average costs decrease by excluding the inventory costs during collecting and distributing containers (TEUs) at the hub ports. Under such conditions, the chain becomes more cost efficient when served by a fleet of mega ships operating at the slow steaming speed (20 kts) (by about 14 per cent). However, the chain becomes less cost efficient (by about 8 per cent) when served by a fleet of mega ships operating at the super slow steaming speed (15 kts). If the loading and unloading of mega ships at the hub ports is speeded up, the chain's inventory costs substantially decrease, resulting in a decrease of the total average costs. Consequently, if all other costs remain unchanged, the chain served by a fleet of mega ships operating at slow and super slow steaming speed(s) becomes much more cost efficient (by 62 per cent and 34 per cent, respectively) than when it is served by a fleet of smaller ships.

Table 4.2 gives the structure of the chain average costs when the inventory costs during collecting/loading and unloading/distribution of containers (TEUs) at the hub ports are included.

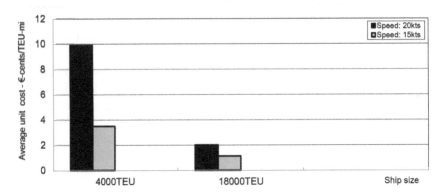

(a) Average transport (ship) operating costs

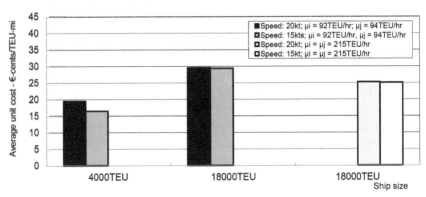

(b) Average chain costs including the inventory costs during collecting/loading and unloading/distributing containers (TEUs)

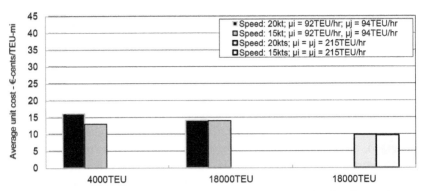

(c) Average chain costs including only the inventory costs during loading and unloading containers (TEUs)

Fig. 4.8: Economic performances of the given supply chain (Janić 2014b).

Table 4.2: Structure of the total costs of the given supply chain
(EC 2009; Janić 2014b)

Operating characteristics	Container ship capacity (TEU)		
	4000	18000	18000
Loading/Unloading rate (TEU/hr)	92/94	92/94	215/215
Operating speed (kt)	20/15	20/15	20/15
Cost component (%)[1]			
Inventory	38/49	85/88	83/86
Handling	12/14	8/8	9/9
Transport	50/37	7/4	8/5

[1]The inventory costs during collecting/loading + unloading/distributing containers (TEU) included.

As can be seen, the share of these (inventory) costs is much lower and the share of transport costs is much higher in the total costs if the chain is served by a fleet of smaller ships compared to when the chain is served by a fleet of mega ships, regardless of their operating speed(s). In any case, reducing the operating speed contributes to the share of inventory costs increasing on account of the share of transport costs. Speeding up the loading and unloading of the mega ships at the hub ports reduces the share of inventory costs very little as compared to when a common loading and unloading speed is applied.

Table 4.3 gives the structure of the chain cost when only the inventory costs during loading and unloading of containers (TEUs) at the hub ports are not taken into account.

Table 4.3: Structure of the total costs of the given supply chain
(EC 2009; Janić 2014b)

Operating characteristics	Container ship capacity (TEU)		
	4000	18000	18000
Loading/Unloading rate (TEU/hr)	92/94	92/94	215/215
Operating speed (kt)	20/15	20/15	20/15
Cost component (%)[1]			
Inventory	24/36	69/75	56/64
Handling	14/18	16/17	24/23
Transport	62/46	15/8	20/13

[1]The inventory cost during loading + unloading containers (TEUs) excluded

As can be seen, by excluding the inventory costs during collecting and distributing containers (TEUs) at both ports, the share of these costs substantially decreases and the share of transport cost increases independent of the class of ship fleet serving the chain. However, the share of the former (inventory) costs remains much higher and the share of the latter (transport) costs remains much lower when the chain is served by a fleet of mega ships compared to when it is served by a smaller fleet. In this case, reducing the ships' operating speed also contributes to increasing the share of inventory costs in the total chain costs.

(a) Environmental performances

The environmental performances of the given supply chain, such as the fuel and emissions of GHG efficiency and use of land/space are shown in Fig. 4.9 (a, b, c), respectively.

Figure 4.9a shows that the fleet of mega ships is between three-and-a-half and four times more fuel efficient than its counterpart of smaller ships under slow and super slow steaming speeds, respectively. In the fleet of smaller ships, changing from the slow (20 kts) to the supper slow (15 kts) steaming speed improves fuel efficiency by about 50 per cent. In the mega fleet, these fuel efficiency improvements amount to about 30 per cent. Figure 4.9b shows very similar relative relationships between the efficiency of emissions of GHG (CO_2) (i.e. EEDI – Energy Efficiency Design Index) of both ship fleets (LR 2011). The fleet of mega ships is again much more efficient, but with lower relative gains achieved by reducing the operating/cruising speed. Figure 4.9c shows that only a single berth is needed at each hub port at both ends of the given chain to accommodate the ship(s) of either class operating under the above-mentioned service frequencies (Fig. 4.7a). However, as intuitively expected, each mega ship(s) occupies about two-and-a-half times larger area of sea near the berth than its smaller counterpart.

(b) Social performances

The social performances reflecting in a certain sense the safety of the given supply chain, such as the costs of the risk of ship loss in an accident are shown in Fig. 4.10.

As mentioned above, these costs are based on the ship's insurance premium and the probability of an accident occurring during the year, causing the ship's loss (the insurance premium for a mega ship is about € 105 million and about € 37 million for a smaller ship (http://www.lloydslist.com/). As can be seen, depending on the operating/cruising speed influencing the required ship fleet size (Fig. 4.7b), the costs of risk of loss of a smaller ship fleet are about 13-14 per cent greater than those of a mega ship fleet. This is due to the fact that despite the insurance premium for the smaller ship(s) being lower (by a factor of about 2.8) compared to that of the mega ship(s), the fleet size of the former is greater than that of the latter by a factor of about 3.2-3.4.

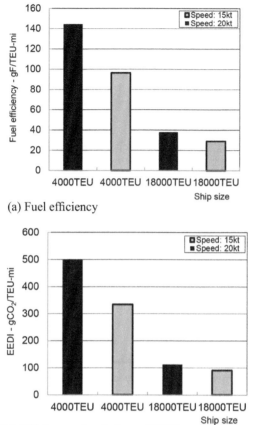

(a) Fuel efficiency

(b) Efficiency of emissions of GHG

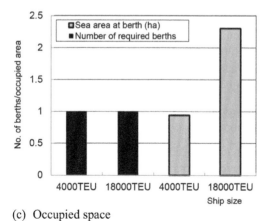

(c) Occupied space

Fig. 4.9: Environmental performances of the given supply chain (Janić 2014b).

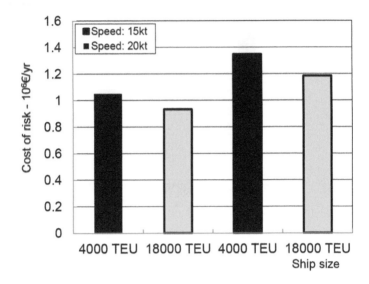

Fig. 4.10: Social performances of the given supply chain –the costs of the risk of ship loss in an accident (Janić 2014b).

4.2.5 Interim Summary

This section has dealt with modelling the operational, economic, environmental and social performances of the supply chain(s) served by different classes of freight vehicles, including mega freight vehicles . In the given context, the supply chain has been considered as a 'transport system'. The models have been applied to an intercontinental supply chain served by liner shipping according to the specified scenarios of exclusively using: (i) nominal container ships (i.e. the 'Panamax' class with a capacity of 4000TEU (Twenty Foot Equivalent Units)), and (ii) mega container ships (i.e. the 'Triple E' class of the capacity of 18000TEU). The results of the application of the models have shown the effects of using a mega container ship fleet on the chain's performances in the given case:

Operational performances: Lower service frequency, smaller required fleet and higher technical productivity;

Economic performances: Significantly lower transport (operational) costs and substantially higher average total costs of the chain due to the dominance of inventory costs;

Environmental performances: Significantly greater relative efficiency of the fuel consumption and related emissions of GHG and larger area at sea, at berths and of the land occupied by the berths needed for accommodation;

however, fewer berths are required due to the lower service frequency despite the longer berth occupancy time; and

Social performances: Lower costs of risk of ship loss in an accident mainly due to the smaller fleet, which consequently results in higher costs of risk of ship loss. Consequently, in general, mega freight transport vehicles can influence the performances of supply chain(s) in relative and absolute terms, both positively and negatively as mentioned above.

4.3 Closely-Spaced Parallel Runways Supported by New Technologies

4.3.1 Background

The runway system capacity of many airports worldwide has reached the saturation point due to continuously growing air transport demand on the one hand and imposing of different constraints in providing adequate capacity to handle such demand, on the other. Some illustrative examples of airports operating at saturation of the runway system capacity during the whole day are London Heathrow airport (United Kingdom) and Dubai International airport (DXB) (United Arab Emirates). London Heathrow (LHR) airport operates two widely-spaced parallel runways in the segregated mode during a limited time of the day mainly due to noise constraints. Dubai International (DXB) airport operates two closely-spaced parallel runways mostly as a single runway without any specific constraints. The former airport is the world's top airport in handling international passengers. The latter airport is one of the fastest growing in the world by developing into the Middle East's strategic hub. Both airports have handled significant proportion of long-distance/intercontinental flights carried out by large/heavy aircraft, including the largest/super heavy A380 aircraft. Under current and prospective conditions characterized by the further growth of air transport demand in terms of the number of passengers, cargo volumes and aircraft operations, both airports have been considering solutions for increasing the runway system capacity as the crucial element for improving the overall operating performances. At Heathrow airport, there has been a longer than 20 years' debate about building a third (parallel) runway (*see* also Chapter 7). At Dubai airport, the new airport (DXC) has been built mainly due to lack of space at the existing DXB airport for building an additional (third) runway. The DXB airport is located almost in the city center with two runways surrounded by passenger and cargo terminals and these are surrounded by other city buildings. Such a development seems to lead to constituting the airport system for Dubai including the existing (DXB) and new (DXC) airport. But in the given context, DXB airport operating two closely-spaced parallel runways has been no exception. Currently, US airports operate 28 pairs of closely-spaced parallel runways. Despite the growing demand, most of these airports are unable to build an additional new runway in order to cope with the imminent growing demand. The main barriers are various spatial, environmental and social constraints, similar to those at the above-mentioned two characteristic airports.

Under such conditions, one short- to medium-term solution for the eventual (marginal) increase in the runway system capacity of closely-spaced parallel runways could include deploying innovative operational procedures supported by new technologies, which have been developing in the scope of the European SEASR (Single European Sky ATM Research) and US NextGen (Next Generation) research and development programs (Erzberger 2004, http://www.sesarju.eu/; http://www.faa.gov/nextgen/).

This section aims at elaborating the effects of some of the above-mentioned new technologies and related operational procedures on the capacity of two closely-spaced parallel runways at the given congested airport (Janić 2016).

4.3.2   The System and Problem—Some New Technologies and Innovative Procedures

Up to now, there have been various short-, medium and long-term solutions for matching the airport airside demand to the runway system capacity at congested airports. These have been applied exclusively and/or in different combinations. For example, short- and medium-term solutions embrace (i) optimum utilization of the existing runway system capacity, (ii) deployment of the above-mentioned innovative operational procedures supported by new technologies, (iii) tactical and strategic air traffic demand management including GHP (Ground Holding Program), and (iv) changing congestion. The long-term solutions include building additional runway(s) at a given airport and/or the airport system and building a completely new airport as in the case of DXB.

The short- to medium-term solution of deployment of innovative operational procedures supported by new technologies seems to be particularly promising in increasing the capacity of two closely-spaced parallel runways (spaced laterally less than 760 m) by enabling safe dependent, i.e. paired ILS/MLS landings and take-offs (Janić 2008, FAA 2013). Such operations could be primarily supported by the WTMA (Wake Turbulence Mitigation for Arrivals) and WTMD (Wake Turbulence Mitigation for Departures) integrated automated system recently deployed at several US airports.

The additional new technologies supporting individual and paired landings could be: ADS-B (Automatic Dependent Surveillance Broadcast) in combination with CDTI (Cockpit Display Traffic Information), SWIM (System Wide Information Management), TFDM (Terminal Flight Data Manager) and TFMS (Traffic Flow Management System), ASDE X (Airport Surface Detection Equipment – Model X) and IDACS (Integrated Departure and Arrival Coordination System), all with the ATC/ATM (Air Traffic Control/Management) ground components and avionics. The above-mentioned technologies, with the exception of the latter two, could also support take-offs (FAA 2013, http://www.faa.gov/nextgen/; http://www.sesarju.eu/).

Specifically, as applied to closely-spaced parallel runways, the WTMA and WTMD systems, providing continuous monitoring and forecasting of the crosswind conditions, enable on-line modification of the existing (no-crosswind) ATC/ATM longitudinal wake-vortex separation rules. In case of landings, this could include applying exclusively or in combination with the existing longitudinal, also diagonal (authorized under FAA Order 7110.308) and/or still not fully authorized vertical separation rules between paired dependent operations. The diagonal separation rules could be applied under conditions when the persistent crosswind is blowing the wakes made by the leading aircraft away from the path of trailing aircraft in the given landing sequences. The vertical separation rules could be applied to the given landing sequences in combination with the constant or a steeper GS (Glide Slope) angle and usually staggered landing thresholds under all weather (crosswind) conditions. They enable the trailing aircraft to remain all the time longitudinally closer to but above the (sinking) wakes of the leading aircraft in the given landing sequence. In addition, the above-mentioned procedures applied to closely-spaced parallel runways appear to be particularly convenient mainly due to avoiding deficiencies of the limited runway length and increased traffic complexity, both as compared to their single-runway counterpart (Janić 2008, 2012, Kolos-Lakatos and Hansman 2013, Tittsworth et al. 2012).

In case of take-offs, the existing or slightly modified ATC/ATM time-based separation rules could be applied under convenient crosswind conditions. The latter implies in particular to successive takes-offs, sequentially always from a different runway while using the lift-off time as a component of the time-based separation rules in combination with diverging trajectories assigned to the successive departure aircraft immediately after take-off.

For mixed operations, an innovative procedure applicable under convenient crosswind conditions could be to allow take-off(s) from a runway different from that of the preceding landing(s), i.e. without the need of waiting for the previously landing aircraft to clear its runway as in the case of a single runway. At the same time, this take-off should be safely longitudinally separated from the following landing at either runways.

4.3.3 Modelling the Capacity of Closely-Spaced Parallel Runways

4.3.3.1 Some Related Research

Modelling of the 'ultimate' and 'practical' capacity of the airport runway systems has occupied researchers, planners and the aviation industry for considerable time. As a result, many analytical and simulation models have been developed. In particular, analytical models usually provide two value parameters for a single runway—one for landing and the other for take-off (Blumstein 1959, Donohue 1999, Gilbo 1993, 1997; Harris 1972, Hockaday and Kanafani 1974, Janic and Tosić 1982, Janić 2006, 2016, Newell 1979). Some other models including the FAA Airport Capacity Model, the LMI Runway Capacity Model and DELAYS as 'Quasi-Analytical Models of Airport Capacity and Delay', based on the analytical single-runway 'ultimate' and 'practical' capacity models, calculate the 'ultimate' capacity coverage curves and associated aircraft delays, both enabling deriving the 'practical' capacity under the given conditions (Gilbo 1993, Newell 1979). Recently, analytical models for estimating the ultimate landing capacity of closely-spaced parallel runways and investigating the effects of innovative operational procedures supported by the new technologies developing within the European SESAR and US NextGen research programs, have been developed (Janić 2008, 2014c, 2015, 2016; http://www.faa.gov/nextgen/, http://www.sesarju.eu/) (*see also* Chapter 3, Section 3.2).

4.3.3.2 Objectives and Assumptions

The objective is to develop a methodology consisting of several models to investigate the effects of the above-mentioned innovative operational procedures supported by new technologies on the potential increase in the current 'ultimate' and 'practical' capacity of two closely-spaced parallel runways at DXB airport. For such a purpose, the latest analytical models for calculating the 'ultimate' capacity and appropriately adapted existing models for calculating 'practical' capacity (based on the steady-state queuing model) of closely-spaced parallel runways are appropriately modified keeping in mind the most recent proved and prospective developments. Such modified models are based on the following assumptions:

- The demand for landings and take-offs on the given two closely-spaced parallel runways is constant during the specified period of time (usually one hour);
- The two closely-spaced parallel runways are used in the mixed mode, depending on the prevailing demand, simultaneously for paired landings, paired take-offs and paired mixed landings/take-offs;

- The aircraft are categorized according to their wake-vortex characteristics mainly depending on the aircraft MTOW (Maximum Take-Off Weight), wing span, while respecting prevailing weather (wind) conditions, with all influencing the approach and landing speed and the runway landing/take-off occupancy time;
- The aircraft landing speeds are constant along the final approach trajectories connecting FAGs and runway landing threshold(s);
- The aircraft strictly follow their prescribed four-dimensional approach/departure trajectories appearing exactly as being expected at particular locations;
- The ATC/ATM minimum longitudinal, diagonal and vertical distance-based separation rules are applied between landings exclusively or in various combinations; the existing and/or modified time-based separation rules are applied between take-offs; and
- The maximum average delay per ACM (Aircraft Movement) is specified in order to derive the 'practical' capacity from the calculated 'ultimate' capacity, using the delay-capacity relationship under steady-state conditions.

4.3.3.3 Structure of the Methodology

(a) 'Ultimate' capacity

Landings and taking-offs

As at the other analytical models of 'ultimate' runway system capacity, the average inter-event time for different combinations of landing and/or taking-off sequences at the corresponding runway thresholds of two closely-spaced parallel runways can be calculated as follows:

$$\bar{t} = \sum_{i/k, j/l} p_{i/k} * \tau_{ik/jl} * p_{j/l} \tag{4.12a}$$

and the landing and/or taking-off capacity (ACM/h):

$$\lambda = T / \bar{t} \tag{4.12b}$$

where

i, j is the leading and trailing aircraft category, respectively, in the landing sequence $(ij)[(i, j) \in N]$;

N is the number of the aircraft wake-vortex categories in the landing and/or departing fleet mix;

k, l is the landing runway of the aircraft (i) and (j), respectively $(k, l = 1, 2)$;

$p_{i/k}, p_{j/l}$ is the proportion of the aircraft category (i) and category (j) in the aircraft fleet mix, which land at or depart from the runway (k) and (l), respectively;

$\tau_{ik/jl}$ is the minimum time between landing or departing of the aircraft of category (i) and (j) at and from the runway (k) and (l), respectively (s); and

T is the time interval for which the capacity is calculated (h).

Total capacity

The total runway system capacity for mixed operations (ACM/h) can be calculated as follows:

$$\lambda = (1 + q_d)\lambda_a \qquad (4.13a)$$

$$\lambda = (1 + q_a)\lambda_d \qquad (4.13b)$$

where

q_d, q_a is the probability of time gaps enabling safe take-offs and/or landings between successive landings and/or take-offs, respectively; and

λ_a, λ_d is the landing and taking-off runway system capacity, respectively (ACM/hr).

Inter-event time between landings

The inter-event time $\tau_{ik/jl}$ in Equation 4.12a for landings is determined for different sequences respecting the aircraft final approach speeds and the ATC/ATM separation rules applied. For example, Figure 4.12(a, b) shows scenarios in which the ATC/ATM minimum vertical separation rules are applied to different landing sequences. The following notation is used:

$\alpha_{i/k}, \alpha_{j/l}$ is the GS (Glide Slope) (i.e. final approach) angle of the leading and trailing aircraft (i) and (j) landing at the runways (k) and (l), respectively (^{0});

$\gamma_{i/k}, \gamma_{j/l}$ is the length of final approach path of the leading aircraft (i) landing at the runway (k) and of the trailing aircraft (j) landing at the runway (l), respectively (nm);

ε_{kl} is the staggered distance between two closely-spaced parallel runways (k) and (l) (nm);

$\delta^l_{ij}; \delta^d_{ij}$ are the ATC/ATM minimum longitudinal and diagonal separation rules, respectively, applied to the aircraft landing sequence (ij) (nm – nautical mile);

d_{kl} is the lateral (right angle) separation of two closely-spaced parallel runways (nm);

H_{ij}^0 are the ATC/ATM minimum vertical separation rules applied to the aircraft landing sequence (ij) (ft – feet));

$v_{i/k}, v_{j/l}$ is the final approach speed of the landing aircraft (i) to the runway (k) and the aircraft (j) to the runway (l), respectively (kt – knot); and

$t_{a/ik}$ is the runway landing occupancy time by the aircraft (i) landing at the runway (k).

(a) Sequence: $v_{i/k} \leq v_{j/l}$

Figure 4.11a(i) shows the case when the leading aircraft (i) lands on the closer runway (k) and the trailing aircraft (j) lands on the staggered runway (l). Figure 4.11a(ii) shows the opposite runway use. In both cases, the minimum separation is established at the moment when the leading aircraft (i) is at its landing threshold.

The inter-arrival time of aircraft (i) and (j) at the landing thresholds of their runways (k) and (l), respectively, when the ATC/ATM horizontal, vertical, or diagonal separation rules are exclusively applied, can be calculated as follows:

$$\tau_{ik/jl} = \max\left\{0; \min\left(\frac{\delta_{l/ij}}{v_{jl}}; \frac{\sqrt{\delta_{d/ij}^2 - d_{kl}^2}}{v_{jl}}; \frac{H_{ij}^0}{v_{jl}tg\alpha_{jl}}\right) \pm \frac{\varepsilon_{kl}}{v_{jl}}\right\} \qquad (4.14a)$$

(b) Sequence: $v_{i/k} > v_{j/l}$

Figure 4.11b(i) shows the case when the leading aircraft (i) lands on the closer runway (k) and the trailing aircraft (j) on the staggered runway (l). Figure 4.12b(ii) shows the opposite runway use. The ATM/ATC minimum vertical separation rules are applied at the moment when the leading aircraft (i) is at its FAG and the trailing aircraft (j) is behind it at a safe vertical (and corresponding longitudinal) distance. The inter-arrival time of the aircraft (i) and (j) at the landing thresholds of their runways (k) and (l) respectively, when different ATC/ATM separation rules—horizontal, vertical, diagonal— are exclusively applied, can be calculated as follows:

$$\tau_{ik/jl} = \frac{\min\left[(H_{ij}^0 \pm \varepsilon_{kl} * tg\alpha_{jl}); \delta_{l/ij}; \sqrt{\delta_{d/ij}^2 - d_{kl}^2}\right] + \gamma_{jl}}{v_{jl}} - \frac{\gamma_{ik}}{v_{ik}} \qquad (4.14b)$$

The term ($\varepsilon_{kl}/v_{j/k}$) in Equation 4.14 (a, b) takes the positive sign ("+") if the leading aircraft (i) lands on the closer and the trailing aircraft (j) on the staggered runway (Fig. 4.11a(i), 4.11b(i)), and the negative sign ("–"), if otherwise (Fig. 4.11a(ii), 4.11b(ii)). If the variable $d_{kl} = 0$, the aircraft (i) and (j) are assumed to land on the same runway at a displaced threshold. If both variables $d_{kl} = 0$ and $\varepsilon_{kl} = 0$ in Equation 4.14 (a, b), both aircraft (i) and (j) land at the same runway threshold.

It should be mentioned that if the aircraft (i) and (j) land on the different runways and if $\varepsilon_{kl} \neq 0$, the runway occupancy time of the leading aircraft (i) in the landing sequence (ij), ($t_{a/ik}$) does not influence the minimum inter-arrival time ($\tau_{ik/jl}$) in Equation 4.14 (a, b).

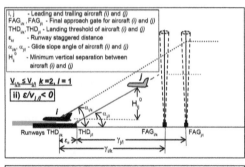

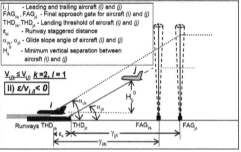

(a) Sequences: $v_{i/k} \leq v_{j/j}$

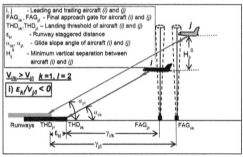

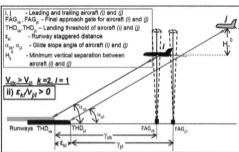

(b) Sequences: $v_{i/k} > v_{j/l}$

Fig. 4.11: Scenarios of landing at closely-spaced parallel runways when the ATC/ATM vertical separation rules are applied (Janić 2008, 2014c, 2015, 2016).

Inter-event time between take-offs

It is assumed that particular take-offs in the given sequences are carried out sequentially always on different runways (k) and (l), respectively. Under conditions without crosswind, the ATC/ATM applies the existing time-based wake-vortex separation rules between successive take-offs. Under conditions of convenient crosswinds, with the support of WTMD, the trailing aircraft in the given take-off sequence can start its take-off from its runway immediately after the leading aircraft has lifted-off from its runway as shown in Fig. 4.12.

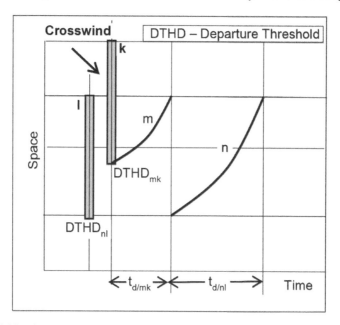

Fig. 4.12: Time-space diagram for the taking-off sequence (m) and (n) from the runways (k) and (l), respectively—convenient crosswinds (Janić 2016).

In order to additionally diminish the impact of wake vortices, the aircraft in the given sequence can be assigned diverging departure trajectories immediately after taking-off (Mayer 2011). Under the above-mentioned conditions, the minimum time between successive take-offs from the closely-spaced parallel runways can be calculated by employing the ATC/ATM time-based separation rules. These depend of the runway occupancy time of the leading aircraft in the given departure sequence, the minimum wake-vortex separation rules (if applicable) and the number of successive departures from the runway of the leading aircraft in the given take-off sequence. Consequently, this minimum time can be estimated as follows:

$$\tau_{d/mk,nl} = [n_{d/k} * \max(t_{d/mk}; \Delta_{d/mn})] \text{ (without crosswind)} \quad (4.15a)$$

and

$$\tau_{d/mk,nl} = n_{d/k} * t_{d/mk} \text{ (with crosswind)} \tag{4.15b}$$

where

m, n is the leading and trailing aircraft in the taking-off sequence [$(m, n) \in N$];

$t_{d/mk}, t_{d/nl}$ is the runway occupancy (lift-off) time of the taking-off aircraft (m) and (n) from the parallel runways (k) and (l), respectively (min.);

$\Delta_{d/mn}$ is the minimum ATC/ATM time-based separation rules between successive taking-off of the aircraft category (m) and (n) (min.); and

$n_{d/k}$ is the number of successive take-offs from the runway (k).

Equation 4.15(a, b) implies that the number of take-offs $n_{d/k}$ is always equal to or greater than 1. Specifically, if $n_{d/k} = 1$, then it is considered to be the take-off of aircraft (m). The above-mentioned Fig. 4.12 shows the time-space diagram when $n_{d/k} = 1$.

Inter-event time between different operations

(a) Take-offs between successive landings

As mentioned above, the aircraft (i) and (j) in the landing sequence (ij) and/or the take-off sequence (mn) are assumed to always land and/or take-off on different runways (k) and (l), respectively, independent of the ATC/ATM separation rules applied. In addition to the previously mentioned, the following notation is used:

$\delta_{da/mj}$ is the minimum longitudinal distance between the taking-off aircraft (m) and the trailing landing aircraft (j) in the landing sequence (ij).

A take-off between any two landings can be carried out in different combinations as follows:

(A) The leading aircraft (i) in the landing sequence (ij) lands on the closer runway $(k = 1)$ and the trailing aircraft (j) on the staggered runway $(l = 2)$ [Fig, 4.11a(i) and 4.11b(i)] while the aircraft (m) take-offs from the same runway where the aircraft (i) landed $(k = 1)$ or from the runway where aircraft (j) is to land $(l = 2)$; and

(B) The leading aircraft (i) lands on the staggered runway $(k = 2)$ and the trailing aircraft (j) on the closer runway $(l = 1)$ [Fig. 4.11a(ii) and 4.11b(ii)] while the aircraft (m) takes-off from the staggered runway $(k = 2)$ where the aircraft (i) landed or the closer runway $(l = 1)$ where the aircraft (j) is to land.

Under conditions of operating closely-spaced parallel runways as a single runway, the previously landed aircraft (i) has to clear its runway and the departing aircraft (m) has to be at the minimum longitudinal distance from the approaching aircraft (j) of $\delta_{da/mj}$ at the moment when starting to take-off, independently of the combination of runways used for landings and take-offs. Figure 4.13 shows the time-space diagram of operating closely-spaced parallel runways under the above-mentioned different (crosswind) conditions.

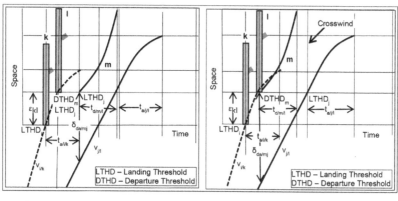

(a) Without crosswind (Case A) (b) With crosswind (Case A)

Fig. 4.13: Time-space diagram for mixed operations at closely-spaced parallel runways under different weather conditions (Janić 2016).

Specifically, Fig. 4.13a shows that the operating Case A is identical to that of a single runway, independently of the crosswind conditions. Figure 4.13b shows the above-mentioned operating Case A under convenient crosswind conditions. With the support of WTMD, WTMA and other above-mentioned technologies, the aircraft (m) can start taking-off from its runway immediately after the landing aircraft (i) touches down on the other runway, implying no waiting of the aircraft (m) for the aircraft (i) to clear its runway. At the same time the approaching aircraft (j) should again be at least at the minimum longitudinal distance $\delta_{da/mj}$ from the aircraft (m) the moment it starts to take-off. A similar time-space diagram as in Fig. 4.13(a, b) can be drawn for the above-mentioned Case B. Consequently, the ATC/ATM minimum time interval enabling (n_d) take-offs between successive landings (i) and (j) under the above-mentioned conditions can be calculated as follows:

$$\tau_{d/ik/jl} = \begin{bmatrix} (n_d - 1) * (t_{a/i/k} + \delta_{ad/mj}/v_{j/l}),\ \text{current and/or under} \\ \text{crosswind if } (i) \text{ and } (m) \text{ use the same runway;} \\ (n_d - 1) * [\max (t_{d/m/k}; \delta_{ad/mj}/v_{j/l})],\ \text{with crosswind} \\ \text{if } (i) \text{ and } (m) \text{ use different runways} \end{bmatrix} \quad (4.16a)$$

where all symbols are analogous to those in the previous equations.

(b) Landings between successive take-offs

A landing can also be carried out between successive take-offs. According to the above-mentioned notation, the landing (*i*) can be realized between two successive take-offs (*m*) and (*n*) departing from the runways (*k*) and (*l*), respectively, if a sufficient time gap is ensured, i.e. the taking-off aircraft (*m*) should, independently of the applied ATC/ATM separation rules, clear its departure runway and the landing aircraft (*i*) should clear its arrival runway, before the successive take-off (*n*) is allowed at either runway. This minimum time gap can be estimated as follows:

$$t_{d/mn} = t_{d/mk} + t_{a/i/k/l} \qquad (4.16b)$$

where all symbols are as in the previous equations.

(b) 'Practical' capacity

When the demand for landings and/or take-offs does not generally exceed the runway system's 'ultimate' capacity over a longer period of time, any ACM delays are stochastic and not particularly long. In such a case, in addition to the 'ultimate' capacity, the specified average delay(s) per ACM can be used for determining the 'practical' capacity to be declared by the airport in terms of the number of slots per hour (or 15 min.) during the day. For such a purpose, the modified expression for the average delay per ACM derived from the steady-state queuing system theory can be used with the following notation (Newell 1979):

λ_p is the 'practical' landing and/or take-off capacity (ACM/hr);

λ_u is the 'ultimate' landing and/or take-off capacity as the reciprocal of the corresponding mean service times [$\lambda_u = 1/\bar{t}$, Equation 4.4 (a, b)] (ACM/hr);

σ is the standard deviation of service time of an arrival and/or of departure (hr^2); and

D^* is the maximum average delay per landing and/or take-off specified for setting up the 'practical' capacity (min.).

The maximum average delay is calculated as follows:

$$D^* = \frac{\lambda_p(\sigma^2 + 1/\lambda_u^2)}{2(1 - \lambda_p/\lambda_u)} \qquad (4.17a)$$

Equation 4.9a is valid if $\lambda_p < \lambda_u$; if it comes closer to one, i.e. if the difference between two capacities decreases, the average delay grows exponentially.

After setting the variable σ = 0 just for the purpose of simplification, the 'practical' capacity can be derived from Equation 4.9a as follows:

$$\lambda_p = \frac{2D^*\lambda_u^2}{2D^*\lambda_u + 1} \qquad (4.17b)$$

where all the symbols are analogous to those as in the previous equations.

4.3.4 Application of the Models

The above-mentioned models of the 'ultimate' and 'practical' capacity are applied to the case of Dubai International airport (DXB), which operates a pair of closely-spaced parallel runways. Since the new airport (DXC) has been built and most of the traffic is likely to move there, the application is only for the purpose of illustrating the generosity of the proposed models to the given and similar cases.

4.3.4.1 Case—Dubai International Airport (DXB)

(a) Demand

The numbers of passengers, volumes of cargo and aircraft movements at Dubai International airport (DXB) have grown tremendously over the past decade (DA 2013). Figure 4.14 (a, b) shows such growth of the number of passengers and ACMs (Aircraft Movement(s)) during the periods 2000-2013 and 2013-2020 (DA 2013).

Figure 4.14a shows the nearly exponential increase in the annual number of passengers from about $12*10^6$ in the year 2000 to about $62*10^6$ in the year 2013. This has been mainly driven by the development of the main airlines— domestic Emirates and its code-sharing partner Qantas, both using the airport as their primary and secondary hub, respectively, of their long-haul hub-and-spoke networks, fly Dubai operating the short-to medium-haul routes of its point-to-point network and more than 130 other international airlines, all serving about 215 destinations worldwide. Simple regression analysis using the data from the period 2000-2013 shows a strong relationship between the annual number of passengers and the above-mentioned main driving forces as follows:

$$PAX_{ap} = 4.90 + 1.48PAX_{al}; R^2 = 0.996$$
$$t-stat \quad (4.776) \quad (18.196) \qquad (4.18)$$

where
PAX_{ap} is the annual number of passengers handled at the airport (10^6); and
PAX_{al} is the annual number of passengers carried by the main airline (Emirates) (10^6).

Figure 4.14b shows that the annual number of ACMs (Aircraft Movement(s)) has similarly grown in line with the growth of the number of

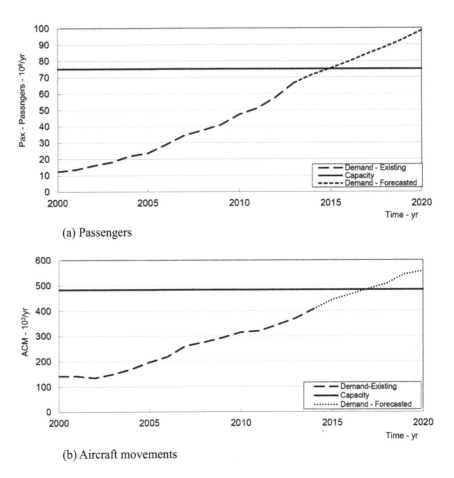

(a) Passengers

(b) Aircraft movements

Fig. 4.14: Development of DXB airport (2000-2013 and 2013-2020) (DA 2013).

passengers, i.e. it has increased from about $141*10^3$ in the year 2000 to about $370*10^3$ in the year 2013. Derived from the previous two, the average number of passengers per aircraft movement has grown more than proportionally with an average of 207 PAX/ACM during the observed period (2000-2013). This is mainly due to the relatively substantial proportion of heavy and supper heavy long-haul aircraft in the airport's fleet mix.

If the above-mentioned developments continue similarly in the future as in the past, i.e. over the period 2013-2020 (both extrapolated by the average past growth rate), the airport's 'practical' annual passenger terminal capacity of $75*10^6$ passengers and the 'practical' runway system capacity of $483*10^3$ aircraft movements will be saturated already in the year 2015 and 2017, respectively, as shown in Fig. 4.14 (a, b).

(b) Terminal airspace, runway system and capacity

The aircraft movements were accommodated in the DXB airport's airside area including terminal airspace, the runway system consisting of two closely-spaced staggered parallel runways, the network of taxiways and 157 apron/gate aircraft parking stands (DA 2013). In particular, the terminal airspace of DXB airport was equipped with four WPs (Way Points) supported by VOR/DMEDXB, all used to define the holding pattern of the arriving aircraft (Jeppsen 2007). Figure 4.15 shows that the holding points are WPs [Way-Point(s)] UKRIM and PEDOW for approaches and landings on RWY (Runway) 12L or 12R, respectively, and WPs SEDPO and LOVOL for approaches and landings on RWY 30L or 30R, respectively.

In the former case, the distance UKRIM - RWY12L is 11.4 nm (nautical mile) and WP PEDOV-RWY12R is 11.5 nm. In the latter case, distances between WPs SEDPO and LOVOL and corresponding RWYs (thresholds) 30L/30R are 11 nm. The approaching and landing aircraft on RWY12R/12L, after leaving the holding pattern, fly between WPs PEDOV and PUDGA, and WPs UKRIM and UMALI, respectively, at a constant altitude of 2000 ft (600 m). The WPs [Way-Point(s)] PUDGA and UMALI represent the FAGs [Final Approach Gate(s)] for starting the final approach and landing, always along the ILS (Instrument Landing System) 3-D defined trajectory. The procedure is similar for approaches and landings on RWYs30L/30R, where WPs MODUS and WP LADGA, respectively, are the FAGs. The holding procedure of 4 min. is performed around all WPs at altitudes between 2000 ft and 4000 ft (Jeppsen 2007).

The length of RWY12L/30R and RWY12R/30L in Fig. 4.15 is 4000 m and 4447 m, respectively, and their width is 60 m. They can accommodate all large/heavy including the largest/super heavy A380 aircraft. The lateral spacing between the two runways is 385 m, i.e. less than 760 m (2500 ft), which categorizes them as closely-spaced parallel runways, currently safely operating as a single runway (DA 2013, Janić 2008). The runways are staggered for 1553 m in courses 12L/30R and 2000 m in courses 12R/30L, respectively.

The 'practical' runway system capacity of DXB airport is specified by the declared number of daily slots for the year 2014—661/661 for landings/taking-offs, respectively. This gives a daily total of 1322 ACM and an annual total of 4,82,530 ACM, if the airport is assumed to operate continuously throughout the year without any constraints affecting its capacity (DA 2013). At the same time, the capacity of the apron/gate complex where the gate/stands are exclusively used by particular aircraft categories amounts to 91 aircraft/h. This gives a total capacity of 2184 aircraft/day and an annual capacity of 7,97,160 aircraft/year, implying continuous operations during the year. The assumption concerning the continuity of airport operations over the year is introduced only for illustrative purposes. Actually, this assumption is highly unrealistic at most airports, including this one (DA 2013, DeNeufville and Odoni 2003).

4.3.4.2 Inputs

The proposed models of 'ultimate' and 'practical' capacity are applied to calculating the runway system capacity of Dubai International airport (DXB). The inputs used are the geometry of the terminal airspace and runway system (Fig. 4.15), characteristics of the current and future aircraft fleet mix, with ATC/ATM current and innovative operational procedures applied to both landings and take-offs (Figs. 4.11-4.13). The characteristics of the terminal airspace and of the runway system are summarized in Table 4.4.

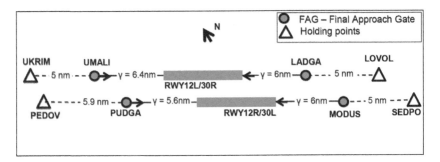

Fig. 4.15: Simplified geometry of the terminal airspace of DXB airport (Jeppsen 2007).

Table 4.4: Characteristics of terminal airspace and runway system at DXB airport (DA 2013)

Runway	Length/ Width (m/m)[1]	Lateral separation d (m)	Staggering distance ε (m)	Length of the final approach path γ (nm)
12L/30R	4000/60	385	1533	6.4/5.6
12R/30L	4447/60	385	2000	6.0/6.0

[1]meter

The characteristics of the current and future aircraft fleet mix are given in Table 4.5.

The aircraft category C/B757 and E/Small are not considered as these do not operate at the airport. The ATC/ATM minimum longitudinal/diagonal distance- and time-based wake-vortex separation rules between landings and taking-offs, respectively, are given in Tables 4.6 and 4.7.

Table 4.5: Characteristics of aircraft fleet at DXB airport (DA 2013)

Aircraft category[1]	Type	Proportion (c/f)[2] (%)	Approach speed[3] (kt)	Runway landing occupancy time (s)	Take-off run (lift-off) time[4] (s)	Runway take-off occupancy time[5] (s)
A/Super Heavy	A380	17/23	145	60	44	60
B/Heavy	A300-600, A330, A340, A350, B747, B767, B777, B787,	69/77	140	60	44	60
D/Large	B737, A320, 321s	14/0	130	55	37	50

[1]RECAT/ICAO categorization; [2]Current/future; [3]Ground speed based on IAS (Indicated Air Speed + headwind of 10 kts); [4]Average (typical) time to lift-off; [5]Time for passing the runway during take-off

Table 4.6: The FAA/RECAT minimum IFR wake-vortex longitudinal separation rules for landings—$\delta_{l/ij}$ and $\delta_{d/ij}$ (nm) (CAA 2014, ICAO 2001, 2008, EEC/FAA 2008, FAA 2012)

A/C sequence	A/Super heavy	B/Heavy	D/Large
A/Super Heavy	2.5[1]	5	7
B/Heavy	2.5	4	5
D/Large	2.5 (1.5)[2]	2.5 (1.5)	2.5 (1.5)

[1]RECAT (Tittsworth et al. 2012); [2]Diagonal separation rules

Table 4.7: The ICAO/FAA minimum wake-vortex time-based separation rules for take-offs—$t_{d/kl/min}$ (min.) (ICAO 2001, FAA 2012, CAA 2014)

A/C sequence	A/Super heavy	B/Heavy	D/Large
A/Super Heavy	2.0	2.0	2.0
B/Heavy	1.5	1.5	2.0
D/Large	1.0	1.0	1.0

In Table 4.6, it is assumed that the ATC/ATM minimum diagonal separation rules applied between paired landings on the closely-spaced parallel runways without any wind conditions/restrictions are $\delta_{d/ij} = 1.5$ nm, if the leading aircraft (*i*) belongs to D/Large and/or E/Small and the trailing aircraft (*j*) to any wake-vortex category. The minimum vertical separation rules applied to any landing sequence are $H_{ij}^0 = 1000$ ft (ft – feet). In addition, the ATC/ATM longitudinal separation rules enabling a take-off between any two landings are $\delta_{d/jk} = 2$ nm (CAA 2014, ICAO 2001, 2008; EEC/FAA, 2008, FAA 2012). Furthermore, the ILS GS (Glide Slope), i.e. final approach and landing angle for all aircraft categories is adopted as: $\alpha = 3°$ (Jeppsen 2007)

4.3.4.3 Scenarios for Calculating Capacity

The 'ultimate' capacity of the runway system at Dubai International airport (DXB) is calculated by using the above-inputs for calculating the scenarios given in Table 4.8.

Table 4.8: Scenarios for calculating the runway system capacity at DXB airport

Capacity	Element	Description
	Runways in use	12L/12R; 30L/30R ($\varepsilon > 0$; $\varepsilon < 0$)
Landings		
	• ATC/ATM minimum separation rules	• Longitudinal FAA/RECAT only • Vertical only • Longitudinal FAA/RECAT + FAA diagonal • Vertical + FAA diagonal
Take-offs		
	• ATC/ATM minimum separation rules	• Current • Weather (crosswind) dependent
Mixed		
	• ATC/ATM minimum separation rules	• Current • Weather (crosswind) dependent

In the scenarios given in Table 4.8, the current and future aircraft fleet mix is considered. The 'practical' capacity is calculated based on the 'ultimate' capacity in the scenarios in Table 4.8.

4.3.4.4 Results

The above-mentioned capacity models are applied to calculating the 'ultimate' and 'practical' capacity using the above-mentioned inputs and scenarios of operating closely-spaced parallel runways at DXB airport. Based on the mentioned 'ultimate' capacities, the 'practical' or 'declared' capacity is additionally calculated and compared to the corresponding current airport-specified capacity values (declared number of slots for the year 2014).

(a) 'Ultimate' capacity

The calculated runway system's 'ultimate' capacity is calculated for the scenarios of using two parallel runways described in Table 4.8 carried out according to the operating procedures shown in Figs. 4.11-4.13 and 4.15. The results show that the landing capacity, independent of the ATC/ATM separation rules applied, is higher if the leading aircraft lands on the staggered and the trailing aircraft on the closer runway than vice versa. Furthermore, this difference increases with increase in the runway staggering distance. The take-off capacity and the capacity for mixed operations remain the same independent of the pattern of runway use. Assuming that different above-mentioned cases of using runways are practiced in equal proportions over a longer period of time, the average runway system capacity, to be used for both planning and operational purposes, can be represented as the capacity envelopes shown in Figs. 4.16, 4.17, 4.18, and 4.19.

Figure 4.16 shows the runway system capacity envelopes when different ATC/ATM separation rules are applied to landings of the current aircraft fleet mix as given in Table 4.5. The capacity envelope of a single runway and the airport-specified 'practical' capacity are also shown as benchmarking cases, i.e. for comparative purposes (DA 2013, Janić 2016).

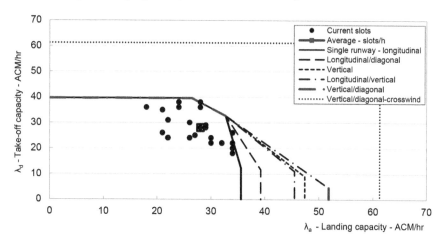

Fig. 4.16: The runway system capacity envelopes at DXB airport when different ATC/ATM separation rules are applied to landings of the current aircraft fleet mix (Janić 2016).

As can be seen and as intuitively expected, the runway system take-off and mixed operation capacity remain the same despite the ATC/ATM separation rules applied between landings. However, the landing capacity appears very sensitive to the type of ATC/ATM separation rules applied. As compared to the landing capacity of a single runway and that of two closely-spaced parallel runways operating as a single runway, when the ATC/ATM current longitudinal distance based-separation rules are applied, the landing capacity increases when paired landings are realized successively on different runways. This increase amounts to 5.3 per cent when longitudinal, 10.3 per cent when mixed longitudinal/diagonal, 33.1 per cent when vertical, 27.8 per cent when mixed longitudinal/vertical, and 45.5 per cent when mixed vertical/diagonal ATC/ATM separation rules are applied. At the same time, with increase in the landing capacity, the take-off capacity decreases. Specifically, under convenient crosswind conditions, the landing, take-off and mixed operation capacity could 'explode', implying carrying out an ACM every minute or even every half-minute (but this is just a hypothetical, hardly realistic, situation in the given case). In addition, the 'ultimate' capacity envelopes lie above the airport-specified 'practical' capacity figures as expected (DA 2013). For example, the average 'ultimate' capacity for mixed operations is about 18 per cent higher than its 'practical' airport-specified counterpart. If the airport operates 24 hours a day all 365 days in the year, the annual 'ultimate' capacity will increase from the current 483 to about 578 thousand ACMs. Consequently, considering the development of demand in Fig. 4.14b, such 'ultimate' capacity will be saturated up to about 96 per cent by the year 2020.

Figure 4.17 shows the capacity envelopes when different ATC/ATM separation rules are applied to landings of the future aircraft fleet mix as given in Table 4.5. The capacity envelope for a single runway serving the current aircraft fleet mix is again provided as a benchmark, i.e. for comparative purposes.

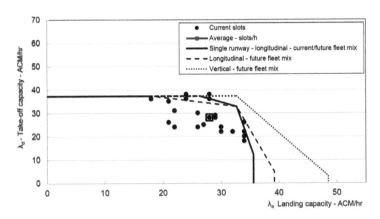

Fig. 4.17: The runway system capacity envelopes at DXB airport when different ATC/ATM separation rules are applied to landings of the future aircraft fleet mix (Janić 2016).

In this case the 'ultimate' take-off capacity for the future aircraft fleet mix is lower than that for the present one by about 6.6 per cent, while the capacity for mixed (50/50 per cent) operations remains the same. The landing capacity of a single runway will increase negligibly, that is, by 1.7 per cent, just due to the lower heterogeneity of the future landing aircraft fleet mix. However, the landing capacity of dual runways will increase by 8.3 per cent if the longitudinal and by 34 per cent if the vertical ATC/ATM separation rules are applied as compared to their single runway counterpart. At the same time, with increase in the landing capacity, its take-off counterpart will decrease. The other above-mentioned combinations of ATC/ATM separation rules between landings are not applicable to the future fleet mix which is expected to exclusively consist of heavy and super heavy aircraft.

Consequently, the effects on the total annual capacity and its saturation until the year 2020 could be similar to those in Fig. 4.16.

Figure 4.18 shows the capacity envelopes when the ATC/ATM vertical separation rules are exclusively applied to landings of the current and future aircraft fleet mix. The capacity envelope for a single runway, current aircraft fleet mix and the ATC/ATM existing wake-vortex longitudinal separation rules is given as a benchmark, i.e. for comparative purposes.

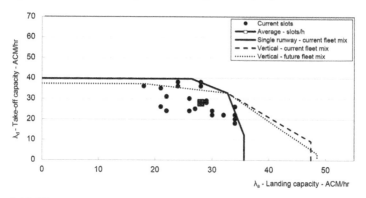

Fig. 4.18: The runway system capacity envelopes at DXB airport when the ATC/ATM vertical separation rules are applied to the current and future aircraft fleet mix (Janić 2016).

As can be seen, the take-off capacity and that for mixed operations are the same as for their single runway counterpart. If ATC/ATM vertical separation rules are applied to the current and future aircraft fleet mix, the corresponding landing capacities will increase by about 31 per cent and 34 per cent, respectively, as compared to their single runway counterpart. Again, its take-off counterpart will decrease. The airport-specified 'practical' capacity is again below the calculated capacity envelopes. The effects of the capacity gains for mixed operations at the annual level have similar effects to those of saturation of the overall runway system capacity, as explained in Figs. 4.16 and 4.17.

(b) 'Practical' capacity

The average number of slots for mixed operations as the 'practical' or 'declared' and calculated 'ultimate' capacity for mixed operations shown in Figs. 4.16-4.18 are used to estimate the specified average delay per an ACM (landing and/or take-off) under the given conditions. Based on Equation 4.9a, this amounts to 4.85 min. per ACM. With increase in the average delay per ACM, which implies deterioration of the specified service quality, the 'practical' capacity could shift closer to its 'ultimate' counterpart. This is illustrated by the relationship between the 'ultimate' and 'practical' landing capacity as shown in Fig. 4.19.

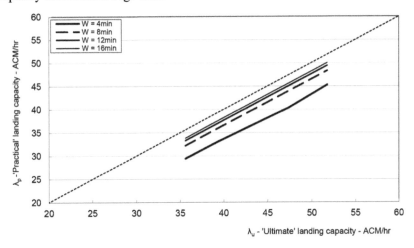

Fig. 4.19: Relationship between 'practical' and 'ultimate' capacity and maximum average delay imposed on ACM (landing) at DXB airport (Janić 2016).

The 'ultimate' landing capacities are based on the above-mentioned application of different ATC/ATM separation rules. As can be seen, the 'practical' capacity increases in line with increase in the 'ultimate' capacity for the given average delay per ACM. In addition, the gap between two capacities decreases with increase in the average delay. This implies that the airport can generally use the average delay as an instrument for increasing the number of declared landing slots but only on account of deteriorating service due to prolonged landing delay(s).

4.3.5 Interim Summary

This section deals with modelling the 'ultimate' and 'practical' capacity of closely-spaced parallel runways considered as operational performances at a congested airport under the given conditions. These conditions are mainly determined by new technologies (under development in European SESAR and US NextGen programs) supporting innovative aircraft landing and take-off procedures. The existing analytical models of both 'ultimate' and 'practical' capacity of closely-spaced parallel runways are case-modified while taking into account the ATC/ATM (Air Traffic Control/Air Traffic Management) current longitudinal and diagonal, and prospective vertical separation rules between landings, time-based separation rules between taking-offs and mixed time-distance based separation rules between mixed operations. In addition, a 'what-if' scenario of operating the runway system under convenient crosswind conditions is taken into consideration. Furthermore, the 'ultimate' and 'practical' capacity are interrelated by means of the average delay per ACM (landing and/or take-off). The models have been applied using the inputs on the terminal airspace and runway system configuration at DXB (Dubai International) airport, the above-mentioned ATC/ATM separation rules and scenarios of their use and characteristics of the current and future aircraft fleet mix.

The results show (i) convenience in application of the models under the given circumstances; (ii) increase in the landing and take-off 'ultimate' capacity as compared to that of a single runway thanks to the paired use of two parallel runways, while performing innovative operational procedures supported by the new technologies in combination with different ATC/ATM longitudinal/diagonal separation rules applied to the current aircraft fleet mix; (iii) substantial increase in the 'ultimate' landing capacity by applying ATC/ATM vertical separation rules to the paired landings of the current and future aircraft fleet mix; (iv) tremendous increase in the landing, take-off and mixed operation 'ultimate' capacity under convenient crosswind conditions as compared to the previous capacity counterparts (taken as a hypothetical case); (v) increase in the 'practical' capacity by increasing the 'ultimate' capacity and balancing between the two by the specified average delay per ACM (Aircraft Movement) (landing and/or take-off); and (vi) the obvious potential of innovative procedures supported by new technologies for increasing the 'ultimate' and consequently 'practical' or 'declared' capacity of closely-spaced parallel runways at the given airport.

4.4 Concluding Remarks

This chapter deals with modelling the effects of innovative and new technologies on the performances of transport systems. The considered systems are the supply chains served by mega vehicles and an airport runway system consisting of two parallel runways, where aircraft operations are supported by new (forthcoming) ATC/ATM technologies, both onboard the aircraft and on the ground. In the former case, the models of the given chain's infrastructural, technical/technological, operational, economic, environmental and social performances are developed and applied to the case of an intercontinental supply chain spreading between Europe and Asia (China), which is served by conventional and mega container ships. These mega container ships are considered as a new technology primarily characterized by their size compared to their conventional counterpart. The results show that in absolute terms, mega ships (vehicles) generally contribute to the deteriorating of the above-mentioned particular performances of the given supply chain under the given conditions. However, in relative terms, they can improve these performances, except for the average inventory costs of goods/freight shipments at begin/origin and end/destination nodes of the chain—the hub ports.

In the case of closely-spaced parallel runways, their 'ultimate' and 'practical' capacity are modelled under conditions when landings are supported by new ATC/ATM technologies, enabling implementation of innovative operational procedures. The models are applied to the closely-spaced parallel runways at Dubai International (DXB) airport. The results show that under specified conditions, both 'ultimate' and 'practical' landing capacity could be increased substantially and the 'ultimate' departure capacity only marginally, compared to the existing practice.

The above-mentioned cases indicate that new technologies, in addition to their obvious advantage in contributing to improved performances of transport systems, can also act in the opposite direction under specific circumstances.

References

AECOM/URS. 2012. *NC Maritime Strategy: Vessel Size vs. Cost*. Prepared for North Carolina Department of Transportation. AECOM Technology Corporation/URS Corporation, Los Angeles, USA.

Allianz. 2013. *Safety and Shipping Review* 2013. London, UK.

Blumstein A. 1959. The Landing Capacity of a Runway. *Operations Research* 7: 52-763.

CAA. 2014. *Wake Turbulence*. Supplementary Instruction (SI) CAP 493 MATS Part 1. Safety and Airspace Regulation Group, Airspace, ATM and Aerodromes, Civil Aviation Authority, London, UK.

Cullinane K., Khanna M. 2000. Economies of Scale in Large Container Ships. *Journal of Transport Economics and Policy* 33 Part 2: 185-208.

DA. 2013. *Connecting the World Today and Tomorrow—Strategic Plan 2020*. Dubai Airports, Dubai, UAE.

Daganzo C. 2005. *Logistics Systems Analysis*. 4th Edition. Springer Berlin Heilderberg, New York.

Davidson N. 2014. *Global Impacts of Ship Size Development and Liner Alliances on Port Planning and Productivity*. IPAH Mid-term Conference for Planning and Investment. April 2014, Sydney, Australia.

DeNeufville, R., Odoni R.A. 2003. *Airport Systems: Planning, Design and Management*. McGraw hill Book Company, New York, USA.

Donohue G.L. 1999. A Simplified Air Transportation Capacity Model. *Journal of ATC*, April-June: 8-15.

EC. 1999. *Freight Logistics and Transport Systems in Europe*. Executive Summary Euro-CASE. European Council of Applied Sciences and Engineering. Paris, France.

EC. 2009. *Terminal Handling Charges During and After the Liner Conference Era*. European Commission, Publication Office of the European Union, Luxembourg.

EEC/FAA. 2008. *Re-Categorization of the Wake Turbulence Separation Minima (RECAT)*. EUROCONTROL, Brussels, Belgium/Federal Aviation Administration, Washington DC, USA.

Erzberger H. 2004. *Transforming the NAS: The Next Generation Air Traffic Control System*. 24 ICAS (International Congress of the Aeronautical Sciences). September. Yokohama, Japan.

Evangelista P., Sweeney P., Ferruzzi G., Carrasco C.J. 2010. Green Supply Chain Initiatives in Transport Logistics Service Industry: An Exploratory Case. *Arrow@ Dit, Professional Journals*, Dublin Institute of Technology, Dublin, Ireland.

FAA. 2012. FAA ORDER JO 7110.65U. *Air Traffic Organization Policy*. Federal Aviation Administration, US Department of Transportation, Washington DC, USA.

FAA. 2013. *NextGen Implementation Plan*. Office of the NextGen, Federal Aviation Administration, US Department of Transportation, Washington DC, USA.

Fauske H., Kollberg M., Dreyer C.H., Bolseth S. 2006. *Criteria for Supply Chain Performance Measurement Systems*. 14 International Annual Euroma Conference, 17 June. Ankara, Turkey.

Forslund H., Jonsson P. 2007. Dyadic Integration of the Process Management. *International Journal of Physical Distribution and Logistics Management* 37: 546-567.

Giannakis M. 2007. Performance Measurement of Supplier Relationships. Supply Chain Management. *An International Journal* 12: 400-411.

Gilbo E.P. 1993. *Airport Capacity: Representation Estimation, Optimization*. IEEE Transactions on Control System Technology 1: 144-153.

Gilbo E.P. 1997. Optimizing Airport Capacity Utilization in Air Traffic Flow Management Subject to Constraints at Arrival and Departure Fixes. *IEEE Transactions on Control Systems Technology* 5: 490-503.

Gunasekaran A., Patel C., McGaughey R.A. 2004. A Framework for Supply Chain Performance Measurement. *International Journal of Production Economics* 87: 333-347.

Hall R.W. 1993. Design for Local Area Freight Networks. *Transportation Research* B 27B: 70-95.

Harris R.M. 1972. Models for Runway Capacity Analysis. *The MITRE Corporation Technical Report: MTR-4102*. Rev. 2, Langley, Virginia, USA.

Hockaday S.L.M., Kanafani A. 1974. Development in Airport Capacity Analysis. *Transportation Research* 8: 171-180.

Huang S.H., Sheoran S.K., Keskar H. 2005. Computer-Assisted Supply Chain Configuration Based on Supply Chain Operations Reference (SCOR) Model. *Computers & Industrial Engineering* 48: 77-394.

ICAO. 2001. *Air Traffic Management. Procedures for Air Navigation Services.* Doc 4444. ATM/501. International Civil Aviation Organization, Montreal, Canada.

ICAO. 2008. *Guidance on A380-800 Wake Vortex Aspects.* Attachment to TEC/OPS/SEP (T11/72) – 06-0320.SLG. International Civil Aviation Organization, Montreal, Canada.

Janić M., Tosić V. 1982. Terminal Airspace Capacity Model. *Transportation Research* A: 253-260.

Janić M., Reggiani A., Nijkamp P. 1999. Sustainability of the European Freight Transport System: Evaluation of the Innovative Bundling Networks. *Transportation Planning and Technology* 23: 129-156.

Janić M. 2005. Modelling Performances of Intermodal Freight Transport Networks. *Logistics and Sustainable Transport* 1: 19-26.

Janić M. 2006: A Model of Ultimate Capacity of Dual-Dependent Parallel Runways. *Transportation Research Record* 1951: 76-85.

Janic M. 2008. Modelling the Capacity of Closely-Spaced Parallel Runways Using Innovative Approach Procedures. *Transportation Research* C 16: 704-730.

Janić M. and Vleugel, J. 2012. Estimating Potential Reductions in Externalities from Rail-Road Substitution in Trans-European Transport Corridors. *Transportation Research* D 17: 154-160.

Janić M. 2014a. *Advanced Transport Systems: Analysis, Modelling and Evaluation of Performances.* Springer, UK.

Janić M. 2014b. *Modelling Performances of the Supply Chain(s) Served by the Mega Freight Transport Vehicles.* Proceedings of ICTTE (International Conference on Traffic and Transport Engineering), 27-28 November 2014, Belgrade, Serbia, 22.

Janić M. 2014c. Modelling the Effects of Different Air Traffic Control (ATC) Operational Procedures, Separation Rules and Service Priority Disciplines on Runway Landing Capacity. *Journal of Advanced Transportation* 48: 556-574.

Janić M. 2016. Modelling Capacity of Closely-Spaced Parallel Runways Supported by New Technologies and Innovative Procedures: The Case of Dubai International Airport. *Journal of Airport Management* 10, 1: 84-106.

Jeppsen. 2007. *OMDB Dubai International.* Jepp View 3.5.2.0. Airport Information, Notebook Info, Jeppsen Sanderson, Inc.

Kolos-Lakatos T., Hansman J.R. 2013. *The Influence of Runway Occupancy Time and Wake Vortex Separation Requirements on Runway Throughput.* MSc Thesis. Report No. ICAT-2013-08. MIT International Center for Air Transportation (ICAT), Department of Aeronautics & Astronautics Massachusetts Institute of Technology, Cambridge, Massachusetts, USA.

Lai K.H., Ngai E.W.T., Cheng T.C.E. 2002. Measures for Evaluating Supply Chain Performance in Transport Logistics. *Transportation Research* E 38: 439-456.

LR. 2011. *Assessment of IMO Mandated Energy Efficiency Measures for International Shipping: Estimated CO_2 Emissions Reduction from Introduction of Mandatory Technical and Operational Energy Efficiency Measures for Ships.* MEPC 63/INF.2 Annex. Lloyd's Register, London, UK.

Mayer H.R., Zondervan J.D., Herndon A.A., Smith T. 2011. *A Standard for Equivalent Lateral Spacing Operations: Parallel and Reduced Divergence Departures.* 9 USA/Europe Air Traffic Management Research and Development Seminar (ATM2011). The MITRE Corporation, McLean, VA, USA.

Mongelluzzo B. 2013. Bigger Ships and Tighter Supply Chains Shine a New Light on Port Productivity and Its Importance to Shippers. *The Journal of Commerce.* www.joc.com: 11-17.

Newell G.F. 1979. Airport Capacity and Delays. *Transportation Science* 13: 201-241.

Olugu U.E., Wong Y.K. 2009. Supply Chain Performances Evaluation: Trends and Challenges. *American Journal of Engineering and Applied Sciences* 1: 202-211.

Otto A., Kotzab H. 2003. Does Supply Chain Management Really Pay? Six Perspectives to Measure the Performance of Managing a Supply Chain. *European Journal of Operational Research* 144: 306-320.

PR. 2011. *Port Vison 2030.* Prepared for the City Council of Rotterdam. Rotterdam, The Netherlands.

REM Associates. 2014. *Methodology of Calculating Inventory Carrying Costs.* REM Associates Management Consultants, Princeton, New Jersey, USA.

Rodrigue P. 2013a. Container Shipping Costs and Cargo Value. *Geography of Transport Systems,* https://people.hofstra.edu/geotrans/eng/ch3en/conc3en/table_containershippingcosts.html

SCG. 2013. Global Supply Chain News: Maersk Triple E Cost Advantages are too Great to Ignore. *Supply Chain Digest,* http://www.scdigest.com/index.php

Simatupang T. M., Sridharan R. 2005. The Collaboration Index: a Measure for Supply Chain Collaboration. *International Journal of Physical Distribution and Logistics Management* 35: 44-62.

Srivastara S.K. 2007. Green Supply-Chain Management: A State-of-the-Art Literature Review. *International Journal of Management Reviews* 9: 53-80.

Stevels A. 2002. Green Supply Chain Management much more than Questionnaires and ISO 14.001. *IEEE Xplore* 96-100.

Stopford M. 2003. *Maritime Economics.* Routledge (Taylor & Francis). London, UK.

Sys C., Blauwens G., Omei E., Van de Vooede E., Witlox F. 2008. In Search of the Link between Ship Size and Operations. *Transportation Planning and Technology* 31: 435-463.

Tittsworth A.J., Lang, R.S., Johnson, J.E. and Barnes, S. 2012. *Federal Aviation Administration Wake Turbulence Program—Recent Highlights.* 57 Air Traffic Control Association (ATCA) Annual Conference & Exposition. 1-3 October. Maryland, USA.

Tseng Y.Y., Yue L.W., Taylor P.M. 2005. *The Role of Transportation in Logistics Chain.* Proceedings of Eastern Asia Society for Transportation Studies 5: 1657-1672.

UNCTAD. 2013. *Review of Maritime Transport 2013.* United Nations Conference on Trade and Development. New York, USA.

Van Woensel T., Vandaele N. 2007. Modelling Traffic Flows with Queueing Models: A Review. *Asia-Pacific Journal of Operational Research* 24: 1-27.

VTI. 2013. *Value of Freight Time Variability Reductions: Results from a Pilot Study for the Swedish Transport Administration.* VTInotat 39A-2013. The Swedish National Road and Transport Research Institute (VTI), Stockholm, Sweden.

WEF. 2009. *Supply Chain De-carbonization: the Role of Logistics and Transport in Reducing Supply Chain Carbon Emissions*. World Economic Forum, Geneva, Switzerland.

Zhang M., Wiegmans B., Tavasszy L.A. 2009. *A Comparative Study on Port Hinterland Intermodal Container Transport: Shanghai and Rotterdam*. 5 Advanced Forum on Transportation of China (AFTC), 17 October, Beijing, China.

http://www.investopedia.com/terms/s/supplychain.asp

http://www.lloydslist.com/

http://www.scdigest.com/ontarget/13-09-12-1.php?cid=7401

http://www.ship.gr/news6/hanjin28.htm

http://www.sea-distances.org/

http://en.wikipedia.org/wiki/Supply_chain

http://en.wikipedia.org/wiki/List_of_largest_container_ships

http://www.worldcontainerindex.com/

http://www.worldslargestship.com/

http://www.sesarju.eu/

http://www.faa.gov/nextgen/

MODELLING TRANSPORT SYSTEMS—III
Resilience

5.1   Introduction

This chapter deals with modelling the resilience of transport systems affected by disruptive events. These events can generally be unpredictable failures of the internal systems' components or external factors, such as bad weather and natural disasters like earthquakes and volcanic eruptions. In general, the impacts of disruptive events deteriorate the planned—usually operational—performances of the affected systems, which further compromise their economic, environmental and social performances (Hosseini et al. 2016). The economic performances are compromised by losses of revenue due to not performing the planned operations, i.e. transport services on the one hand and the additional costs for recovery of the affected services, on the other. The environmental performances can be compromised by causing damage to third parties and/or due to the spillage of fuel and oil from the affected vehicles, which can potentially contaminate the neighboring environment. The social performances can be compromised by traffic incidents and accidents caused by disruptive events, which may result in causalities, injuries and damage to property. Under such circumstances, the resilience of transport systems means their ability to stay operational at the planned or some other specified level during and just after impacts of disruptive events. As in the previous chapters, two cases are elaborated. The first deals with modelling the operational and economic performances of logistics networks operating under regular and irregular conditions, the latter while being exposed to disruptive event(s). The models are applied to a hypothetical logistics network. The other deals with modelling the resilience of an air transport network affected by a large-scale disruptive event of the 'bad weather' type. The models are applied to the part of the east coast of the US air transport network impacted by a large-scale hurricane. In both cases, the primary objective is to estimate the costs associated with the corresponding disruptive events.

Consequently, Section 5.2 describes modelling the resilience of a logistics network (in Chapter 4, such networks are referred to as 'supply chains'). Section 5.3 deals with modelling the resilience of the above-mentioned air transport network. The last section summarizes some concluding remarks.

5.2 Logistics Network Operating Under Regular and Irregular (Disturbing) Conditions

5.2.1 Background

Logistics is usually defined as a set of mutually interrelated organizational and operational activities concerned with handling particular goods from their origins to their destinations efficiently, effectively and safely. The origins of particular goods/freight shipments are the manufacturing plants or the goods/freight distributors. The destinations are either the manufacturing plants if the goods/freight shipments need finalization before being sent to the intermediate storage, or the end users, i.e. retailers and/or consumers. Figure 5.1 shows a simplified self-explanatory scheme of a given logistics network (an alternative term for a 'logistics network' is a 'network of supply chains').

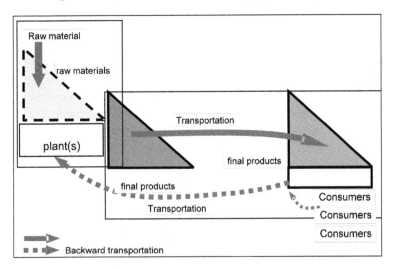

Fig. 5.1: A simplified scheme of a given logistics network (Janić 2009a).

As can be seen, the inventories of raw materials as inputs for manufacturing the given goods/freight shipments may exist at some manufacturing locations. The inventories of the final goods/freight shipments may exist at both the manufacturing and the retailers'/consumers' locations. Under regular operating/market conditions, the inventories at both ends of the network generally decrease over time. They are renewed just before being completely or nearly completely exhausted. In that context, the logistics network links the processes of manufacturing, begin, end, and/or intermediate storage, transport and the final distribution of the goods/freight shipments. The network consists of the chains connecting either the individual pairs or the clusters of manufacturers and consumers of given goods/freight shipments. The main actors in the network are: (i) the manufacturers and the retailers/consumers who are characterized, respectively, by their manufacturing and consuming rates of goods/freight shipments and by their strategies of dealing with their inventories; (ii) the transport operators, who enable the physical transfer of goods/freight shipments between manufacturers and retailers/consumers; (iii) the collectors and distributors of information on the progress of goods/freight shipments through the network (these are usually goods' forwarders); and (iv) the co-coordinators of organizational, physical and communications activities between particular actors.

In terms of the geographical scale, logistics networks can operate over local, national-state, continental and/or intercontinental area(s). They can refer either to particular industries, such as the aircraft industry, the automotive industry and the electronic industry, or to the clusters of different industries (firms) constituting and/or sharing common transport network(s) elements— infrastructure and services.

The operational and economic performances of a given logistics network can be characterized as follows: the production and consumption rate and the level and the cost of inventories of the goods concerned at the manufacturers' and the retailers/consumers' locations; the quantity of goods in motion; the average speed (time), frequency of delivery, the cost of transport and the reliability of deliveries reflecting the vulnerability of a given network to different external and internal disruptions. In general, the above-mentioned performances can be used to synthesize the overall cost as the economic performance of a given logistics network (Janić 2004, 2009a).

This section deals with modelling the operational and economic performances of a given logistics network operating under regular and irregular (disruptive) conditions. The model(s) developed is particularly intended to carry out a sensitivity analysis of the network's economic performances, i.e. the total costs with respect to changes of the most influential parameters, such as the goods/freight shipment delivery frequency and the intensity of impact and duration of the disruptive event. This type of model(s) could eventually contribute to providing some explanation for the rather low market share of the inland non-road transport modes in particular European regions.

5.2.2 The System and Problem—Resilience of Logistics Network(s)

5.2.2.1 Developments

For a long time, the main interest of the majority of manufacturers and suppliers was efficient, effective and safe door-to-door delivery of their goods/freight shipments. Efficiency implies delivery at minimum costs. Effectiveness implies delivery in terms of respecting the specified date and time. Safety implies delivering goods/freight shipments without damage due to known reasons. Logistics networks that fulfill the above-mentioned requirements have gradually developed with the following characteristics: (i) a relatively dispersed concentration of the goods/freight manufacturing, storage and consumption locations; (ii) the use of a different configuration of transport networks for delivery of the goods/freight shipments; (iii) an increase in the volume of direct deliveries with the nominated day and time; (iv) a decrease in the size of shipments and consequently an increase in the delivery frequencies; and (v) an increase in the utilization of transport vehicles. These characteristics have then been supported by the increased use of ICT (Information/Communication Technologies), which have enabled customers to track and trace their shipments through given logistics networks (Groothedde 2005, Zografos and Regan 2004). The main objective of such developments has been to reduce the share of logistics costs in the production costs of particular goods, which in turn has put the freight transport sector and particular transport modes under additional pressure to adequately respond to such requirements in terms of flexibility, availability, quality and cost.[1] Due to generally increasing volumes of goods/freight shipments to be transported on distances up to about 500 km, decrease in the shipment size with a consequent increase in the frequency of goods/freight delivery and shorter lead time(s), the road freight transport is most capable of adequately responding and consequently retaining the dominant market share in these markets in Europe. The increased application of the JIT (Just-in-Time) manufacturing concept, which has generally diminished inventories and increased requirements in terms of the reliability of delivery time (as agreed) and flexibility (the time between the order and delivery), has also contributed to such a development.

---

[1] In Europe, the share of transport costs in the production of different commodities is estimated as follows: retail products (0.7 per cent), petroleum products (3.6-3.8), foodstuff (3.6-3.9), iron and steel (4.5-5.0) and building material (6.4-7.2) (Aberle 2001).

Contrary, over the longer distances with ultimately lower volumes and less time- and frequency-sensitive goods/freight shipments, railways have shown to be the preferred mode (*see* Chapter 3, Section 3.3). Figure 5.2 shows an example of the development of the modal split in freight transport in the EU (European Union) 25 Member States for the period 1995-2011 (EC 2013)

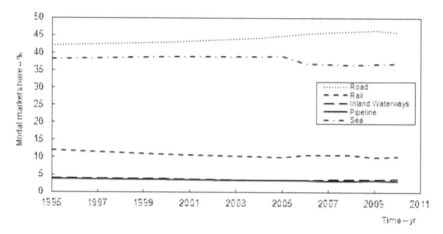

Fig. 5.2: Development of modal split between different freight transport modes in the EU (European Union) (t-km – ton-kilometers) (EC 2013).

As can be seen, during the past decade and a half, road transport has slightly increased its market share in terms of the annual volumes of t-km (ton-kilometers) and consequently maintained its dominant market position during the observed period (42-44 per cent). Sea transport has gained the second largest but rather volatile market share during the observed period (38-39 per cent). Rail transport has gradually lost its market share (from 12 per cent to 10 per cent), and oil pipeline and inland waterways have also slightly lost their market shares (from 3.9 per cent to 3.3 per cent and from 3.7 per cent to about 3.4 per cent, respectively).

5.2.2.2 Trends

Future trends are likely to be based on even more increased requirements for reducing logistics costs. This will happen under conditions of a further increase in the volumes of goods on generally decreasing delivery distances, changes in the structure of particular goods' categories, further diminishing of the shipment size and even more increased users' needs for control over the services they receive (by more intensive use of ICT). In addition, the inventories at the manufacturers will be reduced further through shortening of the manufacturing cycle time. This will continue to drive the transition from PUSH systems driven by the supply of raw materials and final goods to PULL systems, in which the actual demand for particular goods will trigger the upstream manufacturing processes in terms of time and quantities. This will also reduce the inventories at the retailers/consumers' locations. In addition, since the demand becomes known more precisely in advance, it will be increased by direct deliveries without any intermediate storage. Consequently, the inventories will be further shifted from the warehouses of the manufacturers and distribution centers to the rolling stock of particular transport modes. The latter will have to respond appropriately. This in turn will increase the need for cooperation rather than competition, either within the same or between different transport modes.

5.2.2.3 Resilience

The dynamism of a particular logistics network can be defined as the average speed of moving goods through it. This speed is influenced by the production and consumption rates of goods/freight shipments and distance and is particularly dependent on the frequency and speed of transport services. Consequently, those logistics networks with a higher speed of moving goods/freight shipments can be considered as more dynamic than the others (Blumenfeld et al. 1985, Daganzo 1984, 1999). In addition to these advantages, a disadvantage of these logistics networks implicitly assumed to carry goods/freight shipments of higher time-sensitivity is also their higher vulnerability, i.e. a rather weak resilience to both internal and external disruptions. This implies that these disruptions can affect one or more network components and deteriorate the regularity and punctuality of supplying raw materials, thus affecting the manufacturing and consumption rate(s), the transport service frequency and speed, etc. In general, disruptions with a rather modest impact and a shorter duration mainly cause particular transport operations and processes to slow down, and consequently lead to the creation of higher inventory 'buffers' at both ends of the given logistics network (Qi et al. 2004). Disruptions with a severe impact and longer duration usually mean cut-off of the major transport axes and consequently cancellation of the affected transport services. In some cases, alternative, very often less convenient routes and transport modes in terms of the goods/freight shipment transit times and costs, can be used (EC 2002, Zografos and Regan 2004). Consequently, in the former case of mild disruptions, the quality of transport services deteriorates, which in turn causes the overall inventory costs to rise at both manufacturers and retailers/consumers (McCann 2001). In the latter case of severe disruptions, the volumes of goods/freight shipments in transportation decrease due to cancellation of particular transport services, which cause a decrease in the planned utilization of the allocated transport vehicles/fleet(s) on the one hand, and diminish the overall utilization of the related infrastructure due to the cancelled services, on the other (McCann 2001). In addition, the goods/freight shipment production rate at both the manufacturers and further upstream at the suppliers of raw material may be also affected (Thomas and Griffin 1996).

Overall, the scale and scope of impact of a given disruptive event on a given logistics network depend on: (i) the network characteristics in terms of its size, type, the volume of goods/freight shipments and the spatial coverage (regional, interregional); (ii) the goods/freight shipment service time, delivery frequency and inherent vulnerability; (iii) the intensity, scale and duration of the given disruptive event; and (iv) the availability of alternatives to temporarily take over the affected goods/freight flows.

5.2.3 Modelling Resilience of Logistics Network(s)

5.2.3.1 Some Related Research

Extensive research on different operational and economic performance of logistics networks was carried out. This particularly referred to optimizing the coordination of logistics networks, i.e. supply chains defined as the management of material and information flows between the vendors, manufacturing and assembly plants and the distribution centers. The focus was on categories of operational coordination such as Bayer-Vendor, Production-Distribution and Inventory-Distribution. An overview of the achievements in optimizing the costs of such coordination was compiled by Thomas and Griffin (1996). In addition, some research has focused on the choice of the transport mode within given logistics networks/chains. The main criterion included the cost of the entire network/chain consisting of production, storage and shipping costs (Benjamin 1990). In that context, the door-to-door delivery time of goods/freight shipments was of great importance for both shippers and receivers, thus reflecting the potential benefits from transport investments. Nevertheless, the importance of time has always been dependent on the perception of the particular actors involved (Allen et al. 1985, Wigan et al. 2000).

The present modelling is based on ideas and elements of the analytical modelling of logistics networks developed by Blumenfeld et al. (1985), Campbell (1990, 1992), Daganzo (1984, 1999), Daganzo and Newell (1985), Hall (1987, 1993) and Janić et al. (1999). In particular, as a start, the simplistic analytical models dealing with handling, inventory and transportation costs of complex logistics operations that deliver a bulk of goods/freight shipments from the manufacturers to the retailers/consumers either directly or via consolidating terminals as developed by Daganzo (1984, 1999) are used. In these models, detailed data on the particular operations and the associated costs are replaced by their summaries, enabling the use of simple analytical models instead of complex, mainly computer-supported numerical structures.

In addition, some research has also been carried out on the impacts of disruptions due to production and consumption processes on the network cost performance (Dejax 1991, Qi et al. 2004). Therefore, modelling the cost performances of a given logistics network operating under regular and irregular (disruptive) conditions represents a complement to the above-mentioned related research (Janić 2009a).

5.2.3.2 Objectives and Assumptions

Regarding the above-mentioned current and future characteristics of a given logistics network, the main objectives of this research are as follows:

- Developing a methodology consisting of corresponding analytical models to enable estimation of the resilience of a given logistics network in terms of its cost performance, i.e. the total and average cost, while operating under planned-regular and unplanned-irregular (disruptive) conditions;
- Carrying out a sensitivity analysis of resilience, i.e. the above-mentioned costs with respect to changes in the most influential parameters; in the given context, these are assumed to be the configuration and type of transport network and transport mode, the frequency of goods/freight shipment deliveries between the manufacturers and the retailers/suppliers, and the intensity of impact and duration of the disruptive event; and
- Providing explanations on why particular non-road inland freight transport modes (railways, inland waterways, pipelines) are losing their market shares in some European regions despite being stimulated by national and international (EU) policies.

The above-mentioned modelling of the resilience of a given logistics network is based on the following assumptions:

- The number of manufacturers and consumers of given goods/freight shipments is known. The same manufacturer can produce goods/freight shipments for different consumers; the same consumer can receive goods/freight shipments from different manufacturers. This assumption closely reflects the real situation, since the number of both manufacturers and retailers/consumers of given goods/freight shipments in a given region is countable. In addition, most of them use different clients in order to provide the required quantities and reduce higher prices of goods/freight shipments.
- Goods/freight shipments are consolidated into compact forms in terms of size and weight, such as pallets or containers; consequently, they are countable rather than expressed in units of weight or volume. On the one hand, this implies that intermodal transport can be used in a given logistics network by operating different transport modes. On the other hand,

regarding the overall rate of the containerized goods/freight shipments, the quantities of goods in the given network are limited when compared with the total quantities transported in the market.

- The production and consumption rates of given goods/freight shipments at particular manufacturers and consumers-retailers, respectively, are constant. This seems to be realistic under conditions of a relatively stable demand for them at given prices during a given period of time.

- The technological, operational and economic characteristics of particular transport mode(s) operating within a given logistics network in terms of the type and capacity of transport means, frequency of services, speed and the corresponding handling and operational costs, are given. This sounds reasonable when the transport prices offered to the given manufacturers and retailers/consumers reflect the total costs of the particular transport operators.

- Disruptive event(s) can affect a given logistics network by compromising the planned time and punctuality of delivery of goods/freight shipments, either by slowing down the operations and processes in the network or by completely cutting-off the transport services between particular manufacturers and retailers/consumers. The type, intensity and duration of the impact of a given disruptive event are known. In general, disruptive events can be of different types, like the intensity of impact, duration and time of occurrence. Since they are usually unpredictable, their impact and related consequences are unavoidable and the consequences are also unpredictable. Therefore, assuming them to be certain in the modelling of these networks enables the estimation of their impacts using the 'what-if' scenario approach.

5.2.3.3 Structure of the Methodology

(a) Configuration of the network

Logistic networks can have different spatial configurations. In particular, the transport component of these networks can have different configurations, which can be used either exclusively or in different combinations to connect particular manufacturers and retailers/consumers of the goods. The former cases appear convenient for analysis, modelling and comparison of different network configurations in the given context. The latter cases frequently exist in practice. Figure 5.3 (a, b, c) shows a simplified scheme of these particular network configurations for the purpose of their analysis and modelling in the given context.

As can be seen, the manufacturers (white circles denoted by index (i)) are clustered in the 'manufacturer area'. The retailers/consumers [black circles denoted by index (j)] are clustered in the 'consumer area'. These both represent the network nodes where the flows of goods/freight shipments originate and end, respectively. The arrows indicate the direction of movement of these flows (Janić 2007, 2009a). In addition, the particular configurations of the transport networks in Fig. 5.3 (a, b, c) can be as follows:

- Configuration with direct connections of particular pair(s) of manufacturers and retailers/consumers (Fig. 5.3a);
- Configuration with indirect connections of particular pairs of manufacturers and retailers/consumers including one consolidation and one de-consolidation of the flows of goods/freight shipments at the different locations or terminals (Fig. 5.3b); and
- Configuration with indirect connection of particular pairs of manufacturers and retailers/consumers, including one consolidation/deconsolidation of the flows of goods/freight shipments at the same location or terminals (Fig. 5.3c).

One or a few operators of the same or different transport modes might be involved in the above-mentioned transport mode connections as follows:

- The configuration with direct connections implies that the road transport operators provide direct door-to-door connections between particular manufacturers and retailers/consumers. In the past, and still at the present, railways also provide such connections along tracks called industrial tracks, namely between the manufacturers and the doors of retailers/consumers (Fig. 5.3a).
- The configuration with one consolidation and one deconsolidation of goods at different and distant locations (terminals) requires use of at least two different transport modes. Usually, the road transport operators transfer goods from the door(s) of manufacturers to the consolidation terminal (T_1) and then from the deconsolidation terminal (T_2) to the door(s) of consumers-retailers. Any transport mode—road, rail, inland waterway, or air—can be used to operate between terminal (T_1) and (T_2). If rail is used as the main mode, the goods/freight shipments consolidated (packed) into the standardized units—containers, swap-bodies and semi-trailers—will require transshipment (sometimes combined with short-time storage) at the rail/road terminal (T_1) and (T_2). If maritime transport is used as the main mode, the terminals (T_1) and (T_2) will be located in their ports. In that case, the goods/freight shipments packed in maritime containers are collected from and distributed to these port terminals by road, rail, or both. If air transport operates as the main mode, which is the practice of express freight delivery operators like FedEx (Federal Express), UPS (United Parcel Services) and DHL (Deutsche Post AG), terminals (T_1) and (T_2) are the cargo terminals at the goods/freight shipment origin

and destination airports, respectively. The goods/freight shipments are consolidated in boxes of small size and weight (letters and small limited-weight packages) collected within the 'manufacturer' (i.e. the 'sender') area and distributed within the 'consumer' (i.e. the 'receiver') area by road (Fig. 5.3b).

- The configuration with one consolidation/deconsolidation of goods at the same location (terminal) usually requires the use of only one transport mode. It can be either road or rail exclusively. If it is rail, the goods/freight shipments usually consolidated into containers (loading units) are loaded on to the flat wagons at the doors of particular manufacturers; these wagons are assembled into trains and then dispatched to terminal (*T*), which is usually the rail-shunting yard. There, the incoming trains are decomposed and the outgoing trains reassembled and sent to the 'consumer' area(s). After decomposing these trains, the rail wagons are distributed to particular consumers-retailers along the industrial tracks or road transport can be used at both ends of the network. If air transport is used, the ultimate 'manufacturers' and the ultimate 'consumers' of goods/freight shipments are the cargo terminals at local airports. Terminal (*T*) enables the exchange of goods/freight shipments between incoming and outgoing aircraft/flights before they proceed towards their final destination(s). In any case, road transport is used for collection and distribution of goods/freight shipments from and to, respectively, the real 'manufacturers/consumers-retailers', and from and to the local airports (Fig. 5.3c).

The above-mentioned configurations of logistics networks can be identified for specific purposes of their analysis and modelling. In practice, particular manufacturers and retailers/consumers are usually connected by different types of mixed networks. For example, these can often be road networks consisting of elements of the above-mentioned configurations (a) and (c) (*see* Figs. 5.3a and 5.3c).

(b) Model for regular operations

The methodology for estimating the resilience of a given logistics network includes two models—one of its operation under regular and another of its operation under irregular (disruptive) conditions. In the modelling process that follows, the following notation is used:

i, j, k are the index of the given manufacturer, retailer/consumer and items-goods, respectively; they range from 1 to (M), (N), (K);

m_{ijk}, q_{jik} are the production and the consumption rate of the goods/freight shipments (k) at the manufacturer (i) and at the retailer/consumer (j), respectively (ton/TU; TU – Time Unit);

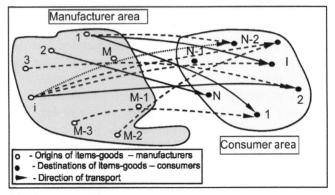

(a) Direct connections

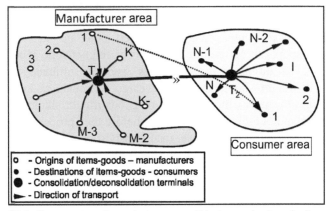

(b) Indirect connections via two consolidation nodes/terminals

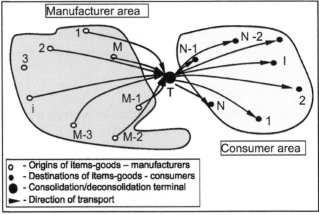

(c) Indirect connections via single consolidation node/terminal

Fig. 5.3: Simplified layouts of particular configurations of logistics networks (Janić 2009).

h_{ijk}, h_{jik}	are the holding costs of goods/freight shipments (k) at the manufacturer (i) and the retailer/consumer (j), respectively (€ or \$US/ton-TU);
p_{jik}	is the value (price) of the unit of quantity of the goods/freight shipments (k) produced by the manufacturer (i) and consumed by the retailer/consumer (j) (€ or \$US/ton);
r_k	is the interest (discount) rate of the unit of the goods/freight shipments (k), implying that its value decreases over time (%);
H_{ijk}, H_{jik}	are the time intervals between successive orders/deliveries of the goods/freight shipments (k) from manufacturer (i) to the retailer/consumer (j), respectively (TU);
τ_{ijk}	is the duration of a disruptive event affecting the transport of goods (k) between the manufacturer (i) and the retailer/consumer (j) (TU);
t_{ijk}	is the average delivery time of the goods/freight shipments (k) from the manufacturer (i) to the retailer/consumer (j) (TU);
T	is the period of time in which the cost of a given logistics network is considered (TU);
α_{ijk}	is the value of the time needed to transport the goods/freight shipments *(k)* between the manufacturer (i) and retailer/consumer (j) (€ or \$US/ton-TU)[2].
d_{ijk}	is the transport distance for goods/freight shipments (k) between the manufacturer (i) and the retailer/consumer (j) (km);
$v_{ijk}(\cdot)$	is the average speed of transfer-transport of the goods/freight shipments (k) between the manufacturing plant (i) and the retailer/consumption plant (j) (km/h);
W_{ijk}	is the anticipated delay while transporting the goods/freight shipments (k) between the manufacturer (i) and the retailer/consumer (j) (TU);
f_{ijk}^{*}	is the frequency of sending the goods/freight shipments (k) directly from the manufacturer (i) to the retailer/consumer (j) (dep/TU);
d_{i1k}, d_{2jk}	are the incoming and outgoing distances of the goods/freight shipments (k) from the manufacturer (i) to the consolidation terminal (T_1) and from the deconsolidation terminal (T_2) to the consumer-retailer (j), respectively (km);

---

[2] As in the case of inventories at both ends of the given logistics network, this value of time may depend on the value of the given item and the interest rate of capital reflecting the value of the item over time.

$v_{i1k}(*), v_{2jk}(*)$ are the average transport speed of the goods/freight shipments (k) along the consolidation/deconsolidation incoming and outgoing distances (d_{i1k}) and (d_{2jk}), respectively (km/h);

W_{i1k}, W_{2jk} are the anticipated delay of the goods (k) incoming at the consolidation terminal (T_1) from the manufacturer (i) and outgoing from the deconsolidation terminal (T_2) to the retailer/consumer (j), respectively (TU);

τ_{i1k}, τ_{2jk} are the average time the goods/freight shipments (k) spend at the consolidation and the deconsolidation terminals (T_1) and (T_2), respectively (TU);

$d_{12k}, v_{12k}(*)$ are the distance and the average speed for the goods/freight shipments (k) sent between the terminals (T_1) and (T_2), respectively (km, km/hr);

W_{12k} is the anticipated delay of the goods (k) while being sent between the terminals (T_1) and (T_2) (TU);

f_{12k} is the frequency of transport services between the terminals (T_1) and (T_2) (dep/TU);

τ_k is the average time, which the goods/freight shipments (k) spend at the consolidation/deconsolidation terminal (T) (TU);

f_{i1k}, f_{1jk} are the incoming frequency from the manufacturer (i) and the outgoing frequency to the retailer/consumer (j) of the goods (k) at the terminal (T), respectively (dep/TU);

p_{ijk} is the price (cost) of transporting the unit of quantity of the goods/freight shipments (k) from the manufacturer (i) to the retailer/consumer (j) (€ or \$US/ton);

a_k is the price (cost) per unit distance of the goods/freight shipments (k) (€ or \$US/km);

b_k is the price (cost) per unit of weight of the goods/freight shipments (k) (€ or \$US/ton); and

S_{ijk} is the weight of a unit of the goods/freight shipments (k) sent from the manufacturer (i) to the retailer/consumer (j) (ton).

The values of the variables (M), (N), and (K) combined with the distances between particular manufacturers and retailers/consumers included in the given logistics network can be used, respectively, as indicators of its size, coverage and the diversity of the goods/freight shipments concerned. The holding cost of the goods/freight shipments (k), (h_{ijk}) and (h_{jik}) comprises the inventory and the warehousing cost. The former cost mainly depends on the value (price) of a single unit of goods/freight shipment (k), (p_{ijk}) and the interest rate (r_k). The latter cost depends on the size of a given quantity of goods/freight shipment (k), i.e. the required space for its storage, packaging, air conditioning, etc. The periods between the successive orders/deliveries of the goods/freight shipments (k), (H_{ijk}) and (H_{jik}), respectively, depend on the requirements of retailers/consumers, manufacturers and transport operators and their capabilities to fulfill these requirements. In contemporary logistics networks, this interval is becoming shorter, i.e. there are more frequent orders/ deliveries of smaller quantities of given goods/freights shipments.

Inventory cost at the manufacturers

The frequency of orders/deliveries of the goods/freight shipments (k) from the manufacturer (i) to the retailer/consumer (j) during the period (T) can be estimated as (Daganzo 1999):

$$f_{ijk} = T/H_{ijk} \tag{5.1}$$

The frequency of transport services can be equal to or lower than the frequency of orders/deliveries (f_{ijk}) in Equation 5.1. This implies diversity of the capabilities of particular transport modes to appropriately respond to the requirements. In addition, the variable T indicates the period of time for which the cost performance of a given network is estimated. It can be a day, week, month, or a year.

The total quantity of goods/freight shipments (k) manufactured between the two successive orders/deliveries, i.e. during the interval (H_{ijk}), can be determined based on Equation 5.1 as follows:

$$Q_{ijk} = m_{ijk} * H_{ijk} = m_{ijk} * (T / f_{ijk}) \tag{5.2}$$

The inventory cost of goods/freight shipments (Q_{ijk}) can be determined, using Equations 5.1 and 5.2 and Fig. 5.4 as follows:

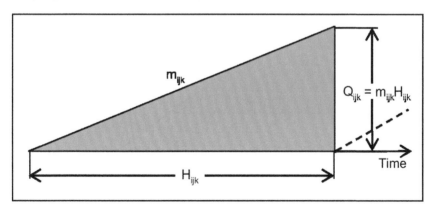

Fig. 5.4: A scheme of the inventories of goods (k) at the manufacturer (i) before being sent to the consumer (j) (Janić 2009a).

$$c_{ijk} = (1/2) * Q_{ijk} * H_{ijk} * h_{ijk} = (1/2) * m_{ijk} * \left(T / f_{ijk}\right)^2 * h_{ijk} \qquad (5.3)$$

The total inventory cost of goods/freight shipments (k) at the manufacturer (i) before being sent to the retailer/consumer (j) during the period (T) can be determined based on Equation 5.3 as follows:

$$C_{ijk} = f_{ijk} * c_{ijk} = (1/2) * m_{ijk} * (T^2 / f_{ijk}) * h_{ijk} \qquad (5.4)$$

Equations 5.2-5.4 imply that the frequency (f_{ijk}) is always positive. If referring to the frequency of services of particular transport modes, it will ultimately be out of the direct control of the users-manufacturers and retailers/consumers, thus giving them a choice of the most convenient transport alternative (Fig. 5.2).

Inventory cost at the consumers

The time interval between the successive arrivals of particular orders of the goods/freight shipments (k) at the retailer/consumer (j) should be approximately the same as their inter-departure interval(s) from the manufacturer (i), i.e. $H_{ijk} = H_{jik}$. The quantity of goods/freight shipments $Q_{ijk} = m_{ijk} * H_{ijk}$ is consumed at the constant rate before ($m_{ijk} < q_{ijk}$), exactly at the time ($m_{ijk} = q_{ijk}$) or after receiving the next order ($m_{ijk} > q_{jik}$). Figure 5.5 shows the case when $m_{ijk} < q_{ijk}$ has two components of the inventory cost: one for the cost of holding inventories by consumers, and the other for the cost of shortage of inventories due to rather too quick consumption.

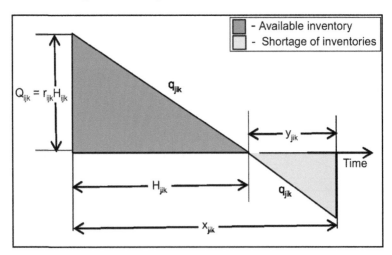

Fig. 5.5: A scheme of the inventories of goods (k) at the consumer (j) after having arrived from the manufacturer (i) (Janić 2009a).

If the deliveries are always on time, i.e. without significant deviation from the schedule, the total cost of the consumed inventories can be estimated as:

$$_1c_{jik} = (1/2)*x_{jik}*m_{ijk}*H_{ijk}*h_{jik} = (1/2)*\left(\frac{m_{ijk}T}{f_{ijk}}\right)^2*\left(\frac{1}{q_{jik}}\right)*h_{jik} \quad (5.5a)$$

Similarly, the total cost of the shortage of inventories can be estimated as:

$$_2c_{jik} = (1/2)*y_{jiki}*q_{jik}*p_{jik} = (1/2)*\left[\left(\frac{T}{f_{ijk}}\right)*\left(1-\frac{m_{ijk}}{q_{jik}}\right)\right]^2*q_{jik}*p_{jik} \quad (5.5b)$$

Combining Equations 5.5 (a, b) gives the total inventory cost of the goods/freight shipment (k), which have arrived from the manufacturer (i) at the retailer/consumer (j). Then, the total inventory cost at the retailer/consumer (j) for the period T can be estimated as:

$$C_{jik} = f_{ijk}*(_1c_{jik}+_2c_{jik}) = (1/2)*\left(\frac{T^2}{f_{ijk}}\right)*$$

$$\left(\frac{m_{ijk}^2}{q_{jik}}*h_{jik}+(q_{jik}-m_{ijk})*p_{jik}\right) \quad (5.5c)$$

In Equation 5 (a, b, c), the frequencies of delivery of goods/freight shipments are again positive and influenced by similar factors as in the case of inventories at the manufacturer(s).

Cost of time while transporting goods/freight shipments

The cost of time taken to transport the goods/freight shipments (k) between the manufacturer (i) and the retailer/consumer (j) over the period (T) can be determined as:

$$C_{ijk} = (m_{ijk}*T)*t_{ijk}*\alpha_{ijk} \quad (5.6)$$

Transport time and transport frequencies

The transport time (t_{ijk}) and the transport frequency (f_{ijk}) in Equations 5.1-5.6 mainly depends on the type of transport network serving the manufacturer (i) and the consumer (j), and the transport modes involved. Referring to Fig. 5.3 (a, b, c), they can be determined as follows:

Direct Transportation

Transport time: The transport time consists of a single component excluding the anticipated delay as follows:

$$t_{ijk} = d_{ijk}/v_{ijk}(d_{ijk})+W_{ijk} \quad (5.7a)$$

The time (t_{ijk}) is actually the stochastic variable with a given probability density function characterized by its mean and standard deviation. The mean mainly depends on the distance and the average speed. The standard deviation mainly depends on the traffic conditions and the other speed-affecting factors along the route, including disruptive events of a relatively milder impact. The standard deviation of (t_{ijk}) compromises the on-time arrival of given goods at the retailer/consumer (j). In order to prevent the shortage of inventories, this retailer/consumer should maintain a buffer of inventories at additional cost[3] (*see* Fig. 5.3a).

Transport frequency: The frequency of sending goods/freight shipments (k) from the manufacturer (i) to the retailer/consumer (j) can be determined as follows:

$$f_{ijk} = f_{ijk}^*$$ (5.7b)

In this case, road transport is presumably used because of its inherent flexibility and ability to respond to the retailers/consumers' and the manufactures' requirements at any time (Fig. 5.3a).

Transport with consolidation/deconsolidation at two different locations/terminals

Transport time: The transport time consists of five components, excluding the anticipated delay(s) as follows:

$$
\begin{aligned}
t_{ijk} = {} & d_{i1k}/v_{i1k}(d_{i1k}) + W_{i1k} + \tau_{i1k} + d_{12k}/v_{12k}(d_{12k}) \quad \text{(5.8a)} \\
& + W_{12k} + \tau_{2jk} + d_{2jk}/v_{2jk}(d_{2jk}) + W_{2jk}
\end{aligned}
$$

The time (t_{ijk}) can also be considered as a stochastic variable, composed of five stochastic components (excluding the anticipated delays). Each component has its probability distribution with the main parameters—mean and standard deviation. Intuitively, one can conclude that this integrated average time might be longer than is the case for direct connections. Certainly, it may likely have a greater standard deviation, consisting of the sum of the standard deviations of the five components (stochastic variables). This may

---

[3] If the time (t_{ijk}) is considered as a stochastic variable with a normal (Gauss) probability distribution, mean $(\overline{t_{ijk}})$ and standard deviation (σt_{ijk}), and if the acceptable risk of a shortage of the goods/freight shipments (k) at the consumption plant (j) is (β_{jk}), the cost of the protective inventories per delivery will be of the order: $(1/2)q_{jik}h_{jik}\left(\sigma_{t_{ijk}}\Phi^{-1}(1-\beta_{jk})\right)^2$. ($F^{-1}$ is the inverse of Laplace function (Winston 1994) and should be added to the cost in Equation 5.5c).

imply the higher transport inventory cost as well as a higher cost of the protective inventories at the retailers/consumers (Winston 1994) (Fig. 5.3b).

Transport frequency: The frequency of sending goods/freight shipments (k) from the manufacturer (i) to the retailer/consumer (j) can be determined as follows:

$$f_{ijk} \equiv f_{12k} \tag{5.8b}$$

The frequency (f_{12}) in Equation 5.8b is usually determined by the schedule of transport modes involved—rail, inland waterways or air. Thus, since many manufacturers and consumers are served at the same time, this frequency is likely to be differently efficient and effective regarding their specific requirements (Fig. 5.3b).

Transport with consolidation/deconsolidation at a common location/terminal

Transport time: The transport time consists of three components excluding the anticipated delays as follows:

$$t_{ijk} = d_{i1k} / v_{i1k} (d_{i1k}) + W_{i1k} + \tau_{1k} + W_{1jk} + d_{2jk} / v_{1jk} (d_{2jk}) \tag{5.9a}$$

The time (t_{ijk}) in Equation 5.9a can be considered similarly as in case (ii) as a stochastic variable composed of three other stochastic variables with a known probability distribution, mean and standard deviation. Intuitively, one can conclude that the standard deviation of the resulting time might be shorter than in case (ii) but longer than in case (i). It again requires a buffer inventory at the retailer/consumer (j), representing additional inventory costs (Fig. 5.3c).

Transport frequency: The frequency of delivering goods (k) between the manufacturer (i) and the retailer/consumer (j) can be determined as:

$$f_{ijk} \equiv \min(f_{i1k}; f_{k1j}) \tag{5.9b}$$

The frequency (f_{ijk}) in Equation 5.9b can again serve a single or a cluster of geographically very close manufacturers and consumers, which again makes it different in terms of convenience for their specific requirements (Fig. 5.3c).

Transport cost

The transport cost implies the cost of the physical movement of the goods/freight shipments (k) from the manufacturer (i) to the retailer/consumer (j). This cost has two aspects: (i) that of the manufacturer and that of the retailer/consumer, in which case the price paid for services is relevant; and (ii) that of transport operators, in which case their operational cost is relevant. In the given case, the former aspect is considered. Consequently, the total transport cost of the service frequencies (f_{ijk}), each carrying the quantity (Q_{ijk}) of the goods/freight shipments (k), during the period (T), can be estimated as follows:

$$C_{ijk}^t = f_{ijk} * Q_{ijk} * P_{ikl} \qquad (5.10a)$$

In Equation 5.10a, the price (P_{ijk}) mainly depends on the distance and size, i.e. the weight or the volume of the order/delivery and thus can be expressed as:

$$P_{ijk} = a_k * d_{ijk} + b_k * Q_{ijk} = a_k * d_{ijk} + b_k * \left(\frac{m_{ijk} T}{f_{ijk}} \right) \qquad (5.10b)$$

$$= a_k * d_{ijk} + b_k * \left(\frac{m_{ijk} * TS_{ijk}}{f_{ijk}} \right)$$

In many cases, the price (P_{ijk}) may include the cost of handling the goods/freight shipments, which refer to operations such as loading, unloading and eventually transshipment between different transport modes at the consolidation /deconsolidation terminals.

(c) Model for irregular (disturbing) operations

Various internal and/or external disruptive events can affect the given logistics network. In the given context, for the chain (ijk), such a disruptive scenario implies cutting-off connections and consequently transportation of the goods/freight shipments (k) from the manufacturer (i) to retailer/consumer (j) for a certain period of time (φ_{ijk}). Under such circumstances, the number of cancelled orders and the number of delivered orders $(f_{ijk/c})$ and (F_{ijk}), respectively, can be estimated as follows:

$$f_{ijk/c} = (\varphi_{ijk} / H_{ijk}) \equiv (\varphi_{ijk} * f_{ijk}) / T \qquad (5.11a)$$

$$F_{ijk} = f_{ijk} - f_{ijk/c} = f_{ijk} - (\varphi_{ijk} * f_{ijk}) / T = f_{ijk} * \left(1 - \frac{\varphi_{ijk}}{T} \right) \qquad (5.11b)$$

In Equation 5.11b, if $\varphi_{ijk} = T$, the network will be completely blocked/closed for any delivery during the entire period (T). Consequently, the inventories of goods/freight shipments (k) at the manufacturer (i) will increase to the level $(m_{ijk} T)$. Otherwise, the retailer/consumer (j) will have to keep a buffer of inventories of about $(q_{ijk} T)$ in order to compensate the shortage of goods/freight shipments (k) during the network breakdown. When $\varphi_{ijk} < T$, the impact of compromised frequency on the cost of the given logistics network can be estimated by replacing the variable (f_{ijk}) in Equation 5.2 with the variable (F_{ijk}) in Equation 5.11b.

(c) Model for the total and average cost

The total cost of the given logistics network can be determined from Equations 5.2-5.11 as follows:

$$C_T = \sum_{ijk} (C_{ijk} + C_{jik} + {}_tC_{ijk} + C_{ijk}^t) \tag{5.12a}$$

The total quantity of goods/freight shipments handled by the network during the period (T) can be determined as:

$$Q_T = \sum_{ijk} m_{ijk} * T \tag{5.12b}$$

Dividing the total cost (Equation 5.12a) by the total quantity of goods/freights shipments (Equation 5.12b) gives the average cost per unit of goods/freight shipment. This might be of interest for comparing the different network configurations, which handle different quantities of various goods using different transport modes operating under either regular or irregular (disruptive) conditions.

5.2.4 Application of the Models

5.2.4.1 Inputs

The above-mentioned inputs are applied to a logistics network, which consists of $M = 70$ manufacturers and $N = 70$ retailers/consumers. They are clustered in the 'manufacturer' and the 'consumption' area, respectively, at an average door-to-door distance of $d = 900$ km. They exchange goods/freight shipments with each other, which makes 4900 possible interactions. In Europe, this may refer to the areas between the Benelux countries (Belgium, the Netherlands, and Luxembourg) and the north of Italy and/or the south of France. The goods/freight shipments are consolidated into pallets. The number to be sent between the two regions during the period $T = 1$ year amounts to $30*10^6$. If they are uniformly distributed, this gives an average flow of 6122 pallets/year between each manufacturer and retailer/consumer. The average weight of a pallet amounts to 0.75 tons and its value $p = 1500$ €/pallet. The interest rate of goods on each pallet is $r = 6.5$ per cent. Consequently, the average value of time of a pallet is estimated to be $\alpha = p*r = 1500 * 0.65 = 92.5$ €/pallet-yr. The average holding cost of a pallet in the inventories at each manufacturer and each retailer/consumer is assumed to be $h = 5$ €/d. The standard deviation of the arrival of pallets at the retailers/consumers is assumed to be $\sigma = 4$ and $\sigma = 12$ hours/delivery, independently of the order/delivery frequency. The acceptable risk of the shortage of goods at each consumer-retailer is assumed to be $\beta = 0.05$.

In terms of spatial configurations, two networks are assumed to exclusively serve the manufacturers and the retailers/consumers: (a) the direct transport network operated by the road transport mode (Fig. 5.3a); and (b) the intermodal rail/road transport network with one consolidation and one deconsolidation, i.e. transshipment of pallets at two intermodal terminals (Fig. 5.3b). In this case, the pallets are additionally consolidated into containers, swap-bodies and semi-trailers. In road transport, the average speed of moving pallets through the network is assumed to be $v_1 = v_2 = 45$ km/h. For intermodal transport, this average speed is assumed to be $v_1 = v_2 = 30$ km/h along a road haulage distance of $d_1 = d_2 = 50$ km at both ends of the network, and v_{12} $= 30$ km/h along a rail haul distance of $d_{12} = 800$ km (EC 2001a, b, 2002). These speeds also include the time taken to pass through the two intermodal terminals. The vehicle carrying capacity is 28 pallets/truck and 1015 pallets/ train.

The average transport cost of a pallet between any pair of manufacturers and consumers/retailers is assumed to be $P = 28.64$ €/pallet in the road transport network and $P = 30.17$ €/pallet in the intermodal transport network (Groothedde 2005, Janić 2007). The impact of the disruptive event of duration causes the cancellation of transport services.

5.2.4.2 Results

The results from the models are obtained by investigating the sensitivity of the network cost performance to changes in the type of transport network (mode) used, the frequency of orders of goods/freight shipments and the intensity of the impact and duration of a disruptive event. The other inputs are considered as parameters and implicitly independent of each other. This particularly relates to the overall quantity of goods/freight shipments in the network, which may generally depend on their price, the quantities carried by different transport modes, which may depend on the level of their competition and the related transport prices, the transport cost, which may change with the quantity of the transported goods/freight shipments, etc. The results are shown in Figs. 5.6 and 5.7. Figure 5.6 shows the dependence of the average cost per item/pallet on the frequency of orders/delivery between an average pair of manufacturers and retailers/consumers.

As can be seen, if road transport is used, the average inventory cost per pallet at both manufacturers and consumers-retailers decreases more than proportionally with increase in delivery frequency. This cost at the consumer-retailer is a bit higher because of maintaining a buffer of inventories. The transport time and moving costs remain constant. Consequently, the total average cost per pallet decreases more than proportionally with increase in the delivery frequency. Under the assumption that intermodal transport performs

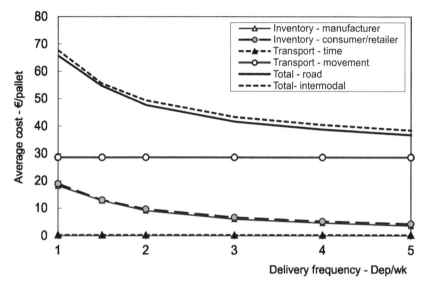

Fig. 5.6: Dependence of the average cost per pallet on the order/delivery
frequencies in the given example (Janić 2009a).

similarly as road transport in terms of the capacity and delivery frequency,
its cost will be higher than the cost of road transport mainly because of the
higher transport costs and higher buffer inventory cost at consumers-retailers.
The above-mentioned generic dependability explains why both manufacturers
and retailers/consumers tend to make more frequent orders/deliveries, which
apply to either the PUSH or the PULL concept—they push inventories from
their stocks into the transport system. The more frequent (smaller) orders/
deliveries require the deployment of a greater number of vehicles. The road
transport mode is often capable of fulfilling such requirements, which may
explain its slight growth in terms of trapping of the market share in Europe.
The intermodal transport usually responds by running daily trains (i.e. up
to five trains per week per operator). This, if combined with the limits of
capacity of each train, the operator's flexibility to appropriately respond and
consequently capture a higher market share(s) is restricted. For example,
when train capacity amounts to one thousand pallets per train, the intermodal
transport can count to catch up about 250 thousand pallets per year, which
is about 0.83 per cent of the total of 30 million pallets in the given example.
If more rail transport operators provide capacity equivalent to ten trains per
week, the market share of intermodal transport would increase to about 8.3
per cent. This reasoning is sensible only if the higher prices of the intermodal
transport are acceptable for particular users—the manufacturers and the
retailers/consumers.

Figure 5.7 shows that the average cost per pallet will increase more than proportionally with increase in the duration of disruption of the above-mentioned logistics network.

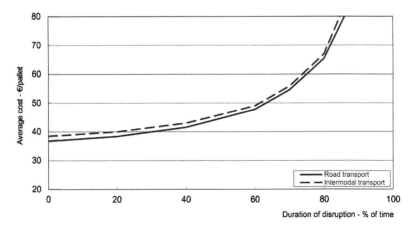

Fig. 5.7: Dependence of the average cost per pallet on the duration of the disruption of the logistics network in the given example (Janić 2009a).

In the given scenario, the disruption is assumed to completely cut-off the transport links between the manufacturers and retailers/consumers for a given period of time. Its duration is varied as a parameter in relative terms. Under such circumstances, combined with the lack of alternative routes, the particular transport services/deliveries will not be carried out, thus causing loss of revenues while reducing the costs of the affected transport operators. In addition, the inventories and the related costs at manufacturers and the shortage of goods/freight shipments and related cost at retailers/consumers will increase. Consequently, the total average cost per unit of the transported quantity of goods/freight shipments will increase more than proportionally with increase in the duration of such a disruptive event. Again, this cost is slightly higher for intermodal than for road transport.

In addition, it should be noted that road transport is less vulnerable to any kind of disruptive event than intermodal transport. For example, if road transport is used, the disruption, unless on a large spatial scale, might affect only individual pairs of manufacturers and retailers/consumers. However, if intermodal transport is used, the disruption of one of the terminals and/or of the rail line between them might ultimately affect almost all manufacturers and all retailers/consumers connected to them. This might consequently increase the total cost of disruption on the intermodal transport network by several times compared with the corresponding cost of using the equivalent road transport network, a fact which does not favor its use in the given context.

5.2.5 Interim Summary

This section deals with modelling the economic performances of two equivalent logistic networks served by the intermodal (rail/road) and road freight transport mode under planned (regular) and unplanned (irregular) or disruptive conditions. In the given case, the networks are considered as 'transport systems', with the economic performances represented by the total costs including the inventory cost of the goods/freight shipments at the manufacturers and the consumers, the costs of goods/freight time while in transportation and the cost of transportation itself. The models have been applied to simplified logistics networks using data from the European freight transport sector. The results show that under regular operating conditions, the average cost per goods/freight shipment (a pallet in this case) decreases more than proportionally at both transport modes with increase in the order/delivery frequency. Under the given conditions, the network served by road appears to be slightly more efficient, i.e. cheaper than its counterpart served by the intermodal (road/rail) transport mode. Under disruptive conditions, the average cost per goods/freight shipment increases more than proportionally with increase in the duration of the disruptive event. Again, this cost is slightly higher for the network served by the intermodal (rail/road) transport mode for the entire duration of the disruptive event.

The results also show that if road transport is exclusively used, the impact of disruptive event(s) might very likely remain relatively limited to the particular manufacturers and retailers/consumers. However, if intermodal transport is used, the impact of disruptive events can affect any of the intermodal terminals and/or rail line(s) connecting them and consequently, many more manufacturers and retailers/consumers. This may act against the more intensive use of intermodal (rail/road) transport mode in the given cases.

5.3 Air Transport Network Affected by a Large-Scale Disruptive Event

5.3.1 Background

Generally, according to the *Oxford Dictionary*, the resilience of an object can be defined as its "ability to recoil or spring back into shape after bending, stretching, or being compressed" (http://complexworld.eu/wiki/Resilience_in_air_transport). In addition, Holling (1973) defines ecological resilience as the ability of a system to absorb changes in state variables, driving variables and parameters and still persist. Furthermore, Holling (1996) and Hollnagel et al. (2006) define engineering resilience as the time required for a system to return to an equilibrium or steady state following a perturbation. Consequently, it can be said that the resilience of a given technical system generally implies its ability to operate under variable and unexpected conditions without substantially compromising its planned performances. As such, resilience can also reflect the robustness of the given system operating under disruptive conditions (Foster 1993).

The above-mentioned concepts and definitions of resilience can also be applied to transport networks comprising nodes, links and the transport services connecting them. The nodes are usually transport terminals as the origins and destinations of transport services serving passenger and/or freight/goods flows. The links are the physical infrastructure (roads, rail lines, air routes, sea routes) stretching between nodes/terminals along which the vehicles perform transport services. While dealing with the resilience of transport networks, deterioration of the planned/scheduled transport services in terms of their delay and cancellation due to the impact of various disruptive events is commonly considered. The scale and scope of such deterioration under the given impact reflect the resilience of the given network. In such a context, the scale of changing resilience after removing (closing) particular nodes (terminals) and/or links (and services) represents the network's friability (Ip and Wang 2011).

The disruptive events generally affecting transport networks can be extremely bad weather (dense fog, heavy rain and/or snowfall, hurricanes, tornadoes, etc.), usually unpredictable catastrophic failures of the transport network components, industrial actions of the transport staff, natural disasters (earthquakes, volcanos, tidal waves), traffic incidents/accidents and terrorist attacks. In some cases, these particular events can be interrelated and occur simultaneously. The commonly affected actors/stakeholders are the network operators, i.e. providers of transport services and their users-passengers and freight/goods shippers/receivers. They are all usually imposed additional costs associated with deteriorated services as well as recovery actions in the aftermath.

An air transport network consisting of airports and airline flights scheduled between them can also be affected by the above-mentioned disruptive events. Their impact adversely affects declared capacity of airports and air routes, consequently leading to long airline flight delays and cancellations.

This section describes a methodology for assessing the resilience and friability of a given air transport network affected by a large-scale disruptive event. In addition, it estimates the consequences for the particular actors/stakeholders involved—airports, airlines and air passengers—which mainly include the costs of long-delayed and cancelled flights. As such, the methodology could be used for both *a prior* and *a posterior* forecasting and assessment of the consequences of particular impacts, respectively, and undertaking the appropriate actions for mitigating them by using the 'what-if' scenario approach. In such a context, the prospectively affected airports, airlines and their passengers need to bear in mind that the time, scale and scope of impacts as the inherent properties of disruptive events cannot be influenced and/or prevented; in contrast, only their consequences can be dealt with (Janić 2015).

5.3.2 The System and Problem—Air Transport Network

5.3.2.1 *Components and Resilience*

An air transport network consists of airports as the network nodes and the air routes stretching between them as the physical links, where airline flights controlled and managed by the ATC/ATM (Air Traffic Control/Air Traffic Management) system are carried out.

(a) Definition

The resilience of an air transport network is defined as its ability to withstand and stay operational at the required level of safety during the impact of a given disruptive event. This definition takes into account only the actions undertaken during the impact of the disruptive event and not the recovery actions in the aftermath (Chen and Miller-Hooks 2012). However, in the much wider context not embraced by the above-mentioned definition, resilience can generally be considered as static and dynamic. The former refers to the air transport network's capability to maintain its planned function during the impact of disruptive events. The latter implies the network's speed of recovery to the desired (specified) state in the aftermath (Chen and Miller-Hooks 2012, Rose 2007). In addition, resilience can be considered in the short-, medium- and long-term periods (Njoka and Raoult 2009).

The actions undertaken particularly during the impact of the disruptive event on an air transport network commonly include significant reduction of the nominal/regular capacity or complete closure of the affected airports (nodes) and air routes (links) between them. This usually causes (rather long) delays and/or cancellations of the affected flights. Based on the nature of air transport operations, the impact of the disruptive event can spread wider to include airports, air routes and flights that would otherwise be unaffected.

(b) Framework

The resilience of a given air transport network can be assessed at three layers as follows:

- The physical layer, which deals with the physical impact on infrastructure —airports, airspace/air routes and ATC/ATM facilities and equipment;

- The service layer, which mainly considers the impact on the air transport service—airline flights; and
- The cognitive layer, which relates to the air passengers' confidence in the affected and subsequently recovered flights (Len et al. 2010).

(c) Tactics and strategies for mitigating losses

The tactics and strategies for mitigating the losses, i.e. the costs of delayed and cancelled flights associated with an affected air transport network can be as follows (Cox et al. 2011):

- Conservation, implying maintaining operation of the network albeit with a reduced number of airline flights (i.e. mainly due their cancellation);
- Relocation, implying repositioning, rescheduling and rerouting some flights and consequently the aircraft fleet required to carry them out;
- Production recapture, implying filling-in several already scheduled and scheduling additional flights after the end of the disruptive event in order to accommodate the remaining passengers of the affected (long-delayed and cancelled) flights; and
- Management effectiveness, referring to the strategies and tactics of restoring the affected flights after the disruptive event.

5.3.2.2 Friability

The particular actors in an air transport network, such as airports, airlines, ATC/ATM service providers, air passengers and/or air cargo shippers/receivers and the authorities at different institutional levels (local, regional, national) are often interested in identifying the least resilient network nodes —airports, links/air routes and airline flights. This actually implies identifying the components whose closure and resulting long flight delays and/or cancellations due to the impact of a given disruptive event, would cause severe reduction in the network's resilience. In practice, these particularly critical elements are known. However, very often, it is rather complex to quantitatively compare their individual importance under given conditions. Consequently, the concept of friability is introduced to enable such quantitative comparison in a systematic way. Therefore, the friability of a given air transport network affected by a given disruptive event is defined as the diminishing rate of its resilience after 'removing' particular components—airports, air routes and/or airline flights. Based on the estimated friability of individual airports and/or links/routes, the friability of the entire air transport network can be estimated (Ip and Wang 2011, Janić 2015).

5.3.2.3 Large-scale Disruptions and Their Consequences

The large-scale disruption of a given air transport network implies that its current operations substantially deviate from the planned ones. These events can be bad weather, natural disasters and failures of the air transport network components, industrial actions of the aviation staff, traffic accidents/incidents and terrorist threats/attacks. Depending on the type, intensity and duration of the disruptive events, their impact can last from a few hours to several days.

(a) Bad weather

Bad weather, such as low clouds, fog and/or heavy rain usually reduces visibility, which can require increase in the ATC/ATM minimum separation rules between landing and taking-off aircraft at the affected airport(s). This inevitably diminishes the corresponding runway system capacities as shown in Fig. 5.8 for the largest European airports. Here, the runway system capacity diminished by between 22 per cent and 48 per cent. If the current demand still remained below such affected capacity, the average delay of arriving aircraft/ flights would increase by between 30 per cent and 90 per cent, respectively (Janić 2005, 2009b, c, 2015; EEC 2005).

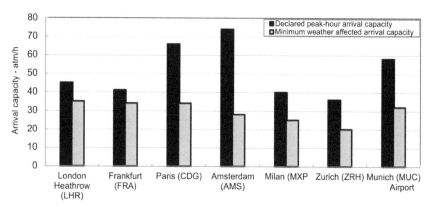

Fig. 5.8: Impact of bad weather on the arrival capacity of select European airports (EEC 2005, Janić 2015).

At US airports, switching from VFR (Visual Flight Rules) to IFR (Instrument Flight Rules) due to bad weather immediately requires an increase in the ATC/ATM separation rules by about 40 per cent, which causes the runway system landing capacity to decrease by about 30 per cent as shown in Fig. 5.9.

Consequently, the average landing delay(s) increase by about 40 per cent (Janić 2005, 2009; FAA 2004). In each of the above-mentioned cases, the affected airports still remain operational. However, hurricanes, thunderstorms and/or heavy snowfall as large-scale disruptive events can cause closure of the affected airports and airspace, thus plummeting their capacity to zero. For example, on 19 January 2013, heavy snowfall caused closure of both runways at London's Heathrow airport before they could be de-iced. This resulted in cancellation of more than 400 of the 1,300 flights scheduled on that day. After reopening the airport, the ATC/ATM separation rules were extended, causing reduction in the runway capacity and consequently increasing the delays in the remaining flights.

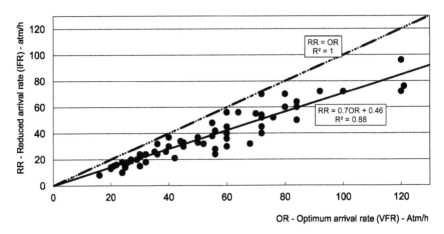

Fig. 5.9: Relationship between the IFR and the VFR arrival capacity at select US airports (FAA 2004, Janić 2015).

(b) Natural disasters

Natural disasters usually damage the infrastructure (airports) of an air transport network causing their closure, and consequently lowering their capacity to zero. For example, frequent earthquakes in Japan (often of a magnitude up to 9.0 on the Richter scale) affect the airports as nodes of the national (and international) air transport network. However, thanks to their adequate design and construction, these airports withstand such impacts and after being temporary closed for a few days, reopen without substantial damage. Natural disastrous events can also cause airspace closure. For example, on 13 April 2010, Iceland's Eyjafjallajokull volcano erupted. Its 11 km-high plume of gases and silicate ash spread over most of Europe. The consequent closing of the airspace between 14 and 24 April caused the cancellation of about two-thirds of European flights and about 180 transatlantic flights in a single day. Both flight delays and cancellations spread far wider—to Canada and Japan. The subsequent opening and closure of the airspace over particular European countries continued until the end of May. The IATA (International Air Transport Association) estimated the total cost of impacts for the global airline industry to the tune of about 1.7 $US billion (http://en.wikipedia.org/wiki/Air_travel_disruption_after_the_2010_Eyjafjallaj_eruption).

(c) Failure of air transport network components

Failure of air transport network components usually occurs at the ATC/ATM and airline facilities, equipment and aircraft. For example, on 26 September 2008, failure of the ATC/ATM central computer caused closure of the airspace across south-east of the UK, thus decreasing its capacity to zero. This impact, which lasted several hours, caused cancellation of 88 flights at five London airports and left about 10,000 passengers stranded. The event also affected flights at Edinburgh, Glasgow, Aberdeen, Cardiff International and Manchester airports. In addition, the most recent failure of the AA (American Airlines) central computer caused the cancellation of more than 700 flights and delayed another 765 flights. Furthermore, on one day in August 2012, the combined reservation system of merged US Continental and United Airlines failed for two hours. On the morning of 16 April 2013, the computer system of US AA failed for several hours. Both events significantly affected the airlines' overall transport capacity.

(d) Industrial action of aviation staff

Industrial action of aviation staff usually causes closure of particular airports and/or airspace as lack of staff virtually halts the aircraft ground servicing and/or the air traffic control tasks, respectively, limits the flight crew and flight attendants to carry out flights, etc. For example, on 11 June 2013, the industrial action (strike) of the French ATC controllers for about two days caused cancellation of 1,800 out of 7,650 flights to/from France, and the delaying, rescheduling and rerouting of many others (http://edition.cnn. com/2013/06/12/business/france-air-traffic-strike).

(e) Traffic incidents/accidents

Traffic incidents/accidents most frequently occur at airports during aircraft landing and take-off, causing temporary closure of the affected airport(s). This brings down their capacities to zero. For example, on 25 February 2009, a Turkish Airlines' B737-800 flying from Istanbul (Turkey) to Amsterdam with 135 persons on board crashed during landing at Amsterdam Schiphol airport in a field approximately 1.5 kilometers north of runway 18R (Polder Baan). The impact caused the death of nine passengers and crew including all the three pilots. The airport was immediately closed for several hours. Some affected flights were diverted to the neighboring airports of Rotterdam and Brussels. After taking care of the people involved and securing the crash site, the airport was gradually reopened (http://airsafe.com/events/models/b737. htm).

(f) Terrorist threats/attacks

Terrorist threats/attacks can impact almost all components of an air transport network. Airports can be blocked and/or aircraft hijacked. Both can substantially directly and/or indirectly reduce the corresponding capacities requiring application of emergency procedures for restoring regular conditions. For example, on 11 September 2001, the U.S. ATC/ATM system managed to land quickly and safely almost 4,500 aircraft that were in the air at the moment the terrorist attacks took place. As a result of the airspace closure over the next few days, all flights within, to and from the US were cancelled (http://usatoday30.usatoday.com/news/sept11/2002-08-12-clearskies_x. htm; (http://en.wikipedia.org/wiki/Closings_and_cancellations_following_ the_September_11_attacks). On 10 August 2006, a terrorist plot aimed at blowing-up aircraft flying between the UK and the US was prevented. Due to immediate closure of the UK airspace, about 2,300 flights were cancelled and others faced long delays over the next seven days. The airline losses of revenue amounted to about EUR 50 million (AEA 2006).

In general, the impacts of the above-mentioned disruptive events are usually of an unpredictable duration and consequences for all the above-mentioned main actors/stakeholders.

5.3.3 Modelling Resilience, Friability and Costs of an Air Transport Network

5.3.3.1 Some Related Research

Research on the resilience and friability of different transport modes/systems such as rail, road and intermodal freight transport networks has been relatively comprehensive. The previous section represents one such example. This research generally includes their definition and their interrelations. In addition, algorithms for optimizing the cost of recovery activities within the specified budget in the aftermath of the given disruptive events have been developed (Berdica 2002, Chen and Miller-Hooks 2012). Furthermore, the framework for evaluating the resilience of the logistics, and both the resilience and friability of the rail transport network, has been defined. This has resulted in developing optimization models and algorithms for allocation of the available resources aimed at guaranteeing security and quality of services in the logistics, and the optimal design of rail networks, based on their resilience and friability (Wang and Ip 2009, Ip and Wang 2011). Specifically, prior research dealing with resilience and friability of air transport networks was recently reviewed. This includes dealing with the topology and dynamics (indirect connectivity and passenger dynamics, air traffic jams and epidemic spreading) of the air transport networks at time-scales ranging from years to days and particularly addressing the resilience of these networks to extreme events (Zanin and Lillo 2013). In addition, some recent research has dealt with defining and understanding the disturbance, resilience and robustness of an ATM (Air Traffic Management) system including development of their qualitative and quantitative measures (EEC 2009, Gluchsenko 2012).

Furthermore, most research has dealt with modelling and estimating the costs of air transport and/or individual airline networks affected by various disruptive events. In such a context, the resilience of a multilayered network consisting of air route networks of each individual airline operating in the given region (European Air Transport Network or ATN) due to failures of particular flights and the consequent need for rescheduling affected passengers in order to reach them to their destinations under given conditions has been modelled and estimated (Cardillo et al. 2013). The performances of communicating networks expressed by the efficiency and the characteristic path length have been modelled to measure the responsiveness of the network to external factors (errors and attacks) (Crucitti et al. 2003). In addition, a wide body of research deals with modelling and estimating the performances of airline networks affected by different types of aircraft failures and the consequent rescheduling of the remaining aircraft to perform the planned flights. This also relates to estimation of the costs of impacts of disruptive events affecting the airline hub airport(s) (Allan et al. 2001, Beatty et al. 1998, Janić 2005, 2009b; Khol et al. 2007, Mayer et al. 1999, Schaefer and Millner 2001, Schavell 2000, Shangyao and Chung-Gee 1997, Welch and Lloyd 2001). Last but not least, the impact of one directly disrupted system/sector on other directly non-disrupted systems/sectors of the national economy measured by the economic losses of all sectors was modelled by developing the concept of inoperability based on the input-output model. This was defined as the level of dysfunction of the affected system, which propagated and consequently affected the other directly non-affected national critical system infrastructure and/or industry sectors. The case study relates to the estimation of impacts of perturbation of air transportation as the primary sector on the other twelve, and then on twenty national-level and then local-level sectors (Santos and Haimes 2004). Subsequently, this approach was extended by development of the inventory DIIM (Dynamic Interoperability Input-Output Model) aimed at assessing the effects of inventories on resilience of the disrupted interdependent systems/economic sectors. The question has been if and how much inventories contribute to delaying inoperability and how the operability of the interdependent systems/sectors is sustained, thus reducing the overall economic losses (Barker and Santos 2010). However, as the authors claim, the proposed modelling approach is not applicable to the service sectors/systems, including transport system/sector where inventories cannot be set up.

5.3.3.2 Objectives and Assumptions

The above-mentioned research does not explicitly consider in the more generic sense the resilience, friability and costs of air transport network(s) affected by large-scale disruptive events, or their relationships. Therefore, this sub-section deals with modelling of resilience, friability and costs of an air transport network and its particular components—airport affected by a given large-scale disruptive event by developing a convenient methodology (Janić 2015). This consists of the corresponding models based on an analogy to the above-mentioned research related to inland transport networks and is to be applied to an *a posterior* real-life case (Chen and Miller-Hooks 2012, Ip and Wang 2011). As such, the methodology can be used by the particular actors/stakeholders involved in air transport network operations on the one hand and researches, on the other, both *a priori* and *a posterior*, as follows:

- Airports can use these models for assessing their convenience while offering their services to airlines as bases or hubs after primarily being considered in terms of their operational and economic advantages;
- Airlines can find these models useful when considering impacts of disruptive events as criteria for designing their schedules in order to mitigate their impact as far as possible;
- ATC/ATM can use these models for designing and assessing the consequences of actions undertaken to manage the air transport operations during disruptive events;
- Air passengers and air cargo shippers can use these models while choosing the air transport mode under given (disruptive) conditions; and
- Researchers can find these models useful while assessing the resilience, friability and costs of transport networks operated by other transport modes.

5.3.3.3 Structure of the Methodology

(a) General

The methodology contains models for quantifying the resilience, friability and costs of an air transport network consisting of (N) airports and air routes spreading between them where different airlines operate their flights during a specified period of time (τ). This period of a few hours to one and/or several days represents the duration of the impact of a given disruptive event. The models imply action for mitigating costs and maintaining the required safety level of operations in the network during the impact of the disruptive event.

(b) Model for estimating resilience

The model for estimating the resilience of a given air transport network is based on the following assumptions:

- Resilience is considered only during the duration of the impact of a given disruptive event; this implies that it does not relate to actions in the aftermath of this event;
- Direct air routes with at least one scheduled flight connect the airports as the network's nodes; if an airport is closed, all incoming and outgoing flights from/to all other airports, respectively, are cancelled, i.e. the connections are cut-off; and
- The number of arriving and departing flights is used for measuring the relative importance, i.e. weight of a particular airport in the network during the impact of a disruptive event (other measures not explicitly considered can be the number of passengers and/or the volume of air cargo).

The model consists of the following components (Janić 2015):

Airport relative importance/weight

The relative importance, i.e. weight of a given airport (*i*) of the air transport network consisting of *N* airports can be estimated as follows:

$$w_i^{\gamma_i}(\tau) = \frac{u_i^{\gamma_i}(\tau)}{\sum\limits_{j=1}^{N} u_j^{\gamma_j}(\tau)} \tag{5.13}$$

where $u_i^{\gamma_i}(\tau)$ is the total number of flights accommodated at the airport (*i*) at the runway system's capacity ratio (γ_i) during time (τ).

The number of flights $u_i^{\gamma_i}(\tau)$ in Equation 5.13 can be determined as follows:

$$u_i^{\gamma_i}(\tau) = \sum\limits_{j=1/\, j\neq i}^{N} \left[f_{ji}^{\gamma_{ai}}(\tau) + f_{ij}^{\gamma_{di}}(\tau) \right] \tag{5.14}$$

where $f_{ji}^{\gamma_{ai}}(\tau)$, $f_{ij}^{\gamma_{di}}(\tau)$ is the number of arriving and departing flights served at the airport (*i*) from the airport (*j*) operating at the runway system arrival and departure capacity ratio (γ_{ai}) and (γ_{di}), respectively, during time (τ).

Equation 5.13 indicates that the weight of a given airport increases in line with the share of its served flights in the total number of flights served in the network during the given period of time under given conditions. These conditions are specified by the current arrival and departure capacity. Equation 5.14 indicates that both arrival and departure flights are taken into account in the weight of a given airport. The ratio $\gamma_{ai}(\tau)$ and $\gamma_{di}(\tau)$ in Equation 5.14 can be determined as: $\gamma_{ai}(\tau) = \mu_{ai}^*(\tau)/\mu_{ai}(\tau)$ and $\gamma_{di}(\tau) = \mu_{di}^*(\tau)/\mu_{di}(\tau)$ where $\mu_{ai}(\tau)$ and $\mu_{di}(\tau)$ are the nominal/regular arrival and departure capacity, respectively, of the runway system at airport (i) during time (τ); $\mu_{ai}^{*}(\tau)$ and $\mu_{di}^{*}(\tau)$ are the arrival and departure capacity, respectively, of the runway system of the airport (i) affected by a disruptive event during time (τ) considered as ultimately irregular and generally different (lower) than the above-mentioned nominal/regular one(s). Thus, depending on the prevailing conditions, the ultimately nominal/regular and affected/irregular capacities can generally take a range of values. Nevertheless, they are generally related as follows: $\mu_{ai}(\tau) \geq \mu_{ai}^{*}(\tau)$ and $\mu_{di}(\tau) \geq \mu_{di}^{*}(\tau)$. In addition, these capacities can also be dependent on each other (Janić 2005, 2009). Consequently, the ratios $[\gamma_{ai}(\tau)]$ and $[\gamma_{di}(\tau)]$ can take any value between 1 and 0. The former implies that the runway system operates under nominal/regular conditions as planned, while the latter implies that the airport is closed for all flights. As such, Equation 5.14 reflects the real circumstances occurring in air transport networks and airports worldwide on an hourly, daily, monthly and seasonally time horizon, when the airports specify the available number of arrival and departure capacities (i.e. slots) depending on the current and prospective short, medium and long-term perceived conditions (hour, day, month, season). In such a case, the most common values of these capacities represent the reference or the nominal state considered for planning purposes when accepting airline requests for slots. However, like any other system, this most common state can be expressed by a single or by a range of values of the arrival and departure capacities of airports included in the network and specified for the given conditions, including those determined by disruptive events. This enables flexibility in specifying the range of reference states of each airport and the entire air transport network, both depending on the prevailing conditions. Consequently, when network resilience is considered, at that time a large severely affected airport can have a lower weight than a smaller much less severely affected one.

Airport self-excluding importance/weight

The self-excluding importance, i.e. weight, of a given airport (i) belonging to the air transport network and consisting of (N) airports implies that its other connected airports do not include it. Thus it can be estimated as follows:

$$\upsilon_i^{\gamma_i}(\tau) = \frac{u_i^{\gamma_i}(\tau)}{\sum\limits_{j=1/\, j\neq i}^{N} u_j^{\gamma_j}(\tau) - u_i^{\gamma_i}(\tau)} \tag{5.15}$$

where all the symbols are as in the previous equations.

Equation 5.15 indicates that the self-excluding weight of a given airport increases more than proportionally with increase in the share of its weight in the given network.

Airport resilience

The resilience of a given airport (i) can be estimated as the sum of the product of all self-excluding importance/weights except the one for its own and the number or proportion of flights carried out as follows:

$$R_i^{\gamma_i}(\tau) = \sum_{j=1/\, j\neq i}^{N} \upsilon_j^{\gamma_j}(\tau) * \left[\delta_{ji}(\tau) * m_{ji}^{\gamma_{ai}}(\tau) + \delta_{ij}(\tau) * m_{ij}^{\gamma_{di}}(\tau) \right] \qquad (5.16)$$

where

$m_{ji}^{\gamma_{ai}}(\tau), m_{ij}^{\gamma_{di}}(\tau)$ is the number of arriving and departing flights at the airport (i) from and to the airport (j), respectively, which operates at the runway system arrival and departure capacity ratio (γ_{ai}) and (γ_{di}), respectively, during time (τ); and

$\delta_{ji}(\tau), \delta_{ij}(\tau)$ is a binary variable taking the value 1 if the airports (j) and (i) and air route between them is operable, and the value 0, otherwise, during time (τ).

The symbols ($m_{ji}^{\gamma_{ai}}(\tau)$) and ($m_{ij}^{\gamma_{di}}(\tau)$) in Equation 5.16 denote realized on-time or delayed, or only on-time flights. In general, on-time flights are those that are exactly on-time or delayed by a maximum of 15 minutes, while delayed flights are those delayed by longer than 15 minutes. Consequently, the total number of scheduled flights in Equation 5.14 under the given conditions can be expressed as: $f_{ji}^{\gamma_{ai}}(\tau) = m_{ji}^{\gamma_{ai}}(\tau) + n_{ji}^{\gamma_{ai}}(\tau)$ and $f_{ij}^{\gamma_{di}}(\tau) = m_{ij}^{\gamma_{di}}(\tau) + n_{ij}^{\gamma_{di}}(\tau)$, where ($n_{ji}^{\gamma_{ai}}(\tau)$) and ($n_{ij}^{\gamma_{di}}(\tau)$) are the number of cancelled arriving and departing flights, respectively. The other symbols are as in the previous equations. The resilience of a given airport in Equation 5.16 is proportional to the sum of the product of the self-excluding weight and the number of actually realized flights to and from each connected airport under the given conditions. In addition, it increases in line with the number of sustained, i.e. actually realized flights.

Air transport network resilience

The resilience of the air transport network consisting of (N) airports can be estimated as the sum of the resilience of each individual airport, based on Equations 5.13 and 5.16 as follows:

$$R^{\gamma}(N,\tau) = \sum_{i=1}^{N} w_i^{\gamma_i}(\tau) * R_i^{\gamma_i}(\tau) \qquad (5.17)$$

where all symbols are as in the previous equations.

Equation 5.17 indicates that the resilience of the air transport network is proportional to the sum of the weighted resilience of each airport belonging to it. Alternatively to Equation 5.17, the resilience of the air transport network consisting of N airports can be measured by an indicator based on the inherent network properties and the set of actions for mitigating costs and maintaining the required safety level of operations. The mitigating actions include delaying, rerouting and/or cancelling flights at the affected airports. In such cases, this indicator can be defined as a proportion or the ratio between the on-time and/or between the actually realized on-time and delayed, and the total number of planned flights during time (τ). In contrast, the proportion of delayed and/or cancelled flights can express the network's vulnerability. Consequently, the indicator of the network's resilience can be specified as follows (Chen and Miller-Hooks 2012, Janić 2015):

$$R^{\gamma}(N,\tau) = \sum_{i=1}^{N} \sum_{j=1/\,j\neq i}^{N} [m_{ij}^{\gamma_{di}}(\tau) + m_{ji}^{\gamma_{di}}(\tau)]/\sum_{i=1}^{N} \sum_{j=1/\,j\neq i}^{N} [f_{ij}^{\gamma_{di}}(\tau) + f_{ji}^{\gamma_{di}}(\tau)] \quad (5.18)$$

where all the symbols are as in the previous equations. The resilience of the given air transport network in Equation 5.18 increases in line with the actually realized and scheduled/planned flights at all its airports under given (disruptive) conditions.

Step-by-step algorithm for estimating the resilience of the air transport network

Step 1: Calculate the weight and self-excluding weight of each airport of the network by Equation 5.13 and Equation 5.15, respectively;

Step 2: Calculate the resilience of each airport of the network by Equation 5.16;

Step 3: Calculate the resilience of the entire air transport network by Equation 5.17 or Equation 5.18; and

Step 4: Repeat Steps 1, 2 and 3, as necessary, if the conditions/impact, configuration and service performance of the network and period of time change.

(c) Model for estimating friability

The model for estimating the friability of an airport and the entire air transport network consisting of (N) airports is based on the assumption that it is possible to quantify their resilience by Equations 5.13-5.17. This model consists of the following components (Janić 2015):

Airport friability

The friability of airport (i) belonging to the air transport network and consisting of N airports can be estimated as follows:

$$F_i^{\gamma_i}(\tau) = R^{\gamma}(N,\tau) - R^{\gamma}(N,\tau/i) \quad (5.19)$$

where $R^\gamma(N, \tau/i)$ is the resilience of the air transport network after removing, i.e. closing airport (i) during time (τ).

The other symbols are analogous to those in Equation 5.17. As can be seen, Equation 5.19 indicates that the friability of a given airport increases in line with resilience of the entire air transport network and decreases as the resilience of the airport itself increases. This implies that the more resilient airports compromise less the overall network resilience.

Maximum friability of an airport

The maximum friability of a given airport (i) as the weakest node of the air transport network consisting of N airports can be estimated as follows:

$$F_i^{\gamma_i}(\tau) = \max[F_i^{\gamma_i}(\tau)/i \in N] \tag{5.20}$$

where all the symbols are as in the previous equations.

Air transport network friability

Friability of the air transport network consisting of N airports while being affected by a disruptive event can be estimated as follows:

$$F^\gamma(N, \tau) = \sum_{i=1}^{N} w_i^{\gamma_i}(\tau) * F_i^{\gamma_i}(\tau) \tag{5.21}$$

where all the symbols are analogous to those in the previous equations.

Among others, Equation 5.21 handles the case of zero friability, i.e. when all airports were removed from the network, i.e. closed. This implies that the network was completely non-functional with a resilience also equal to zero.

Step-by-step algorithm for estimating the friability of the air transport network

Step 1: Calculate the resilience of the air transport network consisting of N airports by Equation 5.17 or Equation 5.18.

Step 2: Calculate the friability of airport (i) of the network by Equation 5.19.

Step 3: Calculate the maximum friability of airport (i) by Equation 5.20.

Step 4: Calculate the friability of the entire air transport network by Equation 5.21.

Step 5: Repeat Steps 1, 2, 3 and 4, if necessary, when the conditions/impact, configuration, the service performance of the network and the period of time change.

(d) Model for estimating the costs

The costs of the air transport network consisting of (N) airports affected by a large-scale disruptive event are represented by the sum of the costs of delayed and cancelled flights as follows (Janić 2015):

$$C^{\gamma}(N,\tau) = \sum_{i=1}^{N} \sum_{j=1/j\equiv i}^{N} \begin{Bmatrix} c_{a/ji}(\tau)*F_{a/ji}[\tau;\gamma_{ai}(\tau)]*d_{a/ji}[\tau;\gamma_{ai}(\tau)]*\psi_{a/ji}[\tau;\gamma_{ai}(\tau)]+ \\ +c_{d/ij}(\tau)*F_{d/ij}[\tau;\gamma_{di}(\tau)]*d_{d/ij}[\tau;\gamma_{di}(\tau)]*\psi_{d/ij}[\tau;\gamma_{di}(\tau)]+ \\ +C_{a/ji}(\tau)*n_{d/ji}[\tau_i,\gamma_{ai}(\tau_i)]+C_{d/ij}(\tau)*n_{d/ij}[\tau,\gamma_{di}(\tau)] \end{Bmatrix} \quad (5.22)$$

where

$c_{a/ji}(\tau), C_{a/ji}(\tau)$ — is the average unit cost of delay and cancellation, respectively, of a flight arriving from airport (j) at airport (i) during time (τ);

$c_{d/ij}(\tau), C_{d/ij}(\tau)$ — is the average unit cost of delay and cancellation, respectively, of a flight departing airport (i) to airport (j) during time (τ);

$F_{a/ji}[\tau;\gamma_{ai}(\tau)], F_{d/ij}[\tau;\gamma_{di}(\tau)]$ — is the number of delayed arriving and departing flights between airports (j) and (i) and vice versa, while operating at the capacity ratio $\gamma_{ai}(\tau)$ and $\gamma_{di}(\tau)$, respectively, during time (τ);

$d_{a/ji}[\tau;\gamma_{ai}(\tau)], d_{d/ij}[\tau;\gamma_{di}(\tau)]$ — is the delay of an arriving and of a departing flight between airports (j) and (i) and vice versa, while operating at the capacity ratio $\gamma_{ai}(\tau)$ and $\gamma_{di}(\tau)$, respectively, during time (τ);

$\psi_{a/ji}[\tau;\gamma_{ai}(\tau)], \psi_{d/ij}[\tau;\gamma_{di}(\tau)]$ — is the delay multiplier of an arriving and of a departing flight between airports (j) and (i) and vice versa, while operating at the capacity ratio $\gamma_{ai}(\tau)$ and $\gamma_{di}(\tau)$, respectively, during time (τ); in general, this reflects the scale of spreading/propagating initial delay(s) from the directly to the other otherwise non-directly affected airports and flights; and

$d_{a/ji}[\tau;\gamma_{ai}(\tau)], d_{d/ij}[\tau;\gamma_{di}(\tau)]$ — is the number of cancelled arriving and departing flights between airports (j) and (i) and vice versa, while operating at the capacity ratio $\gamma_{ai}(\tau)$ and $\gamma_{di}(\tau)$, respectively, during time (τ).

The other symbols are analogous to those in the previous equations.

As a general rule, a flight will be cancelled if the cost of its delay is perceived to be greater than the cost of its cancellation. From Equation 5.22, it follows: $C_{a/ji}(\tau) > c_{a/ji}(\tau)*d_{a/ji}[\tau;\gamma_{ai}(\tau)]$ for an arriving and $C_{d/ij}(\tau) > c_{d/ij}(\tau)*d_{d/ij}[\tau;\gamma_{di}(\tau)]$ for a departing flight.

5.3.4 Application of the Methodology

The above-mentioned models were applied to estimating the resilience, friability and costs of a part of the US air transport network whose 16 airports on the north-east coast were affected by a large-scale disruptive event (Hurricane

Sandy) in October 2012. In addition, the above-mentioned performances were estimated for one of the affected airports—New York (NY) LaGuardia.

5.3.4.2 Inputs

(a) Disruptive event—Hurricane Sandy

The disruptive event, Hurricane Sandy, a very large tropical cyclone, lasted for ten days, i.e. from 21 to 31 October 2012. It struck the Caribbean Islands and then moved towards the north and north-west as shown in Fig. 5.10a. Between 25 and 29 October, the hurricane was moving mainly above the sea almost parallel to the US east coast. Between 28/29 and 31 October it turned to the west towards the coast and further through the continent at a speed of about 20-35 km/h (10-19 kt) (kt – knot (nm/h), nm – nautical mile) (1 nm = 1.852 km). Its surface wind speed reached a maximum of about 180 km/h (97 kt) on 25 October (i.e. after it had just passed Cuba) and about 160 km/h (86 kt) on 29 October (when it strengthened again and turned toward the US north-east cost) as shown in Fig. 5.10b. At the same time, the hurricane's force winds and tropical-storm-force winds were spreading from its center outwards up to 280-300 km (150-160 nm) and 800-900 km (430-490 nm), respectively. This implies that its diameter was almost up to about 1800-2000 km (970-1070 nm). Consequently, the covered/affected area on the ground reached about 2.5-3 million km^2 (1.3-1.6 million square nm). In addition, Fig. 5.10b shows that the hurricane's wind speed was most of the time much higher than the maximum cross-wind speed of about 74 km/h (40 kt) at which most commercial aircraft can safely operate.

(b) The affected air transport network

This hurricane's strong winds accompanied by very high precipitation impacted the following 16 airports of the US air transport network: Atlanta (ATL), Boston (BOS), Baltimore/Washington International (BWI), Washington Ronald Reagan National (DCA), NY Newark Liberty International (EWR), Fort Lauderdale-Hollywood International (FLL), Washington Dulles International (IAD), Jacksonville International (JAX), NY John Fitzgerald Kennedy (JFK), NY LaGuardia (LGA), Orlando International (MCO), Miami International (MIA), Norfolk International (ORF), Philadelphia International (PHL), Providence (PVD), and Raleigh-Durham (RDU). The total daily number of affected scheduled arriving and departing flights to, from and between these airports was between 13,500 and 17,500. This number fluctuated during the disruptive event (http://www.rita.dot.gov/bts/). The measures for reducing the costs and maintaining the required level of safety were adjusted according to the strength of the impact. They consisted of delaying, cancelling and rerouting the directly and potentially affected flights. On the fifth, sixth and seventh day when the disruptive event was closest and its impact the strongest

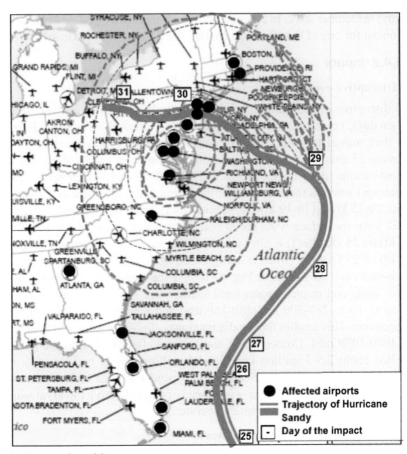

(a) Best track position

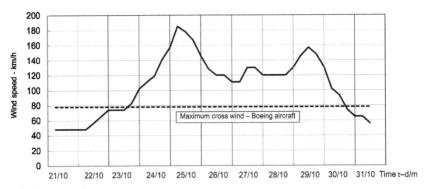

(b) Surface wind speed

Fig. 5.10: Characteristics of the Hurricane Sandy (21 to 31 October 2012). (Blake et al. 2013, Janić 2015, Van Es et al. 2001, http://www.rita.dot.gov/bts)

as shown in Fig. 5.10 (a, b), four, three and two airports were simultaneously closed, respectively.

(c) Cost of delayed and cancelled flights

The average unit cost of delayed and cancelled flights was estimated by combining data from different sources. The average delay of an arriving or a departing flight was estimated to be: $d_{a/ji}(\tau) = d_{d/ij}(\tau) = 53.54$ min. Specifically, this delay for NY La Guardia airport was estimated to be $d_{a/ji}(\tau) = d_{d/ij}(\tau) = 57.51$ min. (these values were reduced by 15 minutes since flights delayed less than 15 minutes are not deemed delayed flights) (http://www.rita.dot. gov/bts/; https://aspm.faa.gov/opsnet/sys/main.asp). The average unit cost of airline delay was adopted to be 75.27 $US/min. This was multiplied by the factor 1.5 in order to take into account the indirect and induced cost impact of air travel delays on the national (US) economy (Janić 2009a). The average cost of passenger time was assumed to be 39.04 $US/h (http://www.airlines. org/Pages/Home.aspx). The average number of passengers per flight in the network of 16 affected airports was calculated to be 121 (http://av-info.faa. gov/). Consequently, the average unit cost of delay is estimated to be: $c_{a/ji}(\tau)$ = $c_{d/ij}(\tau)$ = 192 $US/min. Specifically, for NY LaGuardia airport, the average number of passengers per flight was estimated to be 98 and the average unit cost of delay $c_{a/ji}(\tau) = c_{d/ij}(\tau) = 162$ $US/min. (JEC 2008, PAN&N 2012). In both these cases, the average cost of a cancelled flight carried out by a narrow-body 120/150 seat aircraft is assumed to be $C_{a/ji}(\tau) = C_{d/ij}(\tau) = 21800$ $US/flight. This cost includes the service recovery cost (passenger vouchers, drinks, telephone, hotel), the interline cost (rebooking revenue), the loss of future value (cost of individual passenger delays) and the savings in direct operational costs of the cancelled flight (EEC 2011). The above-mentioned figures possibly indicate that in the affected network and at NY LaGuardia airport, a flight expected to be delayed longer than about two and two-and-a-half hours, respectively, would be cancelled under the given conditions.

5.3.4.3 Results

The results of the application of the proposed methodology to the above-mentioned affected air transport network and the selected airport are shown in Figs. 5.11, 5.12, 5.13, 5.14 and 5.15. In particular, Fig. 5.11 shows the resilience of a part of the US national air transport network with the 16 affected airports estimated by Equations 5.17 and 5.18. In this case, the resilience reflects the network's operational level.

As can be seen, according to Equations 5.17 and 5.18, the resilience of the network reflecting its level of operability during the disruptive event had quite similar dynamics. It decreased with increase in the intensity of impact, which culminated on the fifth day (29 October) when four of the 16 affected airports

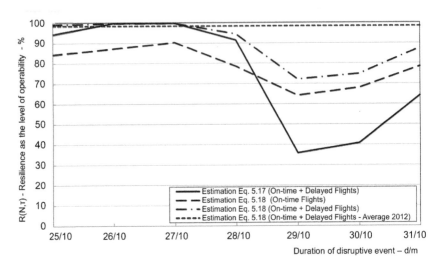

Fig. 5.11: Resilience as the level of operability of the US national air transport network with sixteen affected airports in the given example (Janić 2015).

had to be completely closed and consequently a substantial number of flights cancelled. As the impact subsided later on, i.e. on the sixth and seventh day of the disruptive event, resilience at the level of operability started to increase, thanks to the gradual reopening of the closed airports and resumption of flights. In addition, according to Equation 5.18, the resilience of the network in terms of the 'on-time + delayed flights' was all the time higher than that in terms of 'on-time flights' thanks to the mitigating action of 'delaying flights'. As the impact intensity increased, the effect of this action decreased and that of the action 'cancelling flights' increased. Furthermore, the resilience estimated by Equation 5.17 was lower than that estimated by Equation 5.18 due to differences in the structure of the corresponding models—the former explicitly reflected the impact of closed airports on resilience. Last but not least, most of the time during the disruptive event, the network's resilience as a reflection of the operational level was substantially lower than its average for the entire 2012 (http://www.transtats.bts.gov/). Although substantially compromising the resulting resilience in both cases, the mitigating actions managed to maintain the required level of safety of operations.

Furthermore, the network's resilience, i.e. operational level in Fig. 5.11 was presumed to be influenced by the strength of impact of Hurricane Sandy. In order to estimate the potential nature of such relationship(s), the resilience of the network estimated by Equation 5.18 was taken as the dependent variable ($R(N, \tau)$), and the hurricane's surface wind and distance from the prospectively affected airports as the independent variables (W_R) and (D), respectively. The variable (W_R) is expressed as the ratio between the hurricane's surface wind speed and the maximum allowable cross-wind speed allowing safe departures and arrivals of most commercial aircraft (Fig. 5.10b). The variable *D* expressed in units of distance (nm) actually represents the distance between the center ('eye') of the hurricane and the closest affected airports measured along its path (Fig. 5.10a). The estimated regression relationships for the on-time, realized-on-time, delayed flights, and the rate of flight cancellations are given in Table 5.1. As can be seen, the operational level was compromised as the strength of impact of the disruptive event increased and recovered as the effect of the disruptive event on the affected airports fell. The former had almost the same contribution to the resilience of both groups of flights. The latter contributed to increase in the resilience of the realized flights by about twice that of the on-time flights. This is because, at the same time, the rate of flight cancellations increased with strengthening and decreased with weakening and expelling of the impact from the affected airports.

Figure 5.12 again illustrates the relationship between the resilience estimated by the regression Equation 2 in Table 5.1 and the wind ratio (W_R).

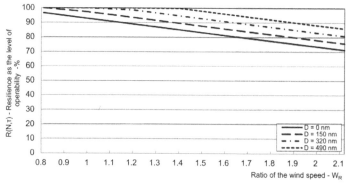

Fig. 5.12: Relationship between the resilience as the level of operability of air transport network and the ratio of the hurricane's and the maximum allowed wind speed in the given example (Compiled from the regression 2 in Table 5.1) (Janić 2015).

As can be seen, the resilience of the air transport network expressed by the realized on-time and delayed flights decrease with increase in the wind speed ratio and the distance of the hurricane's 'eye' and the closest airports of the network. In the worst case defined by the strongest wind at the shortest distance, the resilience dropped to only about 70 per cent.

Figure 5.13 shows the resilience of NY LaGuardia airport as one of 16 affected airports in the given case estimated by Equation 5.18.

Here, the airport's resilience expressed by the proportion of realized flights was higher than that expressed by the proportion of on-time flights. Both generally changed during the disruptive event and gradually decreased with increase in the intensity of its impact. On the fifth day, when the hurricane had

Table 5.1: Relationships between particular categories of resilience and the main characteristics of the disruptive event in the given example (Janić 2015)

Resilience category	Regression equation	R^2	F
1. On-time (departures + arrivals)	$R_{d/o-t}(N, \tau) = 95.104 - 14.395\, W_R + 0.167 D$	0.516	3.067
	t-stat $(3.918)\ (-1.916)\ (2.461)$		
2. Realized (on-time + delayed departures and arrivals)	$R_{d/re}(N, \tau) = 112.200 - 19.335\, W_R + 0.030 D$	0.833	8.761
	t-stat $(5.363)\ (-2.210)\ (4.068)$		
3. Rate of flight cancellation	$r_c(N, \tau) = -15.693\ +22.397\, W_R - 0.033\, D$	0.837	8.776
	t-stat $(-0.575)\ (2.224)\ (-4.075)$		
	$n = 14$; $\quad 0.81 < W_R < 2.12$; $\quad 150 < D < 490$		

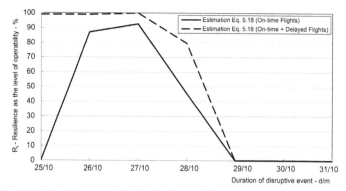

Fig. 5.13: Resilience as the level of operability of NY LaGuardia airport in the given example (Janić 2015).

just arrived and had the strongest impact, the airport was closed and stayed so until the end of the hurricane. Consequently, the airport's resilience, i.e. its level of operability, fell to zero.

Figure 5.14 shows the friability of the NY LaGuardia and Atlanta International airports in the given example.

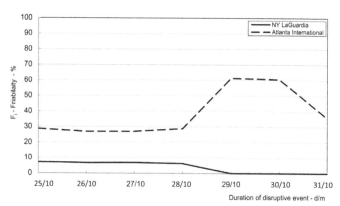

Fig. 5.14: Friability of the selected affected airports in the given example (Janić 2015).

The former airport was selected because it was closed during the last three days of the disruptive event (the main affected airlines were Delta, AirTran and Express Jet with a market share of about 90 per cent). The latter airport was selected as the largest among the affected but still one among the constantly operational airports (the main partially affected airlines were Delta, American and US Airways with a total market share of about 41 per cent) (http://www.transtats.bts.gov/). As can be seen, the friability of Atlanta airport is the highest, indicating that its removal, i.e. closing, as the weakest node under the given conditions would maximally compromise the resilience of the network. In addition, the airport's friability was gradually increased and reached the maximum on the fifth day of the disruptive event, when four other among the 16 affected airports had to be closed. Consequently, closing this additional fifth airport would have had the greatest impact on the remaining resilience of the network. After the subsequent reopening of the closed airports, the given airport's resilience diminished, thus indicating its low influence on the network's overall resilience by its eventual removal, i.e. closure.

The friability of NY LaGuardia airport was much lower than that of Atlanta airport, indicating that its removal, i.e. closure, had a much weaker impact on the resilience of the network. When the airport was closed, its resilience fell to zero and consequently its friability reflected zero resilience.

Figure 5.15 (a, b) reveals the costs of delayed and cancelled flights at all 16 affected airports of the US air transport network and NY LaGuardia airport.

As can be seen, these costs increased in line with strengthening of the impact of Hurricane Sandy. Figure 5.15a shows that during the first four days, the costs of delayed flights in the network were much higher than the costs of cancelled flights. This was in line with their proportions shown in Fig. 5.13. During the last three days of the impact, many more flights were cancelled than delayed. This resulted in increase in the costs of cancelled flights on account of the costs of those delayed. Consequently, the total cumulative costs grew and reached almost $US 500 million at the end of the impact. Figure 5.15b shows that a similar and almost analogous development of the cost and its structure took place at NY La Guardia airport. The exception was the exclusive domination of the costs of cancelled flights in the total costs during the last three days of the impact on account of the airport's closure. The total cumulative costs reached about 10 per cent of the total cumulative network cost at the end of the impact.

5.3.5 Interim Summary

This section deals with modelling the resilience, friability and costs of a given air transport network affected by a large-scale disruptive event. The network considered as the 'transport system' consists of airports as nodes and air routes and flights scheduled between them as the network's links. In such a context, resilience is considered as the network's ability to sustain its planned operations during the impact of a disruptive event, i.e. to retain its planned operational level.

Friability is considered as the rate of reducing resilience because of excluding particular affected airport(s) and/or air route(s) and flights from the network. Costs of delayed and cancelled flights reflect the network's economic performances under the given conditions.

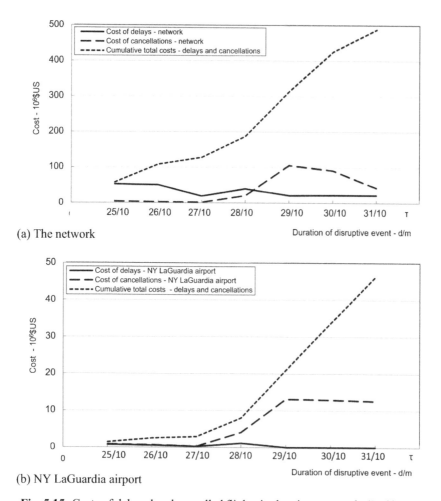

(a) The network

(b) NY LaGuardia airport

Fig. 5.15: Costs of delayed and cancelled flights in the given example (Janić 2015).

The models were applied to a part of the US air transport network in which 16 airports on the north-east of the US were affected by Hurricane Sandy in October 2012. The results indicate that the resilience of the network as well as of the selected airport (NY LaGuardia) were substantially affected by the hurricane. In addition, the friability of the much larger Atlanta airport was greater than that of the smaller NY La Guardia airport, indicating the former airport's greater weakness in case of being removed from the network, i.e. closed under the given conditions.

The costs changed in line with the intensity of impact of the disruptive event—the hurricane during the observed period and of the mitigating actions applied, like delaying and cancelling flights. In terms of both the network and the selected airport (NY LaGuardia), the increasing costs of cancelled and decreasing costs of delayed flights were the highest when the impact of hurricane was the strongest. At the same time, the cumulative network costs reached almost $US 500 *10^6$, about one-tenth of which was shared by the selected (NY LaGuardia) airport by the end of the disruptive event.

5.4 Concluding Remarks

This chapter deals with modelling the resilience of transport systems while focusing on the logistics network operating under regular and irregular (disruptive) conditions and air transport network when affected by a large-scale disruptive event. At the logistics level, the economic performances in terms of the total and average cost per goods/freight shipment consolidated into pallets and delivered between doors of the manufactures and consumers are modelled. The results obtained by applying the models to hypothetical logistics network based on data from European freight transport and logistics sector show the following:

Under regular operating conditions, economies of scale exist, implying decrease in the average cost per freight/goods shipment—pallet—more than proportionally with increase in the transport service frequency by both intermodal rail/road and road transport mode. Under irregular (disruptive) operating conditions, diseconomies of scale exist, implying increase in the average cost per goods/freight shipment—pallet—more than proportionally with the duration of the disruptive event. In both the cases, the average cost of road transport mode is slightly lower than that of its intermodal counterpart.

At the given air transport network, resilience is modelled as the network's ability to sustain its planned operations during the impact of a disruptive event. This ability influences the costs of delayed and cancelled flights, thus reflecting the network (in)-efficient economic performances under the given conditions. The models were applied *a posterior* to the real-life case of the air transport network and two of its airports affected by a large-scale disruptive event (in this case, a hurricane). The results from application of the models showed that larger airport(s), if affected, i.e. closed due to the impact of disruptive event, affects the resilience of the entire network much stronger than the smaller one(s). In addition, the overall costs of delayed and cancelled flights in the network and concerned airports increase with increase in the intensity of the impact of the disruptive event and its duration.

The above-mentioned cases indicate that modelling the resilience of transport systems can be useful for estimating the impact of disruptive events mainly on their operational and consequently economic performances, i.e. the costs imposed on particular actors/stakeholders involved. This can be carried out both *a priori* to the expected impact and *a posterior* after the impact has subsided.

References

Aberle G. 2001. *Globalisierung*. Verkehrsentwicklung und Verkehrskosten (Gießen: Justus-Liebig-Universität), Berlin, Germany.

Allen B.W., Mahmoud M.M., McNeil D. 1985. The Importance of Time in Transit and Reliability of Transit Time for Shippers, Receivers and Carriers. *Transportation Research* B19: 447-456.

AEA. 2006. *Security Threat Level 'Critical': The 10 August Terrorist Alert*. Association of European Airlines, Source 4: 9-10.

Allan S.S., Beesley A.J., Evans E.J., Gaddy G.S. 2001. *Analysis of Delay Causality at Newark International Airport*. 4 USA/Europe Air Traffic Management R&D Seminar, Santa Fe, USA, p. 11.

Barker K., Santos R.J. 2010. Measuring the Efficiency of Inventory with a Dynamic Input-Output Model. *International Journal of Production Economics* 120: 130-143.

Beatty R., Hsu R., Berry L., Rome J. 1998. *Preliminary Evaluation of Flight Delay Propagation Through An Airline Schedule*. 2 USA/Europe Air Traffic Management R&D Seminar, Orlando, USA, p. 9.

Berdica K. 2002. An Introduction to Road Vulnerability: What has been Done is Done and Should be Done. *Transport Policy* 9: 117-127.

Blake S.E., Kimberlain B.T., Berg J.R., Gangialosi P.J., Beren II L.J. 2013. *Tropical Cyclone Report: Hurricane Sandy* (AL182012), 22-31 October 2012. National Hurricane Center, NOAA/National Weather Centre, Miami, Florida, USA.

Benjamin J. 1990. An Analysis of Mode Choice for Shippers in a Constrained Network with Application to Just-in-Time Inventory. *Transportation Research* B 24B: 229-245.

Blumenfeld D.E., Burns L.D., Diltz J.D., Daganzo C.E. 1985. Analyzing Trade-offs between Transportation, Inventory and Production Costs on Freight Networks. *Transportation Research* B 19B: 361-380.

Campbell J.F. 1990. Freight Consolidation and Routing with Transportation Economies of Scale. *Transportation Research* B 24B: 345-361.

Campbell, J.F. 1992. Location-Allocation for Distribution Systems with Transshipments and Transportation Economies of Scale. *Annals of Operations Research* 40: 77-79.

Cardillo A., Zanin M., Gómez-Gardeñes J., Romance M., Del Amo A.J.G., Boccaletti S. 2013. Modelling the Multi-Layer Nature of the European Air Transport Network: Resilience and Passengers Re-scheduling under Random Failures. *The European Physical Journal Special Topics* 215: 23-33.

Chen L., Miller-Hooks E. 2012. Resilience: An Indicator of Recovery Capability in Intermodal Freight Transport. *Transportation Science* 46: 109-123.

Cox A., Prager F., Rose A. 2011. Transportation Security and the Role of Resilience: A Foundation for Operational Metrics. *Transport Policy* 18: 307-317.

Crucitti P., Latora V., Marchiori M., Rapisarda A. 2003. Efficiency of Scale-Free Networks: Error and Attack Tolerance. *Physics A: Statistical Mechanics and its Applications* 320: 622-642.

Daganzo C.F. 1999. *Logistics System Analysis*. 3rd Ed. Springer, Berlin, Germany.

Daganzo C.F. 1984. The Length of Tours in Zones of Different Shapes. *Transportation Research* B 18B: 135-140.

Daganzo C.F., Newell F.G. 1985. Physical Distribution from a Warehouse: Vehicle Coverage and Inventory Levels. *Transportation Research* B 19B: 397-405.

Dejax P. 1991. Goods Transportation by the French National Railway (SNCF): The Measurement and Marketing of Reliability. *Transportation Research* A 25A: 219-225.

EC. 2001a. *Real Cost Reduction of Door-to-Door Intermodal Transport—RECORDIT.* European Commission DG VII, RTD 5 Framework Program. Brussels, Belgium.

EC. 2001b. *Improvement of Pre- and End-Haulage—IMPREND.* European Commission DG VII RTD 4 Framework Program, Brussels, Belgium.

EC. 2002. *EU Intermodal Transport: Key Statistical Data 1992-1999.* Office for Official Publications of European Commission, European Commission, Luxembourg.

EC. 2013. *EU Transport in Figures: Statistical Pocketbook.* European Commission, Luxembourg.

EEC. 2005. *Report on Punctuality of Drivers at Major European Airports.* Prepared for the Performance Review Unit. EUROCONTROL (European Organization for the Safety of Air Navigation) Experimental Centre, Brussels, Belgium.

EEC. 2009. *A White Paper on Resilience Engineering for ATM.* EUROCONTROL (European Organization for the Safety of Air Navigation), Brussels, Belgium.

EEC. 2011. *Standard Inputs for EUROCONTROL Cost Benefit Analysis,* 5th Ed. EUROCONTROL (European Organization for the Safety of Air Navigation), Brussels, Belgium.

FAA. 2004. *Airport Capacity Benchmarking Report 2004.* U.S. Department of Transportation, Federal Aviation Administration, Washington DC, USA.

Foster H. 1993. *Resilience Theory and System Evaluation.* Pp. 35-60. *In:* Wise J.A, Hopkins V.D., Stager P. (Eds). *Verification and Validation of Complex Systems. Human Factor Issues.* NATO Advanced Science Institutes, Series F: Computer and System Sciences. 110. Springer Verlag, New York, USA.

Gluchsenko O. 2012. Definition of Disturbance, Resilience and Robustness in ATM Context. *DLR Report IB 112.* 20012/28DLR Institute for Flight Guidance, Braunschweig, Germany.

Groothedde B. 2005. *Collaborative Logistics and Transportation Networks: A Modelling Approach to Hub Network Design.* PhD Thesis Series. TRAIL, Delft University of Technology, Delft, The Netherlands.

Hall R.W. 1987. Direct Versus Terminal Freight Routing on a Network with Concave Costs. *Transportation Research* B 21B: 287-298.

Hall R.W. 1993. Design for Local Area Freight Networks. *Transportation Research* B 27B: 70-95.

Hollnagel E., Woods D.D., Leveson N. 2006. (Eds). *Resilience Engineering: Concepts and Precepts.* Aldershot, Ashgate, UK.

Holling C.S. 1973. Resilience and Stability of Ecological Systems. *Annual Review of Ecology and Systematics* 4: 1-23.

Holling C.S. 1996. Engineering Resilience versus Ecological Resilience. 31-44. *In:* Schulze, P. (ed.). *Engineering within Ecological Constraints.* National Academy Press, Washington, DC, USA.

Hosseini S., Baker K., Ramirez-Marquez E.J., 2016. A Review of Definitions and Measures of System Resilience. *Reliability Engineering and System Safety* 145: 47-61.

Ip W.H., Wang D. 2011. Resilience and Friability of Transport Networks: Evaluation, Analysis and Optimization. *IEEE Systems Journal* 5: 189-198.

Janić M., Reggiani A., Spicciareli T. 1999. The European Freight Transport System: Theoretical Background of the New Generation Bundling Networks. *Proceedings of the 8th WCTR (World Conference on Transport Research)*, Volume 1: Transport Modes and Systems. Antwerp, Belgium: 421-434.

Janić M. 2004: *Towards Reliable Transport of Goods and Mobility of Persons: The Concept and Prospective Research Agenda.* Position Paper. OTB Research Institute, TU Delft.

Janić M. 2005. Modelling Consequences of Large-Scale Disruptions of an Airline Network. *ASCE Journal of Transportation Engineering* 131: 249-260.

Janić M. 2007. Modelling the Full Cost of an Intermodal and Road Freight Transport Network. *Transportation Research* D12: 33-44.

Janić M. 2009a. Modelling Cost Performance of Logistics Networks Operating Under Regular and Irregular Conditions. *European Journal of Transport and Infrastructure Research* (EJTIR) 9: 100-120.

Janić M. 2009b. Modelling Airport Operations Affected by the Large-Scale Disruption, ASCE. *Journal of Transportation Engineering* 135: 206-216.

Janić M. 2009c. Management of Airside Delays. *Journal of Airport Management* 3: 176-195.

Janić M. 2015. Modelling the Resilience, Friability and Costs of an Air Transport Network Affected by a Large-Scale Disruptive Event. *Transportation Research* A. 71, 1: 1-16.

JEC. 2008. *Your Flight has been Delayed Again: Flight Delays Cost Passengers, Airlines and the US Economy Billions.* Report. The Joint Economic Committee, Washington DC, USA.

Khol N., Larsen A., Larsen J., Ross A., Tiourine S. 2007. Airline Disruption Management—Perspectives, Experiences, and Outlook. *Journal of Air Transport Management* 13: 142-162.

Len G., Abbass H., Curtis N. 2010. *Resilience of Ground Transport Networks: A Case Study of Melbourne.* 33 Australian Transport Research Forum Conference, 27 Sep.-1 Oct, Canberra, Australia.

Mayer E., Rice C., Jaillet P., McNerney M. 1999. *Evaluating the Feasibility of Reliever and Floating Hub Concepts when a Primary Hub Experiences Excessive Delays.* Report. University of Austin, Austin, Texas, USA.

McCann P. 2001. A Proof of the Relationship between Optimal Vehicle Size, Haulage Length and the Structure of Distance-transport Costs. *Transportation Research* A 35A: 671-693.

Njoka W., Raoult E. 2009. *Transportation Resilience: Planning Land Use and Mobility Management for Unpredictable Events.* Briefing Paper, University of Amsterdam, Amsterdam, The Netherlands.

Qi X., Bart J.F., Yu G. 2004. Supply Chain Coordination with Demand Disruptions. *Omega: The International Journal of Management Science* 32: 301-312.

PAN&N. 2012. *Airport Traffic Report.* Aviation Department, the Port Authority of New York & New Jersey, New York, USA.

Rose A. 2007. Economic Resilience to Natural and Man-made Disasters: Multi-disciplinary Origins of Contextual Dimensions. *Environmental Hazards* 7: 383-395.

Santos J.R., Haimes Y.Y. 2004. Modelling the Demand Input-Output (I-O) Interoperability Due to Terrorism of Interconnected Infrastructures. *Risk Analysis* 24: 1437-1451.

Schaefer L., Millner D. 2001. *Flight Delay Propagation Analysis with the Detailed Policy Assessment Tool.* Proceedings of the 2001 IEEE Systems, Man and Cybernetics Conference. Tucson, USA.

Schavell A.Z. 2000. *The Effects of Schedule Disruptions on the Economics of Airline Operations.* 3 USA/Europe Air Traffic Management R&D Seminar. Napoli, Italy: 11.

Shangyao Y., Chung-Gee L. 1997. Airline Scheduling for the Temporary Closure of Airports. *Transportation Science* 31: 72-83.

Thomas D.J., Griffin P.M. 1996. Coordinated Supply Chain Management. *European Journal of Operational Research* 94: 1-15.

Van Es G.W.H., Van der Geest P.J., Nieuwport T.M.H. 2001. *Safety Aspects of Aircraft Operations in Crosswind.* Report NLR-TP-2001-217. National Aerospace Laboratory NLR, Amsterdam. The Netherlands.

Wang D., Ip W.H. 2009. Evaluation and Analysis of Logistics Network Resilience with Application to Aircraft Servicing. *IEEE Systems Journal* 3: 166-173.

Welch D.J., Lloyd T.R. 2001. *Estimating Airport System Delay Performance.* 4 USA/Europe Air Traffic Management R&D Seminar, Santa Fe, USA.

Wigan M., Rockliffe N., Thoresen T., Tsolakis D. 2000. Valuing Long Haul and Metropolitan Travel Time and Reliability. *Journal of Transportation and Statistics* 3: 83-89.

Winston W.L. 1994. *Operations Research: Application and Algorithms.* Duxbury Press, California, USA.

Zanin M., Lillo F. 2013. Modelling the Air Transport with Complex Networks: A Short Review. *The European Physical Journal Special Topics* 215: 5-21.

Zografos G.K., Regan C.A. 2004. Current Challenges for Intermodal Freight Transport and Logistics in Europe and United States. *Transportation Research Record* 1873: 70-78.

https://aspm.faa.gov/opsnet/sys/main.asp

http://av-info.faa.gov/

http://edition.cnn.com/2013/06/12/business/france-air-traffic-strike

http://en.wikipedia.org/wiki/Closings_and_cancellations_following_the_September_11_attacks

http://en.wikipedia.org/wiki/Air_travel_disruption_after_the_2010_Eyjafjallaj_eruption

http://complexworld.eu/wiki/Resilience_in_air_transport

http://usatoday30.usatoday.com/news/sept11/2002-08-12-clearskies_x.htm

http://www.transtats.bts.gov/

http://www.rita.dot.gov/bts/

http://airsafe.com.events/models/b737

http://www.airlines.org/Pages/Home.aspx

PLANNING TRANSPORT SYSTEMS
Infrastructure, Rolling Stock & Planning Process

6.1 Introduction

Planning of transport systems can generally be classified under two categories regarding the time scale—short-term or short-range and long-term or long-range. Their characteristics, including the main objectives and the general nature, differ.

Short-term planning includes projects and measures that can be implemented over a time period of three to five years. These usually do not include substantial investments in transport infrastructure, but pertain to modifications of the existing and introducing new infrastructure lines, networks and related transport services, acquiring new rolling stock(s), introducing innovative pricing and IT system(s), etc. As such, these projects and measures are flexible and based on the current and prospective short-term objectives to be achieved under perceived conditions. For example, these can be deployment of mega container ships in the supply chains (after they have been developed) or introduction of new operational rules and procedures to increase the capacity of runway systems at congested airports, etc.

Long-term planning includes projects and measures to be undertaken and implemented over the period of 10-25 years. In general, these plans include projects of large investments in transport infrastructure and in some cases, development of new rolling stock(s) and other transport service-supporting facilities and equipment. Thus these projects may have significant both global-country or continent and local-regional economic, social and environmental effects. The development of HSR (High Speed Rail) system represents an illustrative case of long-term planning and implementation of transport infrastructure networks (both nodes and links) at the regional, country and continent scale. In Chapter 7, the new runway to an airport system can be considered as an example of long-term planning and implementation of solutions for increasing capacity of important nodes—airports of the existing air transport infrastructure network(s).

In addition to this introductory section, this chapter consists of four other sections—Sections 6.2 and 6.3 describe some characteristics of planning and development of the main components of transport systems – infrastructure and rolling stock, while Section 6.4 focuses on the main steps of the planning process. The last section provides the concluding remarks.

6.2 Infrastructure

Specifically, in the case of transport infrastructure, projects usually require large capital investments and a relatively long time for implementation. One such typical recent and still current example is the development of HSR networks around the world (Japan, Europe, China, USA, etc.) as given in Table 2.1 of Chapter 2, as also the development of the HSR network in China and the related capital investments.

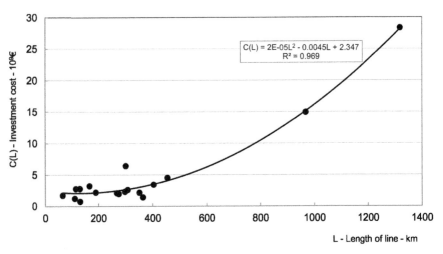

Fig. 6.1: Relationship between the total investment costs and length of HSR (High Speed Rail) line(s) in China (http://en.wikipedia.org/wiki/ High-speed_rail_in_China).

As can be seen, the capital investment costs have increased more than proportionally with the increase in the length of HSR lines/network. In addition, a regression equation is made, indicating the relationship between the capital investments, length of line and designed speed as follows:

$$C(L, V) = -8.268 + 0.018L + 0.024v \qquad (6.1)$$
$$(-2.438)\ (9.282)\ (1.859)$$
$$N = 18;\ R^2 = 0.900;\ F = 37.461$$

The costs $C(L, v)$ are expressed in 10^9 €, the length of line (L) in km and design speed (v) in km/h. The design speed is 250 and 350 km/h. The average cost per unit of length of line is $18 * 10^6$ € and per unit of increasing speed is $24 * 10^6$ €.

Figure 6.2 shows the example of the average unit investment costs in HSR lines in China depending on the length of line.

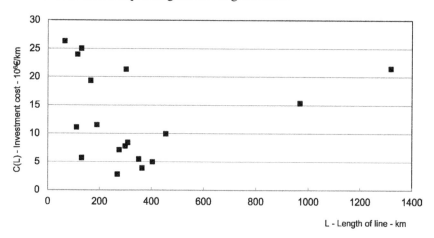

Fig. 6.2: Relationship between the average unit investment cost and length of HSR line(s) in China (http://en.wikipedia.org/wiki/High-speed_rail_in_China).

As can be seen, except for the lines which are longer than about 1000 km, the costs have generally decreased with increase in the line length (up to 400-500 km), thus indicating the existence of economies of scale in the given context. In addition, for comparative purpose, Table 2.5 in Chapter 2 gives examples of the average unit construction cost of HSR lines in certain European countries—both in service and under construction. Furthermore, Fig. 6.3 shows the relationship between the length of HSR line(s) and the time taken for their implementation, i.e. operationalization and beginning commercial/revenue services, in Europe and China. The design speed of the HSR lines in Europe is 200-250 km/h and that in China, 250-350 km/h.

The time of implementation of HSR lines up to 500 km in length is comparable in both regions (two to eight or 10 years) with obvious exceptions in Europe where some lines have taken longer than 20 years to implement. The high dispersion of time in building HSR lines of a similar length in

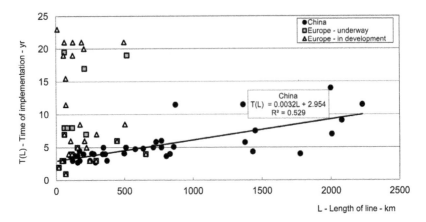

Fig. 6.3: Relationship between the length and time for implementation of HSR line(s)in Europe (period 1990-2025) and China (period 1999-2019) (http://en.wikipedia.org/wiki/High-speed_rail_in_Europe, http://en.wikipedia.org/wiki/High-speed_rail_in_China).

Europe is mainly due to such lines being built separately in each country. Consequently, their time of implementation did not significantly depend on their length. With a few exceptions, it has taken one to 10 years to build lines of a length up to 500-550 km. The projects currently under development is likely share a similar dynamism of implementation. In China, the time of implementation of HSR lines is strongly related to their length. Most lines of a length up to 700-800 km have taken up to six years to be implemented. Lines of 1,500-2,500 km length have taken between five to 15 years for implementation. These examples thus show an inherent difference in the planning and implementation practice and associated processes in two of the world's quite politically and economically different regions. Consequently, the planning, at least in this case, appears to be truly region/country/continent-specific despite the very high similarity, if not identically, of the considered cases.

6.3 Rolling Stock

As far as long-term planning and development of rolling stocks is concerned, some typical examples include development of vehicles: (i) trains; (ii) passenger aircraft; and (iii) container ships—all requiring decades to be fully developed and implemented for commercial operations.

6.3.1 Trains

An example of developing the maximum speed of passenger trains over time is shown in Fig. 6.4 (Boqué 2012).

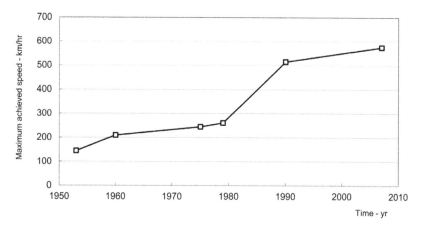

Fig. 6.4: Development of the maximum speed of passenger trains over time (Boqué 2012, Siemens 2008, http://en.wikipedia.org/wiki/Timeline_of_railway_history#20th_century).

It took almost 30 years to increase the maximum speed of passenger trains from 150 to 250/270 km/h; about 10 years to increase this speed to just over 500 km/h, and an additional 20 years to reach speeds of about 570 km/h. It seems that at present, no further increase in this speed is technically/technologically possible.

6.3.2 Commercial Aircraft

Commercial aircraft represent another example of the relatively long time taken in implementation of new models. In general, this time has been different across the range of aircraft size/payload capacity and range. On the one hand, it is mainly influenced by the progress in developing the airframe and engine technologies, while on the other, by the airlines' requirements. Usually, this time includes extensive airframe, engine testing and certification, while the production rates and aircraft life cycles determine the time of aircraft remaining in commercial services. For example, the period of time for developing a medium-sized aircraft typically takes about five to 10 years (that of the currently largest A380 aircraft was 11 years—from 1996 to 2007), while the time of staying in production is 15-20 years, and the aircraft lifetime is 25-35 years—all of which take a total time span of 45-65 years. Figure 6.5 shows an example of development of the payload capacity of cargo commercial aircraft over time.

As will be seen, over the past 50 years, the payload capacity of cargo commercial aircraft has increased ten-fold, starting from milestones like introducing the aircraft DC-6B in the year 1954, then B707-320C, B747-200F, -400F, and -8F, and finally A380F with a payload capacity greater than

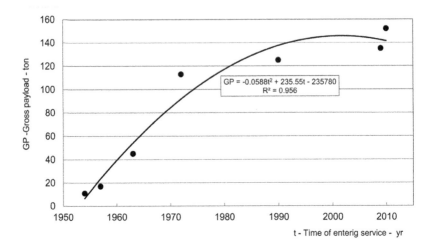

Fig. 6.5: Development of the payload capacity of commercial
cargo aircraft (IIWG 2007).

150 ton in the year 2010. These exclusively cargo versions of aircraft have
better matched the specific needs of air cargo transportation; in particular, the
largest is the Antonov 225 aircraft, a derivative of the Antonov 124 cargo
aircraft, with a payload capacity of 250 ton. However, only a single unit has
been built so far and is hence not represented on the trend line.

6.3.3 Container Ships

An example of the payload capacity of container ships, as shown in Fig. 6.6,
is an additional illustrative example of the time needed for developing rolling
stocks.

Over a period of 30 years (1970-2000), the payload capacity of container
ships increased up to 5,000TEU (Twenty Foot Equivalent Unit), during the
next eight years to 8,000TEU and just in the forthcoming two years sharply
to about 14,000TEU. Then, it took slightly less than 10 years to increase this
capacity to 18,000TEU. It seems very likely that it will take the next seven
to ten years to further increase the payload capacity to 22,000TEU, which
appears to be the limit of the available technology.

The above-mentioned examples indicate that the trends follow the general
shape of the so-called 'logistics curve' representing a relatively low rate of
increase during the starting long period of time, then a sharp increase during a
relatively short period and finally again a lower rate over a reasonably longer
period of time. In a certain sense, the developments particularly over the last
part of the observed period(s) show some signs of exhaustion in the existing
technologies.

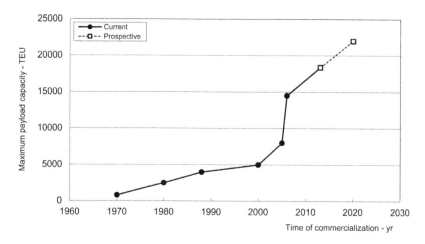

Fig. 6.6: Development of the capacity of container ships over time (Germanischer Lloyd, 2005; Bergmann, 2014; https://www.google.nl/search?q=development+of+container+ships&biw=928&bih=499&tbm=isch&tbo=u&source=univ&sa=X&ei=fzs1VOnPFonUOevTgIgL&ved=0CEQQ7Ak, http://www.dnv.com/industry/maritime/publications/archive/2013/1-2013/index.asp).

6.3.4 Some Effects of Long-Term Planning Projects

Long-term infrastructure projects can impact the surrounding society and environment and therefore need to be carefully evaluated at least at two levels. The first is the so-called multidimensional examination or multidisciplinary approach implying considering their physical/spatial, technical/technological, operational, economic, social, environmental and policy performances; the second includes aspects and preferences of a range of directly and/or indirectly involved actors/stakeholders. These can generally be construction companies and related industries involved in the construction, building and implementation stage, and all the prospective infrastructure and transport operators, authorities at local, national and international level and finally users of these services—generally the passengers and goods/freight shippers and receivers.

Independent of the procedure and methods, evaluation should clearly indicate the benefits and costs of particular projects, i.e. feasibility of investments from the overall social perspective. In addition, the long-term plans and projects are usually scrutinized from time-to-time in order to make the necessary modifications depending on the re-estimated future conditions and circumstances.

6.4 Planning Process

6.4.1 Activities

Planning generally represents one of the basic functions of the modern society. It usually relates to the urban and sub-urban agglomerates including housing, businesses and transport systems, and related supporting systems. Since it interacts directly and/or indirectly with almost all activities of contemporary life, planning of transport systems has to be coordinated with the plans of land use, urban and sub-urban forms and the overall and sometimes specific lifestyle of the population concerned. In particular, transport infrastructure networks consisting of the nodes—passenger and freight terminals—and links—roads, rails, and built inland waterways—need careful planning and spatial integration with other socio-economic activities in order to provide efficient and effective transport services to satisfy the real needs of their users.

In some cases, the long-term planning of transport systems operated by different transport modes assumes a form of integration of their services. Illustrative examples of the infrastructure providing such integration include rail/bus stations in urban and sub-urban areas, rail stations at airports, rail/road/ sea terminals at ports and rail/road/inland waterways intermodal terminals.

The process of planning transport systems is usually country-specific. The common ingredient of this process throughout different countries is the outcome, usually in the format of Master Plan(s) containing physical/ spatial definitions of the systems operated by different transport modes, their components, the prospective layout of infrastructure and specified service networks. The above-mentioned HSR network in Europe developed at the country level (in 1981, the TGV Sud-Est from Paris to Lyon started high-speed rail services in Europe) and China[1] (started in the year 2007), the National System of Interstate Highways in US (adopted by Congress in 1956), the Munich U-Bahn (began operation in 1971) could be considered as some typical examples.

In general, the transport planning process consists of several self-explanatory steps as follows (Vuchic 2004):

- Definition of objectives for the future transport system(s);
- Collection of relevant data about the existing transport systems and their operational and social environment;

---

[1] The HSR (High-Speed Rail) network in China consists of upgraded conventional railways, newly built high-speed passenger designated lines and the world's first high-speed commercial magnetic levitation (Maglev) line (http://en.wikipedia.org/wiki/High-speed_rail_in_China).

- Predicting changes and conditions in future for the selected target years for plans;
- Setting up the evaluation criteria for plans based on the objectives of future transport system(s) defined in the previous step;
- Developing a series of alternative plans depending on the predicted future conditions and meeting the objectives of the future transport systems;
- Testing particular alternative plans regarding their ability to satisfy expected demand and overall impacts on the space (land), society and environment;
- Evaluation of alternative plans considering their above-mentioned multidimensionality and multi-stake holders' involvement, both dictating the evaluation method used; and
- Selection and finalization of the preferred alternative plan and preparing it for implementation.

However, it can be said that technically, the transport planning process generally includes five steps: (i) inventories of current systems and collection of relevant data; (ii) model development to be applied to forecasting demand and supply; (iii) generation of alternate plans; (iv) selection of the preferred plan; and (v) implementation of the preferred plan (Vuchic 2004).

The first step embraces collection of data on the population, economic and social characteristics of the area for which the plan is to be designed, an inventory of land used and physical/spatial and operational performances of existing transport systems and the current volumes of passengers and/or freight carried out by particular transport modes and routes of the corresponding networks.

The second step implies modelling of existing and prospective (passenger and/or freight) transport demand and supply/capacity. In case of planning the infrastructure links between particular regions, this step includes the sub-steps such as (i) generation and attraction of traffic/transport demand flows from/to particular regions, respectively; (ii) distribution of these flows between these regions according to their origins and destinations, (iii) modal choice and modal split; and (iv) assignment of these demand flows to the existing and new infrastructure links and routes of the network(s) operated by different transport modes within the area in question. These sub-steps are carried out based on current and predicted conditions implying analysis, modelling and forecasting of the traffic/transport demand flows on the one hand and planning the supply/capacity performances of particular transport systems and modes, on the other in order to match this demand efficiently, effectively and safely. Then, an evaluation of the updated transport system aimed at assessing to what extent it could cope with the existing and prospective demand is carried out.

The third step consists of generating two types of plans: those containing some short-term improvements of the existing system(s) and those as long-term alternative solutions aimed at satisfying future (forecasted) demand in a sustainable way, i.e., efficiently, effectively and with as low as possible negative impacts on the society and environment.

The fourth step includes identifying the stakeholders involved in evaluation of the above-mentioned long-term alternative plans, selection of evaluation criteria and methods, carrying out the evaluation procedure and adopting the preferred alternative plan(s).

The last step represents implementation of the above-mentioned adopted (preferred) alternative plan under the given conditions.

In the present context, the first step deals with analysis of the existing systems based on the collected data. The second step is particularly interesting and important because it brings modelling into the planning process aimed at setting up the general framework and layout of the new or updated transport system(s) in terms of its physical/spatial scale, components and required resources more consistently, efficiently and effectively.

6.4.2 Transport Demand and Supply/Capacity

Specifically, planning transport demand and supply in the planning context is based on their modelling. In general, it consists of four steps: (i) generation and attraction of traffic/transport demand flows; (ii) distribution of traffic/transport demand flows; (iii) modal choice and modal split; and (iv) assignment of traffic/transport demand on the links and routes of given network(s).

6.4.2.1 Step 1—Generation and Attraction of Demand Flows

The models enable to determine the volumes of traffic/transport demand (trips, passengers and/or quantity of goods/freight), which depart from the particular origins and arrive at particular destinations, respectively, of a given transport system/network under given conditions and over a specified period of time (hour, day, month, year). Different kinds of econometric models taking into account various factors/attributes, i.e. the demand-driving forces, influencing generation and attraction of traffic/transport flows from and to different regions (network nodes) have been used, with many of them in the format of linear or log-linear regression (Florian 2008). For example, for the demand-generating region (node) (i) and/or the demand-attracting region (node) (j), the general expression for the corresponding volumes of demand estimated for a given time period under the given conditions can be as follows:

$$Q_{i0} = f_{i0}(X_{1i}, X_{2i}, .., X_{ki}..., X_{ni})$$

and
$$Q_{i0} = f_{j0}(X_{1j}, X_{2j}, ...X_{lj}, ..., X_{mj}) \qquad (6.2)$$

where

Q_{i0}, Q_{j0} is the volume of demand generated from the region (node) (i) and attracted to the region (node) (j), respectively, at the time (0) representing the beginning of the planning period of (n) years in advance;

X_{ki}, X_{lj} is the demand generating and attracting factor (driving force) in the region (node) (i) based on the node-region (k) and the region-node (j), respecting the region-node (l), respectively, considered as the independent (explanatory) variable ($i, j, k, l \varepsilon N; i \neq k, j \neq l; i \neq j$); and

N is the number of regions (nodes) considered in the network.

Estimating, i.e. calibrating the above-mentioned model(s) was carried out either by using the data from a specified past period of time or by using the cross-sectional data. After indicating the relative importance of particular demand-driving forces (i.e. independent variables Xs), the demand needs to be forecast, since planning is always carried out for some future period—in this case for (n) years in advance. Generally, two approaches are practiced. The first approach implies using the so-called demand expanding factors, i.e. the factors of demand growth (in %) usually appearing in the form $(1 + i)^n$. They have been directly applied to the dependent variable, i.e. the current estimated volumes (Q_i) and (Q_j) in the above-mentioned case as: $Q_{in} = (1 + i)^n Q_{i0}$ and $Q_{jn} = (1 + i)^n Q_{j0}$, where ($i$) is the factor of growth and (n) is the number of time periods (years) in advance from the current period (0). The other approach implies forecasting each of the independent variables (also by means of the growth factors or otherwise), i.e. the demand-driving forces and then using such forecasted values for estimating the future volumes of the demand (Q_{in}) and (Q_{jn}) as dependent variables. Both approaches possess both advantages and disadvantages. The main advantages refer to obtaining the required results fully acceptable under assumption that the conditions, i.e. the main demand-driving forces and character of their influence will remain relatively stable over the future planning period similarly as they used to be in the past. The main disadvantage refers to just the opposite.

6.4.2.2 Step 2—Distribution of Traffic/Transport Demand Flows

The distribution models enable setting up an O-D (Origin-Destination) matrix consisting of the volumes of demand flows between particular origins and destinations, i.e. the network nodes based on the volumes of traffic/transport flows estimated in Step 1. In this case, the gravity model and the entropy model are most frequently used for this purpose (Florian 2008). In addition, the particular O-D demand flows are estimated directly by using the linear and/or log-linear regression form. For example, for a large airport connected to other airports as the main O-Ds for passengers, the log-linear regression equations could be as follows (Janić 2009):

Let (i) and (j) be two mutually connected airports (regions) ($j = 1, 2, 3,...,$ N). In some cases, the airport (region) (j) may refer to the larger geographical (metropolitan) area with a few 'clustered' airports. The model of passenger demand between the airports (i) and (j) based on the regression technique may have the following form:

$$Q_{ij} = a_0 \, (GDP_i * GDP_j)^{a_1} * (Y_{ij} * L_{ij})^{a_2} * T_{ij}^{a_3} * C_{ij}^{a_4} * P_{ij}^{a_5} + \sum_{k=1}^{K} b_k D_k \quad (6.3)$$

where

Q_{ij}	is the passenger demand between airports (i) and (j) (pass/yr);
GDP_i, GDP_j	is GDP (Gross Domestic Product) of regions (i) and (j), respectively (billions € or \$US);
Y_{ij}	is the average yield between airports (i) and (j) (usually in ¢/RPK; ¢/RPM (¢ – cent; RPK – Revenue Passenger Kilometer; RPM – Revenue Passenger Mile);
L_{ij}	is the two-way shortest distance between airports (i) and (j) (km or nm);
T_{ij}	is the total travel time between airports (i) and (j) (hour);
C_{ij}	is the transport capacity supply factor between airports (i) and (j) (the number of flights or seats in single or both directions during a given period of time (days and weeks);
P_{ij}	is the competitive power of airport (i) regarding the region-market (j) as compared to the power of other airports in the same and/or nearby region(s), i.e. proportion of the flight frequencies (seats) between airport (i) and region (j) in the total number of frequencies (seats) between all competing airports and region (j);
D_k	is a dummy variable taking the value '1' if a disruptive event of type k has considerably affected the passenger demand and the value '0' otherwise (terrorism, epidemic diseases, regional wars, etc. are considered disruptive events); and
a_i, b_k	is the coefficient to be calculated by estimating the regression model (i = 1-6; k = 1-K).

Equation 6.3 implies that the above-mentioned main forces drive the passenger demand at the given airport (i). On short- and medium-haul distances (L_{ij}), the variables (T_{ij}), (C_{ij}) and (P_{ij}) need to be modified in order to take into account the potential competition from the surface transport modes, if available. The independent variables (GDP_i) and (GDP_j) relate to the annual GDP of the corresponding regions surrounding the airports (i) and (j),

respectively. Their product reduces the number of variables and characterizes the mutual 'attractiveness' of the given regions. Under the conditions of economic progress, these variables increase over time and seemingly contribute positively to the volumes of passenger demand and vice versa. This implies a positive elasticity, i.e. the coefficient (a_1). Under certain circumstances, these variables can be combined or replaced with 'derivatives', such as PCI (Per Capita Income), population, employment, and/or trade, investments, tourism and exchange of other services between the airport and other regions.

The variable (Y_{ij}) (yield) relates to the weighted average yield of all airlines operating between the airports (i) and (j) during the given period of time (for example, one year). Its product with the variable (L_{ij})—two-way travel distance—gives the average return airfare (AF_{ij}). The variable (L_{ij}) represents the shortest (great circle) distance between the given pair of airports. In general, when the airfare as the 'travel resistance factor' increases, the number of passengers generally decrease and vice versa, which at least in theory reflects the negative elasticity, i.e. the coefficient (a_2).

The variable (T_{ij}) expresses the travel time between airports (i) and (j). It includes the non-stop flying time and eventually the schedule delay, i.e. the waiting time for convenient departure at the airport (i). Since being based on the shortest distance, which usually does not change over time, this variable can appear influential only if a change in aircraft technology takes place (for example, if faster regional jets replace slower turbo-props) and/or if the number of flights changes significantly. Otherwise, it is not particularly relevant. In any case, the passenger demand tends to increase with decrease in the distance, i.e. shortening of the travel time implies a negative elasticity, i.e. the coefficient (a_3).

The variable (C_{ij}) reflects the supply of transport capacity usually expressed by the number of seats (flight frequency times the average aircraft size or seat capacity) or only the flight frequency offered between given airports (i) and (j) by all airlines. Since the supply of capacity (at a reasonable price) is intended not only to satisfy but also to stimulate demand, the latter is theoretically expected to rise with increase in this capacity and vice versa. This generally implies a positive elasticity, i.e. the coefficient (a_4).

The variable (P_{ij}) represents the competitive position (i.e. 'power') of the airport (i) in comparison to other relatively close competing airports in the same or neighboring regions. All considered airports are connected to very similar destination(s) (j). In such a case, this variable refers to the generalized access cost, airfare and departure frequency at the airport (i) as compared to that of other competing airports. In general, the higher competitive power attracts higher passenger demand, thus reflecting the positive elasticity (a_5).

Dummy variables (D_k) $(k = 1, 2, 3,...)$ reflect the impact of various events affecting the airport demand. Generally, such an impact on demand implies a negative elasticity, i.e. the coefficients (b_k).

The proposed causal model suggests that theoretically, passenger demand at a given airport (i) will increase with increase in the exogenous demand-driving forces (regional socio-economic factors) and supply of transport capacity, and decrease with increase in the 'travel resistance' factors, such as the generalized travel cost consisting of the 'door-to-door' out of pocket cost of which the airfare is a dominant part and the cost of travel time. In addition, this demand will increase with strengthening of the airport's competitive position (power). This is mainly inline as can be intuitively expected. The model can be estimated by using the cross-sectional data for particular O-D markets (ij) $(j = 1,..., n)$ for the period of time of at least 12 to 15 past years.

Table 6.1 gives an example of the application of the above-mentioned simplified log-linear regression model for estimating the annual O-D (Origin-Destination) passenger demand flows between Amsterdam Schiphol airport and the other world's regions (market clusters) it is connected to. The relevant data for the 1992-2004 period have been used (Janić 2009). In this case the above-mentioned Equation 6.3 has been modified by specifying the following variables: (Q_{hi}): the annual number of O-D (Origin-Destination) passengers (dependent variable); (GDP_h), (GDP_i): Gross Domestic Product of the area around the given(hub) airport and the region (i) it connects, respectively; (AF_{hi}): the average fare per passenger as the product of the average yield per passenger and the shortest distance between the given (hub) airport and the main airport in the region (i) it connects; (S_{hi}), (S_{lc}): the number of seats offered by the conventional ('legacy') and LCCs (Low Cost Carrier(s)), respectively between the given (hub) airport (h) and the region (i) it connects; (FDI_i): Foreign Direct Investments in the region (i) connected to the given (hub) airport (h); $(D1)$: Dummy variable intended to describe the influence of the disruptive event(s) during the observed period (independent variables).

The coefficients of the particular regression equation indicate that the passenger demand flows are positively influenced by the general socio-economic factors, such as GDP (Gross Domestic Product) at both ends of the particular markets/routes and FDI (Foreign Direct Investments) at the end of the given market/route, and negatively by air fares, which is as expected. In particular, in Europe, this demand has been additionally driven by the seats offered by both conventional ('legacy') airlines and LCCs (Low Cost Carriers). The t-statistics of particular coefficients indicate importance of the selected independent variables (i.e. demand-influencing factors) in all markets/routes. In addition, R^2 statistics indicate that the above-mentioned influence is relatively strong. As well, the entire equations in the given case have been important as indicated by F statistics.

Table 6.1: An example of modelling O-D (Origin-Destination) passenger demand at Amsterdam Schiphol Airport (Janić 2009)

O-D Market cluster	Regression model-equation
The Netherlands—EU (15 until 2004 and 25 Member States after)	$Q_{h1} = 4613.18 \ (\text{GDP}_h * \text{GDP}_1)^{0.082} \text{AF}_{h1}^{-0.889} \text{S}_{h1}^{0.770} \text{S}_{1c}^{-2.370}$ $\qquad(2.990) \qquad\qquad (3.851) \ (-2.961) \ (3.772) \ (-2.970)$ $\qquad R^2 = 0.893; \ F = 16.749; \ N = 13$
The Netherlands—Rest of Europe (22 until 2004 and 12 central and eastern European, and Mediterranean countries after)	$Q_{h2} = 4841.82(\text{GDP}_h * \text{GDP}_2)^{0.356} \text{AF}_{h2}^{-0.880}$ $\qquad(2.618) \qquad\qquad\qquad (2.998) \qquad (-2.875)$ $\qquad R^2 = 0.569; \ F = 8.905; \ N = 13$
The Netherlands—North America (US east coast and central area)	$Q_{h3} = 1640.59(\text{GDP}_h * \text{GDP}_3)^{0.483} \text{AF}_{h3}^{-0.831} - 0.059\text{D1}$ $\qquad(1.501) \qquad\qquad\qquad (3.316) \ (-1.292) \ (-1.336)$ $\qquad R^2 = 0.481; \ F = 4.712; \ N = 13$
The Netherlands—Latin America (Brazil)	$Q_{h4} = 592.93(\text{GDP}_h * \text{GDP}_4)^{0.404} \text{AF}_{h4}^{-0.651}$ $\qquad(6.405) \qquad\qquad\qquad (7.010) \quad (-7.164)$ $\qquad R^2 = 0.899; \ F = 54.405; \ N = 13$
The Netherlands—Africa (Kenya, South Africa)	$Q_{h5} = 693.426 \ \text{FDI}_{h5}^{0.597} \text{AF}_{h5}^{-0.329}$ $\qquad(8.474) \qquad (15.416) \quad (-3.330)$ $\qquad R^2 = 0.981; \ F = 319.258; \ N = 13$
The Netherlands—Middle East (United Arab Emirates, Israel)	$Q_{h6} = 0.561(\text{GDP}_h * \text{GDP}_6)^{0.695} \text{AF}_{h1}^{0.105}$ $\qquad(0.175) \qquad\qquad\qquad (3.439) \ (-0.221)$ $\qquad R^2 = 0.729; \ F = 17.120; \ N = 13$
The Netherlands—Asia (China, Japan, Korea, Malaysia, Singapore)	$Q_{h7} = 173.38(\text{GDP}_h * \text{GDP}_7)^{0.784} \text{AF}_{h7}^{-1.281}$ $\qquad(1.841) \qquad\qquad\qquad (3.993) \ (-5.898)$ $\qquad R^2 = 0.761; \ F = 20.098; \ N = 13$

6.4.2.3 Modal Choice and Modal Split (Step 3)

(a) Modal choice

In the existing transport service networks operated by different transport modes, which serve particular O-D (Origin-Destination) of passenger and/ or goods/freight demand flows, a choice of transport mode and consequent modal split between existing (and at some time forthcoming still not-existing) transport modes needs to be carried out. In such a case, the choice is based on the implicitly considered characteristics/attributes of demand flows on the one hand and the explicitly considered characteristics of transport services by particular modes, on the other. The latter are most often represented by service quality containing attributes, such as service accessibility, time, speed, reliability and punctuality and cost usually considered as the price, i.e. door-to-door expenses for users. The above-mentioned characteristics of both users and suppliers of transport services is expressed by the (dis)utility function for each transport mode serving particular O-D pairs, thus enabling estimation of the probability of choosing a given mode under the given conditions. The most frequently used models of modal choice, in addition to that based on 'all-or-nothing' or 'uniform distribution' principle, include the diversion and logit model. Both usually quantify the probability of choice of a given mode based on its utility function as compared to those of the other modes involved as perceived by prospective users.

Diversion model

The diversion model implies that the attractiveness of a given transport mode expressed by its (dis)utility function increases in proportion to a decrease in the attractiveness of other potentially competing modes expressed by their (dis)utility functions. For example, for two competing freight transport modes, the probability of choice of one of them can be expressed as follows:

$$p_{l/ij}\left[U_{l/ij}(d_{ij}), d_{ij}\right] = \frac{U_{m/ij}(d_{ij})}{U_{l/ij}(d_{ij}) + U_{m/ij}(d_{ij})} \tag{6.4a}$$

where

$p_{l/ij}[U_{l/ij}(d_{ij}), d_{ij}]$ is the probability of choosing the transport mode (l) for delivering goods/freight shipments on the O-D distance (d_{ij});

$U_{l/ij}(d_{ij})$ is the dis(utility) function of the transport mode (l) delivering goods/freight shipments on the O-D distance (d_{ij}); and

$U_{m/ij}(d_{ij})$ is the dis(utility) function of the transport mode (m) delivering goods/freight shipments on the O-D distance (d_{ij}).

The dis(utility) functions can be expressed by interrelating particular (dis) utility components, such as the total door-to-door transport cost (operational + external), the cost of time of goods/freight shipments being transported, the reliability and punctuality of services, etc. Consequently, if these components increase at one transport mode, its counterparts are expected to gain the market share and vice versa. In this case, these are the basic principles of modal choice and modal split philosophy.

Logit model

As at the passenger transport modes, the choice of freight transport mode is usually the subject of many random influencing factors. The main reason is that the freight shippers (and sometimes receivers) usually evaluate differently the utility they obtain by opting for particular available transport modes. In many cases, shippers/receivers may not have complete information on the attributes constituting the utility of the available modes and their services. This also relates to precise knowledge of the actual costs of these modes. Furthermore, some shippers do not opt for a certain mode even though it would be very logical to do so because of its advantages when compared to the others. Under such conditions, the logit model can be used for estimating the modal choice. Its main structure is represented by the probability of choosing the transport mode/alternative (l_l) among (N) possible modes/alternatives for delivering the goods/freight shipments between the given O-D pair (ij) at the distance (d_{ij}) as follows:

$$p_{l/ij}\left[U_{l/ij}(d_{ij}), d_{ij}\right] = \frac{e^{U_{l/ij}(d_{ij})}}{\sum\limits_{k=1}^{N} e^{U_{k/ij}(d_{ij})}} \qquad (6.4b)$$

where

$p_{l/ij}[U_{l/ij}(d_{ij}), d_{ij}]$ is the probability of choosing the transport mode (l) for delivering goods/freight shipments on the O-D distance (d_{ij}); and

$U_{l/ij}(d_{ij})$ is the dis(utility) function of the transport mode (l) while delivering the goods/freight shipments on the O-D distance (d_{ij}).

Usually, the (dis)utility function $U_{l/ij}(d_{ij})$, implicitly including the different (dis)utility components, has a linear form. In some cases, when this function is not linear, the least-squares-based regression technique is used for estimating its parameters (i.e. coefficients). In such a case, the dis(utility) function is based on the aggregated data implying that all freight shippers/receivers use the same dis-utility components while choosing the relevant transport mode, thus assuming that the variables of dis(utility) function are equal for them all. Consequently, the probability of choosing one among the available transport modes using the aggregated data is equal for all shippers/receivers.

(b) Modal split

The above-mentioned models of modal choice can be used for determining the total volume(s) of particular O-D demand flows attracted by the given transport mode(s). Then, given the characteristics of transport vehicles and their load factors, the service frequency by particular modes serving demand between particular O-D pairs can be estimated. This service frequency for the mode (*l*) can be determined based on Equations 6.4a, as follows:

$$f_{l/ij}(T,d_{ij}) = \frac{p_{l/ij}\left[U_{l/ij}(d_{ij}),d_{ij}\right]*Q_{ij}(T,d_{ij})}{CV_l(d_{ij})*\lambda_l(d_{ij})}, for\ (i,j)\in N; i\neq j \quad (6.5)$$

where

$f_{l/ij}(T,d_{ij})$	is the service frequency of transport mode (*l*) on the O-D distance (d_{ij}) during time (*T*);
$Q_{ij}(T,d_{ij})$	is the total volume of goods/freight shipments to be transported on the O-D distance (d_{ij}) during time (*T*);
$CV_l(d_{ij})$	is the average capacity of a vehicle operated by mode (*l*) on the O-D distance (d_{ij}); and
$\lambda_l(d_{ij})$	is the average load factor per vehicle, i.e. service, operated by transport mode (*l*) on the O-D distance (d_{ij}).

(c) An example

Input—The network geographical/spatial configuration

Application of the above-mentioned modal choice and modal split models is illustrated using the example of the road and intermodal rail/road transport network as shown in Fig. 6.7.

The road network is fully operational, while the rail/road transport network is expected to be set up and as such begin to be a competitor to road transport (Janić 2014a). As can be seen, this network consists of eight nodes—the intermodal rail/road terminals and seven routes—the rail and road links/routes—connecting them. In addition, the intermodal terminals and their surroundings represent the ultimate O-D (Origins and Destinations) or the beginning and end terminals of goods/freight flows representing the satisfied demand in this case.

Input—Length of links/routes of the network

The rail-based distances between particular begin and end terminals as the O-D (Origin-Destination) of goods/freight flows in the given network are given in Table 6.2. Their road counterparts are assumed to be approximately the same.

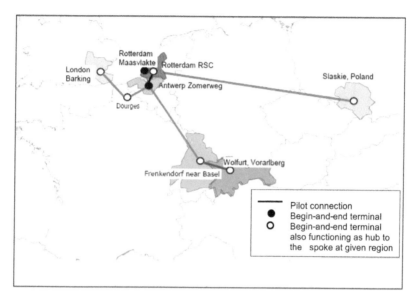

Fig. 6.7: Simplified scheme of the intermodal rail/road freight transport service network in the given example (Janić 2014a).

Input—Goods/freight demand flows

The goods/freight demand flows are consolidated into containers of the size of TEU (Twenty Foot Equivalent Unit). Their volumes exclusively transported by road transport mode in the year 2010 between particular O-D (Origin-Destination) regions surrounding the nodes/terminals of the given intermodal rail/road freight transport network are given in Table 6.3.

It is assumed that when the intermodal rail/road freight transport services start, they will be able to attract some of the above-mentioned flows from the road mode thanks to their competitiveness in terms of the operating door-to-door delivery cost.

Input—Operating costs as (dis)utility functions

The operating costs of the road and intermodal rail/road freight transport services are estimated for each O-D pair of nodes/terminals of the given network. For road transport, the costs are estimated, based on the existing services. They include the cost of loading at the door of shipper(s), the cost of transport by road between the shipper's and receiver's door and the cost of unloading at the door of receiver(s). For intermodal rail/road transport, the costs are estimated for the existing services taking place in the networks with the similar characteristics as the one in the given case. They consist of the loading and transport cost by road from the shipper to the origin intermodal (rail/road) terminal, the cost of loading at the origin terminal, the transport

Table 6.2: Rail-based distances between particular begin and end terminals in the given network (km) (EC 2014, Janić 2014a)

Begin/End terminal	London Barking	Dourges	Antwerp Zomerweg	Rotterdam	Frenkerdorn near Basel	Wolfurt, Vorarlberg	Slaskie Poland
London Barking		250	430	530	1,160	1,320	1,660
Dourges	250		180	280	910	1,070	1,410
Antwerp Zomerweg	430	180		100	730	890	1,230
Rotterdam	530	280	100		830	990	1,130
Frenkerdorn near Basel	1,160	910	730	830		160	1,960
Wolfurt, Vorarlberg	1,320	1,070	890	990	160		2,020
Slaskie, Poland	1,660	1,410	1,230	1,130	1,960	2,020	

Table 6.3: The volumes of goods/freight flow demand transported by the road transport mode between the regions surrounding the begin and end terminals of the given network (TEUs) (EC 2014, Janić 2014a)

Begin/End terminal	London Barking	Dourges	Antwerp Zomerweg	Rotterdam	Frenkerdorn near Basel	Wolfurt, Vorarlberg	Slaskie Poland
London Barking		23,103	29,572	15,431	11,683	5,880	5,905
Dourges	25,664		201,970	15,395			
Antwerp Zomerweg	41,852	308,093		305,460	67,738	9,996	20,298
Rotterdam	24,425	58,773	446,026		48,708	23,078	19,034
Frenkerdorn near Basel	17,598		63,791	25,081			
Wolfurt, Vorarlberg	3,184		8,507	7,646			
Slaskie, Poland	9,876		21,481	16,462			

cost by rail between the origin and destination intermodal terminal, the cost of transshipments (if applicable) at the intermediate (rail/rail) terminals and the transport cost by road between the destination intermodal terminal and the receiver of goods/freight shipment(s), including the cost of its unloading there. An example of the above-mentioned costs for one of the nodes, Antwerp, and its inbound and outbound services in dependence of the O-D door-to-door distance are shown in Fig. 6.8 (a, b) expressed in €/TEU.

As can be seen, as interrelated with the door-to-door distance, the costs for both modes appear to be very similar for both inbound and outbound transport services. By using the least-square regression technique, it is seen that this cost for the road transport mode linearly increases with increase in the door-to-door distance(s). For the intermodal rail/road transport mode, this cost also linearly increases with increase in the door-to-door distance but with higher variations around the average value(s) mainly conditioned by their above-mentioned inherent structure. It may be noted that the rail transport costs are the lowest in the given case, almost 80-100 per cent lower than those of the intermodal rail/road transport along the given range of delivery distances. This is mainly caused by the relatively high road costs and the cost of loading/unloading and transshipments. The cost-distance relationships estimated for other O-D pairs (routes/corridors) of the given network are given in Table 6.4 as (dis)utility functions used in the modal shift models (Equation 6.4 a, b).

Results—modal choice

The results of the prospective modal choice within the given network are obtained from the above-mentioned diversion and logit model and expressed in terms of the probability of choice, i.e. the relative market share of both the road and intermodal rail/road transport mode. For such a purpose, by dividing the cost-distance functions in Table 6.5 by door-to-door distance (d), the average cost in terms of €/TEU-km is obtained and used for the modal choice estimation by means of Equation 6.4 (a, b). An example of the resulted modal shift for the inbound and outbound transport for the Antwerp route/corridor in Table 6.4 is shown in Fig. 6.9 (a, b).

Here it is seen that the market share of the intermodal rail/road transport mode tends to increase with increase in door-to-door distance at a decreasing rate at either modal shift model. In addition, at shorter distances (up to about 400 km), the diversion model provides higher values of the market shares than the logit model, but the differences decrease and disappear with increase in the distance, i.e. at those longer than 400 km. Nevertheless, both models indicate that both modes competing under given conditions can count on a reasonable market share along a range of the door-to-door distances (100-1800 km). Both models are also used for the purpose of mutual comparison.

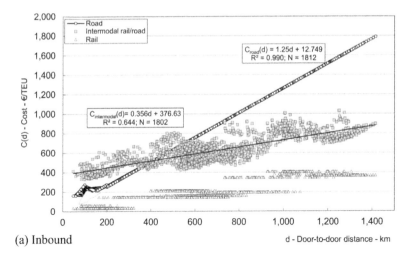

(a) Inbound

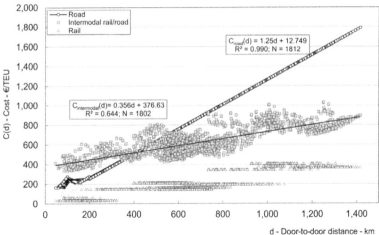

(b) Outbound

Fig. 6.8: Relationship between the operating cost per TEU and its door-to-door delivery distance for the road, intermodal rail/road and rail transport mode in the case of Antwerp (EC 2014, Janić 2014a).

Results—modal split

Based on the rail-based distances in Table 6.2, the volumes of goods/freight flows transported by road in Table 6.3, operating costs in Table 6.4 and the probabilities of modal choice (Fig. 6.9), the O/D demand flows attracted by the intermodal rail/road transport mode from its road counterpart are estimated and given in Table 6.5.

Table 6.4: Operational costs of the road and intermodal rail/road transport mode on the particular routes/corridors of the given network as (dis)utility functions (Janić 2014a)

Route/Corridor	Transport mode	
	Intermodal rail/road	Road
	Costs as (dis)utility function (€/TEU)	
• Antwerp		
• Inbound	$C_{intermodal}(d) = 0.356d + 376.63$ $R^2 = 0.644; N = 1,802$	$C_{road}(d) = 1.25d + 12.749$ $R^2 = 0.990; N = 1,812$
• Outbound	$C_{intermodal}(d) = 0.357d + 376.84$ $R^2 = 0.644; n = 1,802$	$C_{road}(d) = 1.25d + 12.744$ $R^2 = 0.999; N = 1,802$
• Rotterdam		
• Inbound	$C_{intermodal} = 0.359d + 370.25$ $R^2 = 0.693; N = 1,690$	$C_{road} = 1.25d + 12.713$ $R^2 = 0.999; N = 1,690$
• Outbound	$C_{intermodal} = 0.354d + 377.77$ $R^2 = 0.666; N = 1,680$	$C_{road} = 1.25d + 12.732$ $R^2 = 0.999; N = 1,680$
• London		
• Inbound	$C_{intermodal}(d) = 0.334d + 399.3$ $R^2 = 0.809; N = 1,237$	$C_{road}(d) = 1.265d + 0.417$ $R^2 = 1.0; N = 1,237$
• Outbound	$C_{intermodal}(d) = 0.333d + 399.25$ $R^2 = 0.812; N = 1,207$	$C_{road}(d) = 1.265d + 0.418$ $R^2 = 1.000; N = 1,207$

d – distance (km)

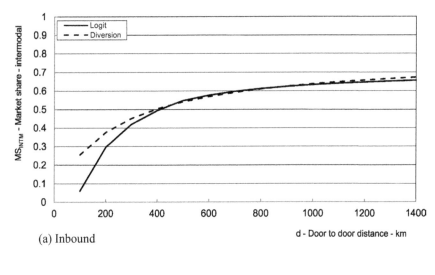

(a) Inbound

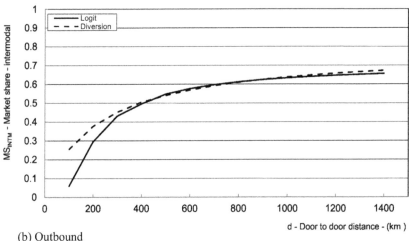

(b) Outbound

Fig. 6.9: Relationship between the market share of the intermodal rail/road transport mode and door-to-door distance (the complement to 1 or 100 per cent is the market share of road) in the case of Antwerp (Janić 2014a).

Results—the service frequencies of intermodal freight trains

The service frequencies of intermodal trains serving the attracted goods/ freight flows between particular O-D terminals of the given network are estimated using the following assumptions:

- The train services are set up for the time period of $T = 1$ week. Each service/ train between each pair of O-D terminals has a fixed configuration. The train is composed of 20 four-axle flat cars, each with a payload capacity

Table 6.5: The estimated market shares and corresponding volumes of goods/freight flows between particular O-D terminals of the given network attracted by the intermodal rail/road transport mode (%; TEU) (Janić 2014a)

Begin/End terminal	London Barking	Dourges	Antwerp Zomerweg	Rotterdam	Frenkerdorn near Basel	Wolfurt, Vorarlberg	Slaskie Poland
London Barking		34/7055	50/14,786	54/8,333	66/7,711	67/3,940	68/3,801
Dourges	35/8,977		24/48,472	40/6,158			
Antwerp Zomerweg	50/20,925	24/73942		5/15,273	60/40,643	63/6,297	65/19,194
Rotterdam	53/12,945	37/22040	6/26,762		62/30,199	64/14,770	65/12,277
Frenkerdorn near Basel	64/11,263		60/38,275	61/15,425			
Wolfurt, Vorarlberg	65/2,097		62/5,274	64/4,893			
Slaskie, Poland	67/6,617		66/14,177	65/10,618			

of 60 ton and tare weight of 22 ton. Since one TEU weighs up to 24 ton, each flat car is assumed to carry 2 TEU, which gives the gross weight of 70 ton/car ($2*24 + 22 = 70$) and 1400 ton/train. Including the weight of the (multi-system) locomotive (typically 87 ton), the total gross weight of each train reaches 1487 ton (Siemens 2008). In addition, the total payload of each train is 40 TEU/train ($2*20 = 40$). Consequently, the train load factor in terms of used space is $\lambda_v = 1.0$ and in terms of weight $\lambda_w = 0.8$.

The resulting train service frequencies, market share and prospective volumes of goods/freight shipments on particular routes of the given network are given in Tables 6.6 and 6.7.

The above-mentioned results indicate that launching services by the new transport mode in addition to the existing ones carried out by the other competitive transport mode (in this case intermodal rail/road vs. road freight) could have a reasonable potential to change the structure of the market currently dominated by a single mode (road).

6.4.2.4 Assignment of Traffic/Transport Demand (Step 4)

In the above-mentioned example, traffic assignment, route assignment, or route choice has been implicitly carried out since the particular O-D routes/corridors for the intermodal rail/road transport services supposed to operate at the given service frequencies are already specified. These routes or paths through the network are assumed to be optimal considering the generalized travel costs consisting of operational and goods/freight time cost. In addition, each link and route of the network has sufficient capacity to accommodate new services, i.e. the rail operators as the new market entrants are able to easily obtain the required slots for launching new services. Thus, the traffic pattern in the network is driven mainly by users— shippers and receivers of the goods/freight shipments—and the intermodal rail/road transport service providers satisfying their needs. In such a way, the all-or-nothing technique of traffic assignment is implicitly applied, assuming that generally all of the train traffic between particular O-D terminals takes the shortest paths (with respect to the generalized cost) and without capacity constraints along particular links and along particular routes of the network. Consequently, updating the intermodal rail/road transport infrastructure, including the intermodal terminals is not needed, at least for the expected volumes of O-D goods/freight demand flows and related vehicle traffic flows serving them under the given conditions.

In general, at other types of transport networks such as urban, sub-urban and intercity rail, road and airline, the transport services driven by the demand flows in the form of either individual service or service networks take place on the corresponding transport infrastructure networks.

Table 6.6: The intermodal rail/road service frequency[1] between particular O-D terminals of the given network (train services/week) (Janić 2014a)

Begin/End terminal	London Barking	Dourges	Antwerp Zomerweg	Rotterdam	Frenkerdorn near Basel	Wolfurt, Vorarlberg	Slaskie Poland
London Barking		3	7	4	4	2	2
Dourges	4		23	3			
Antwerp Zomerweg	10	36		7	20	3	9
Rotterdam	6	11	13		15	7	6
Frenkerdorn near Basel	5		18	7			
Wolfurt, Vorarlberg	1		3	2			
Slaskie, Poland	3		7	5			

[1] Based on the train's composition: 20 flat cars/train × 2 TEU/flat car = 40 TEU/train

Table 6.7: Road and intermodal rail/road volumes of goods/freight shipments, market shares, and service frequency on the routes of the given network (Janić 2014a)

Route	Road TEU/yr (2010)	Intermodal Market Share (%)	Intermodal TEU/yr (2010)	Frequency (Trains/ week)[1]
London Barking – Dourges	91574	34	31135	15
Dourges – London Barking	122599	34.5	42297	20
Dourges – Antwerp Zomerweg	270441	28	75723	36
Antwerp Zomerweg – Dourges	405028	29	117458	56
Antwerp Zomerweg – Rotterdam Maaswlakte/RSC	385467	6	23128	11
Rotterdam Maaswlakte/RSC – Antwerp Zomerweg	186353	6	11181	5
Antwerp Zomerweg – Frenkerdorn near Basel	197083	60.5	119235	57
Frenkerdorn near Basel – Antwerp Zomerweg	125987	60	75592	36
Frenkerdorn near Basel – Wolfurt, Vorarlberg	38954	20	7798	4
Wolfurt, Vorarlberg – Frenkerdorn near Basel	19339	18	3481	2
Rotterdam Maaswlakte/RSC – Slaskie, Poland	45237	64	28952	14
Slaskie, Poland – Rotterdam Maaswlakte/RSC	47819	65	31082	15

[1] Based on the train's composition: 20 flat cars/train × 2TEU/flat car = 40 TEU/train

However, the classical flow assignment problem implies the so-called network equilibrium problem arising in congested networks when the demand flows tend to move between particular O-Ds along the least-cost (time) path(s). This can also be considered as an economic equilibrium problem. In such a case, the demand flows are considered as potential consumers of transport services, whereas the transport service network is considered as the provider of the capacity for matching this demand at prices reflecting the demand's perceived travel costs. Equilibrium occurs when the volume(s) of demand between particular O-Ds equalize with those driven by the market price (Nagurney, 2002).

In general, the main objective of the traffic assignment process is to reproduce the pattern of the transport service network represented by movement of vehicles on the transport infrastructure network under conditions of satisfying the current and expected demand. Under such conditions, the components of generous traffic assignment procedures are as follows (Thomas 1991):

- Estimating the volume of traffic on the links of the network;
- Using the estimated travel costs between particular O-Ds of demand flows in distributing them over the links and routes of the network;
- Obtaining the aggregate measures of the network performances, such as the total vehicular flows, total distance covered by vehicles, the total system travel time, etc.;
- Estimating particular O-D travel times (costs) for a given level of demand, if appropriate; and
- Obtaining the link flows and identifying particularly congested links of the road and/or rail network.

Different types of traffic assignment models have been developed, such as all-or-nothing assignment, incremental assignment, capacity-restraint assignment, user equilibrium assignment, stochastic user equilibrium assignment, system optimum assignment, etc. (Thomas 1991).

Specifically, for planning the urban and sub-urban road (and rail) networks, two types of traffic assignment models are commonly used—the user equilibrium and the system optimization assignment model. The user equilibrium assignment model provides the optimum based on Wardrop's first principle. This states that no car driver can unilaterally reduce his/her travel costs by shifting to another route. If it is assumed that the car drivers have perfect knowledge of their travel costs on a network and choose the best route according to Wardrop's first principle, this leads to the deterministic user equilibrium elaborated as the nonlinear mathematical optimization problem. The system optimization assignment model provides the optimum assignment based on Wardrop's second principle. This states that car drivers cooperate with each other in order to minimize the total system travel time (cost). In such a case, the assignment model enables minimization of congestion when the car drivers are told which routes to take. As such, although this appears not to be a realistic model reflecting the driver's actual behavior, it is useful for planners aiming at minimizing the travel costs and thus achieving an optimum social equilibrium (Thomas 1991). The generic analytical structure of the system optimal assignment model is as follows:

Minimize:
$$C(X) = \sum_a x_a t_a(x_a)$$
(6.6a)

Subject to:
$$\sum_k f_k^{rs} = q_{rs}, \quad \forall r,s$$
(6.6b)

$$x_a = \sum_{r,s,k} \delta_{a,k}^{rs} * f_k^{rs}, \quad \forall a$$

$$f_k^{rs} \geq 0, \quad \forall k,r,s$$

$$x_a \geq 0, \quad \forall a \in A$$

where

A	is the number of links in the network;
x_a	is the equilibrium traffic/transport flow on the link (a) of the network ($a \in A$);
$t_a(x_a)$	is the travel time (cost) on the link (a) of the network, dependent on the level of flow there;
f_k^{rs}	is the traffic/transport flow on the path (k) connecting the O-D pair (r) and (s) matched by the transport capacity by vehicles operated by a given mode on the O-D distance (d_{ij}); and
q_{rs}	is the volume of traffic/transport demand between O-D pair (r) and (s).

In this model, the prime function to be minimized represents the total time (cost) of equilibrium traffic/transport flows in the given network. This time (cost) is an increasing function of flow(s) on particular links, thus making the objective function usually nonlinear. The constraints generally include the flow conservation equations and the non-negativity constraints. Specifically, in Equation 6.6b, the first set of constraints specifies that the traffic/transport flows on all paths connecting particular O-D pair(s) should be equal to the total demand flows (rate) between them. The second set of constraints provides that the equilibrium flows on each link of the network should be equal to the sum of flows on all paths containing them, while connecting particular O-D pairs. The third and fourth set of constraints guarantees non-negativity of traffic/transport flows on the links and paths of the networks.

6.4.2.5 Evaluation of Transport Plans

Evaluation of transport plans usually implies the choice of one among several alternatives developed in the scope of the given solution(s). In practice, the most commonly used include the EAT (Economic Analysis Technique) or BAU (Business As Usual) method. The single-objective EAT or BAU method evaluates particular alternatives on the basis of their 'revenues' and 'costs' during a specified period of time, which is commonly their 'life-cycle'. The outputs are expressed exclusively in monetary terms, such as NPV (Net Present Value), BCR (Benefit-Cost Ratio) and IRR (Internal Rate of Return) (Giuliano 1985, Tabucanon and Mo-Lee 1995). Since the complexity of the DM processes has increased over time due to increase in the number of alternatives in the scope of particular solutions, the number of (usually conflicting) attributes/ criteria per alternative and the number of actors whose (very often diverse and conflicting) points of view need to be taken into account, MCDM (Multi-Criteria Decision Making) or MADM (Multi-Attribute Decision Making) methods are recommended as more convenient tools for looking for the preferable among several alternatives. This implies that MCDM methods discretely consider a usually limited number of alternatives requiring inter- and intra-attribute comparisons involving implicitly or explicitly trade-offs (Zanakis et al. 1998). Some academic research and professional practical and successful applications of the MCDM (Multi-Criteria Decision Making) or MCA (Multi-Criteria Analysis) methods include SAW (Simple Additive Weighting), AHP (Analytical Hierarchy Process), TOPSIS (Technique for Order Preference by Similarity to Ideal Solution), ELECTRE (ELimination Et Choix Traduisant la REalité (ELimination and Choice Expressing Reality)), PROMETEE (Preference Ranking Organization METHod for Enrichment of Evaluations) and many other methods with their modifications (Hwang and Yoon 1981, Janić 2014b, Sauian 2010). Specifically, applying the MCDM methods to evaluation of transport plans/projects in combination with the EAT (Economic Analysis Technique) or BAU (Business As Usual) method has been a matter of wide professional and academic interest (Brucker et al. 2011, Schutte and Brits 2012). An illustrative professional interest was articulated in the European research and development COST 328 Action where the strongest research-based recommendations for using the MCDM instead of the 'pure monetary' EAT or BAU method was made. For example, MCA (Multi-Criteria Analysis) was proposed as a convenient approach for evaluating the projects in the scope of the TENs [Trans-European Transport Network(s)] (EC, 1998). Some academic applications of MCDM methods, such as SAW, TOPSIS and AHP have included the views of particular actors/ stakeholders in the process of planning transport corridors (Bethany et al. 2011), evaluation of their performances (Ding et al. 2008), evaluation of performances of general logistic systems (Sawicka et al. 2010), innovative freight bundling networks in Europe (Janić et al. 1999), and HS (High Speed) transport technologies (Janić 2003, Janić 2014b).

In general, each MCDM process consists of the following steps:

- Identifying the DM (Decision-Maker);
- Defining the objectives of the DM, which include selecting the preferred (i.e. the 'best') among the available alternatives in the scope of the given (required) solution;
- Defining and quantifying the evaluation attributes/criteria of particular alternatives usually reflecting their performances of relevance for the DM;
- Selecting the MCDM method(s);
- Applying the selected MCDM and ranking the considered alternatives; and
- Selecting and then undertaking steps towards implementing the preferred alternative.

However, after the first three above-mentioned steps were successfully carried out, the DM was confronted with the dilemma/question of selecting the most appropriate MCDM method for the given problem. Most authors mention that the main criteria for choosing the method is its validity, implying that the method that is very likely able to reflect (intuitively) the expected outcome(s) should be chosen. In addition, practitioners prefer simple, transparent, easily understandable and applicable methods. In most cases, the methods have to deal with the choice of the preferred among the few alternatives, each having a much greater number of evaluation attributes/criteria. In addition, the experience in applying MCDM methods so far has shown that under the given circumstances, different methods have produced different results (in about 40 per cent of cases). In combination with their complexity and producing different outcomes, some of these have confused their users. Consequently, the MCDM methods should function in the context of the given DSS [Decision Support System(s)] as an aid for users to learn about the problem and its possible solutions in order to reach the ultimate decision. They can be considered as decision aids rather than as decision-making tools. *A posterior* robustness analysis needs always to follow after using any of these models (Zanakis et al. 1998). More details on application of the above-mentioned MCDM methods have already been and will be provided in the forthcoming Chapter 7.

6.5 Concluding Remarks

This section elaborates the main principles of planning transport systems. In the given context, this is considered as the medium- to long-term planning of transport infrastructure and rolling stock. The former is related (again) to HSR networks. The latter includes analysis of the development of train speed and the capacity of commercial aircraft and container ships over a long-term period of time. This illustrates the general direction of developing rolling stock over time—generally on the one hand by planning and developing greater capacity and on the other, by planning and developing higher technical and operating speeds. During particular stages of development, capacity and speed are positively correlated, but with emerging mega container ships and aircraft, this positive correlation has vanished. At trains, the speed and capacity are slightly correlated over time. Then, the known four-step model usually applied to planning transport infrastructure is presented with some illustrative cases for some steps. The first step has only been described. Step 2 is illustrated by the case of distribution of passenger demand flows from a given large airport towards the airports in the continental and intercontinental regions it is connected to. Some elements of Step 1 of the model are contained herein. Step 3 is illustrated by analysis of modal shift and modal split between the rail and road freight transport mode in the given network where both modes compete with each other. Step 4—traffic assignment—is described more generally, as it can be said that Step 3 already contains some elements thereof. Finally, a description of the possible evaluation of transport plans is provided by mentioning the different evaluation methods; in particular, those based on the MCDM approach, which are elaborated in Chapter 7 by means of three representative cases.

References

Bergmann J. 2014. *Future Development of Ultra Large Container Ships: Where Are the Limits?* DNV-GL Group Maritime, Høvik, Norway, www.dnvgl.com

Bethany S., Holland H.J., Noberga A.A.R., O'Hara G.C. 2011. Using Multi-Criteria Decision Making to Highlight Stakeholders' Values in the Corridor Planning Process. *The Journal of Transport and Land Use* 4: 105-118.

Boqué J.R.M. 2012. High-Speed Rail: Economic Evaluation, Decision-Making and Financing. *Master-Arbeit*. Technische Universität Dresden, Fakultät Verkehrswissenschaften "Friedrich Institut für Gestaltung von Bahnanlagen", Dresden, Germany.

Brucker D.K., Macharis C., Verbeke A. 2011. Multi-Criteria Analysis in Transport Project Evaluation: An Institutional Approach. *European Transport\Trasporti Europei* 7: 3-24.

Ding Y., Yuan Z., Li Y. 2008. Performance Evaluation Model For Transportation Corridor Based on Fuzzy AHP Approach. *Fuzzy Systems and Knowledge Discovery (FSKD '08.)*. Fifth International Conference on Fuzzy Systems and Knowledge Discovery. IEEE Computer Society. Los Alamitos, California CA, USA 3: 608-612.

EC. 1998. *Integrated Strategic Infrastructure Networks in Europe*. Final Report on the Action COST 328 EUR 18165. European Commission, Luxembourg.

EC. 2014. *Twin Hub—Intermodal Rail Freight Twin Hub Network North-West Europe*. INTERREGIVB North-West Europe Program. Lille, France, www. twinhubnetwork.eu

Florian M. 2008. *Models and Software for Urban and Regional Transportation Planning: The Contribution of the Center for Research in Transportation*. CIRRELT, Centre on Enterprise Networks, Logistics and Transportation, University of Montreal, Montreal, Canada.

Germanischer Lloyd. 2005. *Trends in the Development of Container Vessels*. Presentation. National Technical University of Athens, Athens, Greece.

Giuliano G. 1985. *A Multicriteria Method for Transportation Investment Planning*. Transportation Research A 19A: 29-41.

Hwang L.C., Yoon K. 1981. *Multi Attribute Decision-Making: Methods and Applications*. Lecture Series in Economics and Mathematical Systems, Springer-Verlag, Berlin, Germany.

IIWG. 2007. *Commercial Aircraft Design Characteristics—Trends and Growth Projections*. Fifth Revision R1. International Industry Working Group, International Air Transport Association, Montreal, Canada.

Janić M., Regglani A., Nijkamp P. 1999. *Sustainability of the European Freight Transport System: Evaluation of Innovative Bundling Networks*. Transportation Planning and Technology 23: 129-156.

Janić M. 2003. *Multi Criteria Evaluation of High-Speed Rail, Transrapid Maglev and Air Passenger Transport in Europe*. Transportation Planning and Technology 26: 491-512.

Janić M. 2009. *Airport Analysis, Planning, and Design*. Nova Science Publishers Inc, New York, USA.

Janić M. 2014a. *Modal Shift Analysis. Deliverable of Project Twin Hub—Intermodal Rail Freight Twin Hub Network in North-West Europe*. INTERREGIVB North West Europe Programme, Lille, France, www.twinhubnetwork.eu

Janić M. 2014b. *Advanced Transport Systems: Analysis, Modelling and Evaluation of Performances*. Springer, UK.

Nagurney A. 2002. *Traffic Network Equilibrium*. Isenberg School of Management University of Massachusetts, Massachusetts, USA.

Sauian S.M. 2010. MCDM: *A Practical Approach in Making Meaningful Decisions*. Proceedings of the Regional Conference on Statistical Sciences 2010 (RCSS'10), June 2010: 139-146.

Sawicka H., Weglinski S., Witort P. 2010. Application of Multiple Criteria Decision Methods in Logistics Systems. *Electronic Scientific Journal of Logistics* 6: 99-109.

Schutte I.C., Brits A. 2012. Prioritizing Transport Infrastructure Projects: Towards a Multi-Criterion Analysis. *Southern African Business Review* 16: 97-112.

Siemens 2008. *Electric Locomotives—Reference List*. Siemens AG Transportation Systems Locomotives, Erlangen, Germany.

Tabucanon T.M., Mo Lee H. 1995. Multiple Criteria Evaluation of Transport System Improvements. *Journal of Advanced Transportation* 29: 127-143.

Thomas R. 1991. *Traffic Assignment Techniques*. Avebury Technical Publication, England, UK.

Zanakis H.S., Solomon A., Wishart N., Dublish S. 1998. Multi-Attribute Decision Making: A Simulation Comparison of Selected Methods. *European Journal of Operations Research* 107: 507-529.

UIC. 2014. *High Speed Lines in the World*. Updated 1st September 2014. UIC High Speed Department, International Union of Railways, Paris, France.

Vuchic R.V. 2004. Urban Transit: Operations, Planning, and Economics. John Wiley & Sons, New Jersey, New York, USA.

http://en.wikipedia.org/wiki/High-speed_rail_in_Europe.

http://en.wikipedia.org/wiki/High-speed_rail_in_China.

http://en.wikipedia.org/wiki/Timeline_of_railway_history#20th_century.

https://www.google.nl/search?q=development+of+container+ships&biw=928&bih=499&tbm=isch&tbo=u&source=univ&sa=X&ei=fzs1VOnPFonUOevTgIgL&ved=0CEQQ7Ak

http://www.dnv.com/industry/maritime/publications/archive/2013/1-2013/index.asp

EVALUATION OF TRANSPORT SYSTEMS
Methodology & Cases

7.1 Introduction

Policy/decision-makers worldwide dealing with decisions on implementing solutions for satisfying the growing demand for transporting passengers and goods/freight shipments regularly confront the problem of choosing what is the best among several alternatives. The main problem has always been setting up and estimating the attributes of particular alternatives, converting them into evaluation criteria and then determining their relative importance, i.e. weights in the given context. Currently, due to increasing efforts for facilitating the more sustainable medium- to long-term development of transport systems, the number of evaluation attributes/criteria has increased alongside the increase in the number of actors/stakeholders involved in the evaluation process, each usually coming up with specific weights for particular criteria. Consequently, the search is on for more systematic methodologies that could support and improve the efficiency and effectiveness of the decision-making process, i.e. choosing the best among the several alternative solutions available for the given transport system. As mentioned in Chapter 6, various single and multi-criteria evaluation methods can be used for evaluation of alternative solutions in transport projects and plans. The most well-known single-criterion method that has been used for a long time is CBA (Cost-Benefit Analysis). However, due to the increasing complexity and sensitivity of the evaluation process and its outcome, different MCDM (Multi Criteria Decision Making) methods have been developed and applied either exclusively or in parallel to CBA. The MCDM methods whose application is presented in this chapter are SAW (Simple Additive Weighting), TOPSIS (Technique for Order Preference by Similarity to the Ideal Solution), and AHP (Analytic Hierarchy Process) method and the entropy method for estimating the relative importance, i.e. weights of particular attributes/criteria of the selected alternatives for the DM (Decision Maker), which are described in Section 7.2. In addition, Sections 7.3, 7.4 and 7.5 elaborate three cases of application in the above-mentioned MCDM methods. Section 7.3 deals with selection of a new hub for an airline. Section 7.4 describes selection of an airport within a given airport system consisting of a few airports where a new runway is added in order to increase the airport's and system's airside (runway) capacity. Section 7.5 deals with selecting one among the two freight transport corridors competing under given conditions. The last section is devoted to the concluding remarks.

7.2 The Evaluation Methodology

In this case, the evaluation methodology includes the MCDM methods and the method for determining the relative importance, i.e. weight of particular attributes, or the evaluation criteria. The three discrete MCDM methods considered are SAW (Simple Additive Weighting), TOPSIS (Technique for Order Preference by Similarity to the Ideal Solution) and AHP (Analytic Hierarchy Process) method (Hwang and Yoon 1981, Saaty 1980, Winston 1994). These methods are shown to be popular and widely used by researchers. Essentially, each one reflects a different approach in solving a given 'discrete MCDM problem of choice of the best among several pre-selected alternatives'. All the methods require a pre-selection of a countable number of alternatives and use of a countable number of quantifiable (conflicting and non-commensurable) attributes or criteria of their performances. The attributes/criteria may mean the 'costs and benefits' for a DM. In such a case, a larger outcome always means greater preference for the 'benefit' and less for the 'cost' attribute/criterion. After inter- and intra-comparison of alternatives with respect to a given set of attributes (criteria) of their performance, the implicit/explicit trade-offs are established and used for ranking the alternatives (Zanakis et al. 1998).

The SAW method is the simplest and clearest method used as a benchmark for comparison of the results obtained from this and other discrete MCDM methods applied to the same problem. The TOPSIS method possesses a unique (specific) but also very logical way of approaching discrete MCDM problems. However, it is computationally more complex than the SAW method. The AHP method is specific due to the certain 'freedom' of a DM to express its preference for particular attributes/criteria by using the original AHP measurement scale.

The SAW and TOPSIS methods require quantification of attributes/ criteria of performance of particular alternatives. The weights used to express the relative importance of these attributes/criteria can be determined either analytically or set empirically by a DM. The last method, AHP, does not require such explicit quantification of attributes/criteria, but it does need specific hierarchical structuring of the MCDM problem. In addition, the method itself generates the weights for criteria by using the AHP measurement scale according to a specified procedure. Under such circumstances, the comparison of results from such different methods applied to the same problem turns out to be very interesting and challenging from both academic and practical perspectives. The subsequent sub-sections describe the basic structure of the three MCDM methods and the procedures for assigning the weights to the evaluation attributes/criteria.

7.2.1 The SAW Method

The SAW (Simple Additive Weighting) method consists of quantification of the values of attributes/criteria for each alternative, construction of the Decision-Matrix (A) containing these values, derivation of the normalized decision-matrix (R), assigning the importance (weights) to the criteria and calculation of the overall score for each alternative. Then, the alternative with the highest score is selected as the preferable or the best one. The analytical structure of the SAW method for (N) alternatives and (M) attributes/criteria can be summarized as follows:

$$S_i = \sum_{j=1}^{M} w_j r_{ij} \quad \text{for } i = 1,\ 2,..,N \tag{7.1}$$

where

S_i is the overall score of i-th alternative;

w_j is the importance (weight) of j-th criterion;

r_{ij} is the normalized rating of i-th alternative on j-th criterion, which computed as $r_{ij} = x_{ij}/(\max_i x_{ij})$ for the 'benefit' and $r_{ij} = (1/x_{ij})/[\max_i/x_{ij})]$ for the 'cost' criterion represents an element of the normalized matrix R;

x_{ij} is an element of the Decision-Matrix A, which represents the 'original' value of j-th criterion of i-th alternative;

N is the number of alternatives; and

M is the number of attributes/criteria.

7.2.2 The TOPSIS Method

The TOPSIS (Technique for Order Preference by Similarity to the Ideal Solution) method, at the first stage, consists of composition of the Decision-Matrix (A) with the values of attributes (criteria) and construction of the normalized-decision matrix (R) based upon the matrix (A). The elements of matrix (R) are computed as: $r_{ij} = x_{ij}/(\sum_{i=1}^{M} x_{ij}^2)^{1/2}$, where ($x_{ij}$) is the value of *j*-th criterion of *i*-th alternative and is, as in Equation 7.1, an element of the Decision Matrix (A). The weighted-normalized decision matrix is obtained by using the normalized decision matrix R and the weights assigned to the criteria as $V[v_{ij}] = [w_j * r_{ij}]$. At the second stage, the ideal (fictitious best) solution A^+ and the negative-ideal (fictitious worst) solution A^- are determined, respectively, as follows:

$$A^+ = \left\{ \left(\max_i v_{ij} \mid j \in J_1 \right); \left(\min_i v_{ij} \mid j \in J_2 \right) \mid i = 1, 2, .., N \right\} = \left\{ v_1^+, v_2^+, ..., v_j^+, ..., v_M^+ \right\}$$

(7.2a)

$$A^- = \left\{ \left(\min_i v_{ij} \mid j \in J_1 \right); \left(\max_i v_{ij} \mid j \in J_2 \right) \mid i = 1, 2, .., N \right\} = \left\{ v_1^-, v_2^-, ..., v_j^-, ..., v_M^- \right\}$$

(7.2b)

where (J_1) is associated with the 'benefit' and (J_2) with the 'cost' criteria.

Consequently, the Euclidean distance of each alternative from the overall ideal and negative-ideal solution is determined, respectively, as follows:

$$S_i^+ = \left[\sum_{j=1}^{M} (v_{ij} - v_j^*) \right]^{1/2} \text{ and } S_i^- = \left[\sum_{j=1}^{M} (v_{ij} - v_j^-)^2 \right]^{1/2} \text{ for } i = 1, 2, .., N \quad (7.3)$$

where all the symbols are as in the previous equations.

The relative closeness of each alternative to the ideal solution is computed as the ratio $C_i^+ = S_i^- / (S_i^+ + S_i^-)$ for $i = 1, 2, .., N$. Finally, the alternative with the highest value of (C_i^+) is selected as the best one (Hwang and Yoon 1981, Zanakis et al. 1998).

7.2.3 The AHP Method

The AHP (Analytic Hierarchy Process) method consists of three steps: decomposition of the problem, comparative judgement and synthesis of priorities (Saaty 1980, Winston 1994).

(a) Decomposition of the problem

This deals with a hierarchical schematic representation of the overall objective and decision alternatives.

(b) Comparative judgement

This includes formation of pairwise matrices and their comparison at two levels: (1) the level at which all the alternatives are compared with respect to each criterion and (2) the level at which the criteria are compared with respect to the overall objective. The following sub-steps are performed:

At level 1, a pairwise comparison matrix with quadratic shape (A_{NXN}) is formed where N corresponds to the number of alternatives. The number matrices of type A are equivalent to the number of criteria M. An element of matrix (A), (a_{ij}) may be assigned any value from the AHP original measurement scale containing the integers from one to nine. The particular number, usually selected by a DM, is used to express the relative importance of each particular criterion being compared across particular alternatives. The following condition should always be fulfilled $a_{ij} = 1/a_{ji}$ if $i \neq j$ and $a_{ij} = 1$. Then, the normalized matrix (A_{norm}) is obtained by dividing each element of matrix (A) in column (i) by the sum of all elements in the same column (i) as follows $r_{ij} = a_{ij} / \sum_{i=1}^{N} a_{ij}$ where $i = 1, 2, ..., N$. Next, the matrix of weights (W) is computed. For example, the weight for i-th row of the matrix (W), (w_i) is determined as the average of elements in row (i) of the matrix (A_{norm}) as follows:

$$w_i = (1/N) \sum_{j=1}^{N} r_{ij} \quad \text{for } i = 1, 2,.., N \tag{7.4}$$

A similar procedure is carried out at level 2 with the matrix of criteria (C), which has dimensions equivalent to the number of criteria. At level 1, checking of the consistency of the DM's comparisons is carried out by computing the matrix $B = Aw^T$ and the value: $P = (1/N) \sum_{i=1}^{N} b_i / w_i^T$, where (b_i) is i-th element of matrix (B) and (w_i^T) is i-th element of matrix (W^T). Then, the Consistency Index (CI) is computed as $CI = (P - N)/(N - 1)$ and compared with the Random Index (RI) [The Random Index (RI) for given (N) is provided by the AHP method].

At level 2, matrix (C) instead of matrix (A) is used to perform the above calculations.

If the condition $CI/RI \leq 0.10$ is fulfilled, the synthesis of priorities is carried out by computing the overall score for each alternative (S_i) as follows (Saaty 1980, Winston 1994):

$$S_i = \sum_{j=1}^{M} w_j v_{ij} \quad \text{for } i = 1, 2,.., N \tag{7.5}$$

where

v_{ij} is the element of a priority vector of i-th alternative on j-th criterion.

Finally, the alternative with the highest overall score is selected as the preferable one. In case the required condition is not fulfilled, the procedure of forming the related pairwise comparison matrices should be repeated.

7.2.4 The Relative Importance (Weight) for Attributes/Criteria

The relative importance (weight) of attributes/criteria can be determined by using different procedures. Broadly analytical, simulation and empirical (heuristic) procedures can be distinguished.

Some of the analytical procedures, which can be used in the applications of the SAW and TOPSIS methods, are the right given value, row and column geometric means, simple raw average, mean transformation method and the entropy method. The meaning of the first four methods is relatively clear. The last method (entropy) is often recommended as a convenient tool to be applied when eliminating criteria with similar values, highlighting the importance of criteria with higher differences in their values and when a DM has no reasons to prefer one criterion to others (Hwang and Yoon 1981, Zanakis et al. 1998).

The entropy idea has played an important role as a concept in physics and in social sciences. In particular, entropy has been widely used in the information theory as a measure of uncertainty of a discrete probability density function as follows (Hwang and Yoon 1981, Straja 2000):

$$S(P_1, P_2, \dots P_n) = -k \sum_{i=1}^{n} p_i \ln(p_i) \tag{7.6a}$$

where
p_i is a probability of i-th outcome; and
k is a constant.

Under the conditions of highest uncertainty, when all probabilities are equal, the entropy function $S(p_1...p_i...p_n)$ will reach its maximum. Since the Decision-Matrix contains a certain amount of information for a set of alternatives and attributes/criteria, the entropy concept can analogously be used to assess the contrasts between the values of attributes/criteria at particular alternatives. According to the entropy idea, for example, if the values of particular criterion are very similar or even the same for given alternatives, entropy will be higher and thus the weight assigned to such criterion is smaller. This is likely the case when the criterion should be eliminated due to its lack of relevance. However, if the values of a given criterion are more different across particular alternatives, their corresponding entropy will be smaller and the weight assigned to such criterion higher.

Let a set of alternatives A_i ($i = 1, 2, 3, ..., N$) be evaluated according to (X_j) criteria ($j = 1, 2, 3, ..., M$). Let (X_{ij}) be the outcome of i-th alternative with respect to j-th criterion and an element of the Decision-Matrix A. Let (p_{ij}) be determined as follows (Hwang and Yoon 1981, Straja 2000):

$$p_{ij} = \frac{X_{ij}}{\sum_{i=1}^{N} X_{ij}}, \text{ for } i \in N; j \in M \qquad (7.6b)$$

The entropy of attribute/criterion (j), (E_j) for (N) alternatives can be expressed as follows:

$$E_j = -1/\ln(M) \sum_{i=1}^{N} p_{ij} \ln(p_{ij}) \quad \text{for } j \in M \qquad (7.6c)$$

where the term $[-1/\ln(M)]$ provides the condition $0 \le E_j \le 1$ to be fulfilled.

If the DM has no reason to prefer one criterion above the others, the weight of criteria (X_j) (w_j) can be determined as follows (Hwang and Yoon 1981):

$$w_j = (1 - E_j) / \sum_{j=1}^{M} (1 - E_j) \qquad (7.6d)$$

where all the symbols are as in the previous equations.

In addition, simulation can be used to determine the weights for attributes/criteria by generating them from the given distribution, whose shape may be dependent on the purpose. For example, in case of 'no distribution', all the weights can be equalized to indicate the same importance of particular criteria. The 'uniform distribution' is used to reflect an indecisive or uninformed DM. Other distributions can also be used, depending on the type and preferences of the DM. This procedure can be used at SAW, TOPSIS and AHP to assign the weight to the attributes/criteria.

In addition, the empirical (heuristic) procedure, including the judgement of a DM on the weights for attributes/criteria, can also be applied. In such case, the assignment of weights is based on the experience (heuristic) or specific preferences of the DM intended to 'justify' *a priori* preference. This procedure can be used at both SAW and TOPSIS method 'as imposed'. At AHP method, it can be combined with the AHP's measurement scale, which offers a flexible but consistent choice of weights for attributes/criteria.

7.3 New Hub Airport for an Airline

7.3.1 Background

The European, like any other air transport system, consists of airports, ATC (Air Traffic Control), i.e. aviation infrastructure and airlines. Before market liberalization, which took place as a gradual process between 1987 and 1997, the intensity of flying between particular airports within particular EU

(European Union) member states had been regulated by numerous inter-state and inter-airline bilateral agreements (Button et al. 1998, Button and Stough 2000, Button and Swan 1991, ICAO 1988, OECD 1988). Consequently, most EU airlines, particularly the national 'flags carriers', built relatively strong 'star-shaped' or 'radial' air route networks around their national hubs—usually the biggest national airports. The allowed routes and agreed flight frequencies have considerably influenced the spatial layout of the airline networks within the EU.

During the post-liberalization period, capturing 'strategic' market positions by using advantages of the obtained freedoms of liberalized market became an important policy strategy of many EU airlines (Stasinopoulos 1992, 1993). Some of them, particularly those from the European periphery, both 'flag carriers' and regional airlines, intended to strengthen their presence in the 'core' area of Europe[1] while some airlines from the 'core' tried to move in the opposite direction. In both cases, in addition to contracting alliances of different types, setting up a new hub airport abroad, i.e. outside the domestic market (in addition to the old—national—one), was considered as a feasible option.

This section describes an application of the Multiple-Criteria Decision-Making (MCDM) approach to the problem of selecting a new hub airport for a hypothetical EU airline. The application contains detailed calculations of particular phases of the evaluation process purely for illustrative purposes (Janić and Reggiani 2002).

7.3.2 The System and Problem—Selection of New Hub Airport by an Airline

Liberalization of the EU aviation market removed the institutional barriers that had hindered freedom and flexibility of air transport operations between particular member states. Consequently, 'free operations' in terms of flight frequencies, fares and entering/leaving the market took place with expectations to instigate competition within the industry, diminish airfares and improve the overall quality of service for both passengers and freight. In parallel, privatization of airlines and airports was carried out as an additional (and complementary) activity with the same purpose, i.e. to improve the overall efficiency and effectiveness of the entire sector and its particular components—airlines and airports. Confronted with the new challenges and conditions, some EU airlines generally used one or few options for maintaining

---

[1] For a long time, the central parts of France and Germany, south part of England, Belgium, The Netherlands and North Italy have been recognised as the 'core' area, which has generated about 35 per cent of the total European air traffic (IFAPA 1988).

their existing or taking a new strategic position in the EU aviation market as follows:

- Abandoning existing ('classical') agreements with other EU airlines and re-designing bilateral and multi-lateral agreements with non-EU airlines, both on the continent and abroad;
- Maintaining the existing and contacting new alliances with both European and non-European partners; and
- Looking for a new hub airport at demand-attractive, i.e. 'strategic' locations within the EU, preferably within its 'core' area, either individually or in the scope of an alliance partnership.

7.3.2.1 Bilateral Agreements

After liberalization, EU airlines abandoned the bilateral agreements concluded between themselves, while retaining and modifying most of the bilateral agreements with other non-EU and non-European partners. These agreements were modified mostly in terms of the increased flexibility in flight frequencies and setting up airfares (Stainland 1998). The existing agreements were expected to be further 'softened' or even completely abandoned by the implementation of the various 'open-skies' initiatives[2] between the EU and the rest of the world.

7.3.2.2 Airline Alliances

Airline alliances of type 'corporate mergers', 'marketing agreements' and 'strong alliances involving holding of stakes/equities by a merger in the partner(s)' have been practiced by particular EU airlines for a long time (Button et al. 1998, Oum et al. 2000, Tretheway 1990). The number and diversity of alliances increased particularly after liberalization of the EU aviation market both at the EU airlines and the most important EU airports with domination of those of the 'marketing agreement' type (Janić 1997, Oum et al. 2000, Panmure WLB 2000, RBI 1995/1999). In general, alliances brought both advantages and disadvantages to the EU airlines. An apparent advantage was the overall improvement in utilization of the airline fleet, which was achieved through complementarity of services and co-operation instead of competition based on 'code-sharing' agreements and balanced schedules on common routes. In addition, the alliances helped many EU airlines, particularly the 'flag carriers', to maintain a dominant position at their main hubs (Burghouwt

---

[2] An 'open skies' agreement may contain all (or most) elements of the completely liberalized aviation market of the partnering countries. For example, according to the US Department of Transportation, 12 European countries already had 'open skies' agreement with the US: The Netherlands, Switzerland, Sweden, Norway, Luxembourg, Iceland, Finland, Denmark, Belgium, Austria, Czech Republic and Germany (Stainland 1998).

et al. 2002). The disadvantage was the unavoidable competition between different (global) alliances. Also, the users—the air passengers experienced both advantages and disadvantages. The apparent advantages consisted of an improved quality of service through increased flight frequencies (i.e. flight concentration on particular routes), increased diversity of destinations-markets, more reliable and efficient transfer of passengers and freight between the alliance's (i.e. code sharing) flights and obtained benefits from FFPs (Frequent Flyer Programs). The evident disadvantage was to maintain the relatively high and diverse airfares throughout the EU market, primarily due to a lack of sufficient competition (Bailey et al. 1985, Button et al. 1998, IFAPA 1988, Janić 1997, RBI, 1995/1999).

7.3.2.3 New Hub Airport

Particular EU airlines considered setting up a new hub airport abroad (i.e. in another member state) as a feasible option to, in addition to the strengthening of their global market position in the EU, diminish the latent risk of failure of the convenient alliance(s). There is evidence of such practices taking place on both the national and international (EU) scene. For instance, on the international scene, Iberia, which operated the national hub Madrid-Barajas airport, considered either Frankfurt-Main or Amsterdam-Schiphol airport as a new – second hub. Finnair, whose hub was Helsinki Vantaa airport, considered Stockholm Arlanda airport as a potential new hub. Both SAS, which had already operated three hubs (Copenhagen Kastrup, Stockholm Arlanda and Oslo Fornebu) and KLM, looked for a new hub (Berechman and de Wit 1996). After Alitalia had moved its hub (two-thirds of the European routes) from Rome-Leonardo da Vinci Airport to Milan Malpensa Airport at the end of 1998, KLM also considered this airport as a potentially new hub through a prospective alliance with Alitalia (AW 2000). Recently, British Airways tried to negotiate an alliance with KLM, but at the same time looked at Brussels International airport as a potentially new hub abroad, particularly after the collapse of the Belgian 'flag' Sabena. (The airline's well-established hub was London Heathrow airport and until recently, London Gatwick airport.) In addition, one of the European LCC (Low Cost Carriers)—Virgin-Express was considering Paris Charles de Gaulle airport (Paris) as an additional hub. The airline's hub had already been Brussels International airport, where its market position was strengthening after the collapse of Sabena in 2001 (http://www.airwise.com/). Another LCC, Ryanair, selected Charleroi airport near Brussels as its fourth 'hub' in addition to three others—London Stansted, Dublin and Shannon (http://www.ryanair.com). On the national (domestic) scene, in addition to East Midlands airport, British Midlands set up an additional (European) hub at London Heathrow airport and an intercontinental hub at

Manchester airport. Lufthansa located its second national hub at Munich airport in addition to the one at Frankfurt-Maine airport.

Bearing in mind the described real-life developments in this section, a hub is broadly considered as an airport at which an airline may have a 'base' for its fleet. From there it may carry out more or less either frequent 'point-to-point' or 'hub-and-spoke' operations. The latter may have spatial but not necessarily the temporal component in terms of 'waving' of incoming and outgoing flights (Burghouwt et al. 2002).

7.3.3 Evaluation of the New Hub Airport

7.3.3.1 Some Related Research

Research dealing with the selection of a new hub facility has always been closely interrelated to the problem of development and operation of hub-and-spoke transport networks. Research has been carried out in fields, such as operations research, spatial planning and economics. Real-life attainments in both passenger and freight transport followed (Aykin 1995).

Operational researchers have mostly dealt with determining the route structure and location of one and/or few hubs that minimizes the total network cost for a transport operator. In such a context, the single hub location problem was always converted into the 'classical' Weber's least-cost location one. The optimal location of two or more hubs emerged as a much more complex problem, which usually required development of complex algorithms based on heuristics and mathematical programming techniques (Adler and Berecham 2001, Aykin 1995, O'Kelly 1986).

Economists mostly applied regression model(s) for studying hub-and-spoke networks and their influence on the operators' and users' welfare (Morrison and Winston 1994). In most cases, a hub-and-spoke network was considered as a given entity in which the problem of 'hub location' did not exist at all. It was assumed that a hub should be located 'logically' at a central location in relation to other nodes/airports of the network and have a significant proportion of local traffic (Bailey et al. 1985). Berechman and de Wit (1996) developed a simulation model to optimally locate the hub airport for a hypothetical West European airline. The 'airline profit' earned by operating the network established around a preselected hub was used as an exclusive decision-making criterion. Then, Adler and Berecham (2001) developed an algorithm for optimizing two hub-and-spoke airline networks operated in a deregulated market. The algorithm maximized the airline profits under the given constraints. Evidently, most of the above studies were based on the optimization of hub location and associated networks by using a single criterion representing the network operator's costs, revenues or profits. Thus explicit evidence indicating that some airlines have used this or a similar procedure to deal with the problem of hub location is lacking. Hence, this seems to remain a matter in the domain of researchers. However, bearing in mind that the particular airline has demonstrated high flexibility in using different operations research techniques at both the tactical and strategic level (Yu 1998), it is realistic to expect that someday it may become a practitioner, more interested in the proposed approach (Janić and Reggiani 2002).

7.3.3.2 Objectives and Assumptions

The objectives of this section are to develop the methodology for evaluation of a potentially new hub airport for an (EU) airline. The methodology is based on generation of alternatives (i.e. candidate airports), selection and quantification of attributes (criteria) of each alternative, assignment of weights to particular attributes (criteria) reflecting their relative importance for the Decision-Maker (DM) and selection of the optimal (preferable) alternative. As such, the methodology is based on the following assumptions:

- The airline in question intends to establish a new hub in addition to the existing one.
- The candidate airports considered as alternatives exclude the one the airline already operates as the primary hub.
- The candidate airports considered as alternatives are characterized by the relevant indicators and measures of performances, which the airline uses as evaluation attributes/criteria in the DM (Decision-Making) process.
- The particular attributes/criteria are quantifiable under the given conditions.
- The airline as the DM uses the MCDM methods for selecting the new hub airport among the *a priori* considered candidates/alternatives.

7.3.3.3 Structure of the Methodology

The evaluation methodology for selection of the new hub by an airline consists of three classes of models: one for estimating attributes/criteria of particular candidate/alternative airports, the MCDM methods and the methods for estimating the relative importance, i.e. weight of particular attributes/criteria mentioned above.

(a) Definition of attributes/criteria

A hypothetical EU airline (the DM) is assumed to consider several candidate airports as alternatives for the potential location of its new hub. The airline defines *a priori* a set of attributes/criteria reflecting the performances of the considered airports of interest. In general, these attributes/criteria are as follows:

- Strength of candidate/alternative airport(s) to generate air transport demand;

- Operational and economic characteristics of candidate/alternative airport(s);
- Airline operating costs; and
- Environmental constraints at candidate/alternative airport(s).

Strength of candidate/alternative airport(s) to generate air transport demand: This includes socio-economic indicators of the airport catchment area, such as *GDP* (Gross Domestic Product), or combined *Population* and *PCI* (Per Capita Income). In addition, some surrogates, such as attractiveness of the region and/or city (urban agglomeration) in terms of business and tourism may also be taken into account.

Gross Domestic Product (GDP) is shown to be the main driving force behind growth of commercial air transport in many countries and regions including those served by the airport concerned. In such a context, growth of *GDP* is always expected to generate growth of air transport demand and vice versa, at both macro (the country) and micro (the region and airport) scale. Consequently, at the micro scale, airports located in the regions (countries) with higher *GDP* are always shown to be more attractive for airlines.

Population traditionally reflects an inherent 'strength' of a region (or the country) as a 'source' of potential air transport demand. However, this attribute should be used carefully and selectively. For example, in regions served by one airport, it seems clear that the whole population is expected to use this single airport. In regions or large urban agglomerations served by several airports, the population uses particular airports depending on their current convenience. Therefore, an adjustment of the size of population expected to use the candidate airport should be carried out. Under such circumstances, without taking into account the competition, which may already exist at the intended location for the new hub, such modified attributes may be used to roughly indicate the potential market size for the airline looking for a new hub. In addition, Per Capita Income (PCI) of a region can be used as an indicator of the market 'strength' in terms of the 'purchasing power' of the local population. In general, the regions with higher PCI are always considered to be more lucrative air-transport markets, independent of the structure of activities and type of preferred trips. In many cases, Population and PCI are considered together instead of GDP. Consequently, the airports serving more inhabited regions with higher PCI are always considered, independent of their number and (market) relationships (co-operation/competition), as stronger generators of air transport demand and thus as more attractive for setting up a new airline business.

Operational and economic characteristics of candidate/alternative airport(s): These embrace attributes such as the airport size, quality of airport surface access, quality of service of airport airside area and cost of airport service.

The airport size reflects the importance of an airport at local (regional), national (country) and global (international) scale. Generally, a larger airport always looks more attractive and more promising for starting a new airline business than the smaller since there it always looks easier to gain prospective commercially feasible demand, either through competition or co-operation with airlines already there.

The quality of surface access reflects efficiency and effectiveness of passenger access to an airport by using the airport surface access systems. In such a context, all airports are assumed to be accessible by individual modes, such as car or taxi. However, the availability, efficiency and effectiveness of public transport, such as rail and bus system, may significantly vary. Generally, airports with a greater number of more efficient (faster/cheaper) and effective (frequent/punctual/reliable) surface public transport systems are always preferred by both passengers and airlines (Ashford 1988). Specifically, the number of public transport systems serving particular airports may emerge as a relevant attribute for evaluation if it significantly differs from other alternatives. For example, the quality of access is not the same at an airport with or without rail connections.

The quality of service of the airport landside area includes the overall quality of the aviation product provided to passengers by an airport while being in the airport terminal. This may include components of the quality of service, such as queuing and waiting at different service counters, safety and security, reliability of inter-flight connections, the risk of losing or damage to baggage and the overall internal cleanness. This attribute (criterion) is preferred to be as high as possible and important for evaluation, particularly in cases when the airports themselves look after the above elements of the quality of service. However, if airlines take care of these elements, as well as when the alternative airports offer very similar conditions, this attribute (criterion) appears to be less relevant (CAA 2000, Bowen and Headly 2002).

The quality of service of airport airside area includes attributes, such as the volume utilization and distribution of the airport airside capacity among the airlines operated there. Indirectly, these attributes reflect the ease for an airline as a new entrant to get the desired number of landing and departure slots at a preferable time. Generally, at airports with a greater but lower utilized capacity, the entrance and setting up of the desired network of routes and services is easier and thus this location is always considered more attractive. The distribution of capacity (i.e. the available slots) among airlines already operated at the airport in question indicates a level of 'market deregulation', including the incumbent's (and its alliance's) relative 'market strength'. Consequently, if the slots are distributed more evenly among the airlines not being the alliance's partners, the airport market is considered to be more 'liberal-deregulated' and the incumbents' influence on slot allocation weaker. This may make the new entry much easier and consequently the airport more attractive. In addition, the average delay per aircraft operation caused by the airport can be used as an attribute of the airside quality of service. The values of this attribute are preferred to be as small as possible (Burghouwt et al. 2002, EEC 2002, Janić 1997).

The cost of airport service includes passenger tax, airline landing fee or both. Actually this cost reflects a rate charged by an airport for a service, i.e. for serving a unit of air transport demand, either passenger or aircraft. According to the business policy of many airlines, and particularly LCCs, to keep the operational costs under strict control, the average cost of service may appear as an important factor while considering an airport as a new hub. In general, bigger and privatized, more efficient airports as well as the smaller regional airports struggling to attract more air transport demand by offering cheaper services are generally considered by most airlines to be more attractive (Doganis 1992).

Airline operating costs: These costs consist of the total expenses imposed on an airline while operating the 'renovated' hub-and-spoke network, which contains the new hub. They depend on internal and external factors. The internal factors include the size of the airline network expressed by the number of airports and routes, flight frequencies on particular routes, types (capacity) of aircraft engaged, the airline routing strategy to incorporate the selected airport in the existing network and fixed cost of setting up a new hub at the preselected airport. The external factors include the prices of inputs, such as, labor, energy-fuel and capital. The airline operating costs generally increase with increase in both internal and external factors and vice versa and they are preferred to be as low as possible for the new hub (Aykin 1995, Janić 2001).

Environmental constraints at candidate airport(s): These constraints may exist at particular airports in terms of aircraft noise, air pollution and land take. These constraints may work as a 'deterring factor' while considering an airport as a candidate for a new hub due to several reasons. First, they could significantly affect the intended volume of operations. Second, they may be completely unacceptable for airlines using the 'old-technology' aircraft in terms of noise and air pollution. Finally, congested airports, without prospective option(s) for expansion due to constraints in land, are always considered as less attractive locations for launching a prospective airline business. In general, the airports with a smaller number of less strict environmental constraints are always preferable. Consequently, twelve attributes/criteria of performances are defined as relevant for evaluation of the location of a new hub for an airline as follows:

- Population
- Per Capita Income
- Airport size
- Generalized airport access cost
- Quality of passenger service in an airport terminal
- Airline cost of operating the 'renovated' air route network
- Average cost of airport service
- Airport capacity
- The incumbent's market share
- Utilization of airport capacity
- Airport-induced delay
- Environmental constraints

Generally, some of the above attributes/criteria may be dependent on each other. For example, the attribute 'airport size' is dependent on 'population' and 'PCI'. This is particularly the case at airports with a large proportion of terminating traffic and vice versa. In addition, the attribute 'airport size' may also depend on the airport location in the airline and air transport route network, in which case the transit/transfer traffic generated by the airline itself may have a significant proportion in the total airport traffic. The attribute 'generalized airport access cost' reflecting the availability, efficiency and effectiveness of the airport surface access modes may be dependent on the 'airport size'. The attribute 'airport capacity' is mainly correlated with the 'airport size' and vice versa. The attribute 'average cost of airport service' may also be dependent on the 'airport size' and vice versa. The attribute 'airport-induced delay' may be dependent on the 'airport size' and 'airport capacity'. However, such overall interdependence between particular attributes/criteria does not preclude their consideration by a DM, both individually and independently. This may be an argument in favor of the application of the above-mentioned MCDM methods. In addition, such an approach allows the airline as the DM to be selective and flexible in selecting particular attributes/criteria and setting up their values.

(b) Models for estimating attributes/criteria

The particular attributes/criteria of airport performance can be quantified by using different methods. For example, some of them, such as 'Per Capita Income' and 'airport size' can simply be extracted from the corresponding database(s). 'Population' can also be extracted from the appropriate databases, but in most cases it needs additional modification respecting the allocation to particular airports. The attributes such as 'airport capacity' and 'environmental constraints' can be obtained from the given airport and ATC (Air Traffic Control). The values of other attributes/criteria, such as 'generalized access cost', 'reliability of passenger and baggage handling', 'airline operating cost of 'renovated' air route network', 'average cost of airport service', 'incumbent's market share', 'utilization of airport capacity' and 'airport induced delays' can be compiled from the corresponding databases. Consequently, the following attributes/criteria of performances are modelled:

Generalized airport access cost: This cost includes both the passengers' out-of-pocket costs paid for travel and the cost of their time while being within particular surface access systems. The 'time of being within the system' includes 'defer' time, which is dependent on the departure frequency and 'in-vehicle' time, which is dependent on the average running speed and distance between an airport and its catchment area. The value of passenger time may be dependent on the type of travel (business, leisure) and characteristic of passenger (sex, age, etc.) (Janić 2001). In general, this cost can be estimated as follows:

$$c_g = p(d) + \alpha T(d) \tag{7.7}$$

where

$p(d)$ is a fare paid by the passenger for travelling to/from an airport by one of the available surface public airport access systems (€/km);

d is the average travel distance between an airport and its catchment area (km);

α is the average value of passenger time while being within a given airport surface access system; this value may be dependent on type of passengers (leisure, business) and type of journeys (domestic, international) (€/hr-pass); and

$T(d)$ is the perceived travel time on the distance (d) between a given airport and its catchment area $(T(d) = s + d/v(d)$, where (s) is a 'slack' or 'defer' time dependent on the departure frequency of a given access system and $(v(d))$ is the system average speed on distance $(d))$ (km).

Quality of service in an airport terminal: This can be expressed by the average passenger delay while getting the basic service within the terminal (Janić 2001). Another measure may be reliability of service expressed by a proportion of miss-connecting flights or miss-handled/damaged baggage during a given period (month, year). The values of this attribute can be obtained from the airport airlines and dedicated consumers' reports (Bowen and Headly 2002).

Airline cost of operating the 'renovated' air route network: The cost of operating 'renovated' hub-and-spoke network can be estimated for a given network configuration (size, structure-two hubs) and traffic scenario determined by the flight frequencies on particular routes, aircraft types (size) and the average cost per unit of airline output—passenger-kilometer. Thus, the total operating cost of an airline two-hub-and-spoke network for case when k-th alternative airport is considered as the new-second hub is estimated as follows (O'Kelly 1986, Aykin 1995).

$$
C_T(k) =
\begin{bmatrix}
\displaystyle\sum_{i=1}^{P-1}\sum_{j=1}^{P-1} Q_{ij}\left(c_{i\,h1}l_{i\,h1} + c_{h1\,j}l_{h1\,j}\right)\Big|_{i\neq j;\ ij\in P; k\in K} + \\[2mm]
\displaystyle\sum_{i=1}^{P+Q}\sum_{j=1}^{P+Q} Q_{ij}\left(c_{i\,h1}l_{i\,h1} + c_{h1\,hk}l_{h1\,hk} + c_{hk\,j}l_{hk\,j}\right)\Big|_{i\in P;\ j\in Q;\ k\in K} + \\[2mm]
\displaystyle\sum_{i=1}^{Q-1}\sum_{j=1}^{Q-1} Q_{ij}\left(c_{ihk}l_{i\,hk} + c_{hkj}l_{hk\,j}\right)\Big|_{i\neq j;\ ij\in Q;\ k\in K} + C_k
\end{bmatrix}
\qquad (7.8)
$$

where

P	is the number of spokes assigned to existing hub (h_l);
Q	is the number of spokes assigned to the new hub (h_k)($k = 1, 2,...K$);
K	is the number of preselected alternative airports for a new hub ($K \in P+Q$);
Q_{ij}	is a passenger flow between spokes (i) and (j) (pass);
c_{ihl}, c_{hlj}	is the average cost per unit of passenger flow while connecting the spokes (i) and (j) with existing hub (h_l) (€ or \$US/pass);
$c_{hl,hk}$	is the average unit cost of passenger flow while connecting existing hub (h_l) to the new one (h_k) (€ or \$US/pass);
c_{ihk}, c_{hkj}	is the average cost per unit of passenger flow while connecting spokes (i) and (j) to the new hub (h_k) (€ or \$US/pass);
l_{ihl}, l_{hlj}	is the length of a route connecting existing hub (h_l) to the spokes (i) and (j), respectively (km);
$l_{hl\,hk}$	is length of a route connecting existing hub (h_l) to the new hub (h_k) (km);
l_{ihk}, l_{hkj}	is length of a route connecting the new hub (h_k) to the spokes (i) and (j), respectively (km); and
C_k	is the fixed cost for location of the new hub at a preselected airport k (€ or \$US) ($k \in K$).

Equation 7.8 is modified according to the specific conditions under which the location of the existing hub is fixed and the location of the new-additional hub is alternatively chosen from a given set of alternatives. It consists of four components: the cost of connecting the existing hub with the associated spokes; the costs of connecting the spokes assigned to different hubs; the cost of connecting the new hub to the assigned spokes; and the airline fixed costs

needed to set up the new hub. The first component is not directly dependent on the location of the new hub while the following three components are. For each location (k) ($k \in K$), each component of Equation 7.8 is computed for a given 'strict' routing policy, O-D matrix of the given passenger flows, the airline unit cost per passenger-kilometer and the route length.

Average cost of airport service: In most cases, this cost can be obtained by using convenient modelling techniques. In such a context, regression analysis is frequently applied to estimate the relationship between this cost (dependent variable) and the volume of airport output (independent variable).

The incumbent's market share: The incumbent's market share can be estimated for a given airport by dividing the total number of incumbent's incoming and outgoing flights by the total number of incoming and outgoing flights carried out by all airlines during the given period of time (hr, d, mon, yr). This should include use of the aircraft of a comparable seat capacity.

Utilization of airport capacity: Utilization of the given airport capacity can be expressed as the ratio between the actual number of aircraft movements and the airport capacity.[3]

Airport-induced delay: The airport-induced delay can be obtained from the airport and air traffic control reports. However, sometimes it is very difficult to extract the portion of such induced delay from the available aggregate figures.

(a) Evaluation methods

Three above-mentioned MCDM methods—SAW, TOPSIS and AHP—are considered to be applied to the given case of selection of the new hub airport for an airline. They are supposed to use the estimated/quantified particular attributes as criteria. The relative importance, i.e. weight of particular criteria is estimated by means of three scenarios: Scenario (a)—equal weights; Scenario (b)—weights generated from the uniform distribution [0, 1] by simulation; and Scenario (c)—weights estimated from the above-mentioned entropy method (SAW and TOPSIS) and by own weighting procedure (AHP).

7.3.4 Application of the Evaluation Methodology

7.3.4.1 Geographical Scope

Application of three proposed MCDM methods is carried out under the assumption that a hypothetical EU airline already operated a network with one hub located, as an example, coincidentally at Rome-Leonardo da Vinci airport

---

[3] The airport capacity is usually defined as the maximum number of aircraft movements accommodated at the airport during the given period of time (one hour) under given conditions (Janić 2001).

(Italy). Evidently, such a geographical position at the European periphery relative to its 'core' area makes the airline intentions to look for a new-additional hub sensible. Seven airport alternatives are preselected as potential locations for a new hub as follows: Brussels—A_1, Paris (Charles de Gaulle-CDG) — A_2, Frankfurt Main—A_3, Düsseldorf—A_4, Amsterdam Schiphol—A_5, London Heathrow—A_6 and Milan Malpensa—A_7. Six of the above-mentioned airports are located inside and the seventh one at the edge of the core area. These were shown to be the most attractive airports with prospective lucrative markets for the European, both continental and intercontinental, traffic. However, these are also the most congested European airports where the incumbents and their alliance partners (with the exception of Brussels international airport after the collapse of Sabena) have the majority of slots. In general, some evidence indicates that bilateral agreements related to the intercontinental services were the main reasons why the incumbents still strongly held on to these airports as their national hubs (Burghouwt et al. 2002; Panamure WLB 2000). Under such circumstances, setting up a new additional hub at some of these airports would be a very difficult or impossible task. Therefore, the presented numerical example intends to illustrate how the MCDM evaluation of these seven airports could be carried out and test the convenience and consistency of the proposed methods for prospective academic and eventual practical use.

7.3.4.2 Inputs

In order to apply the SAW and TOPSIS methods, the values of relevant attributes are defined for each of the seven preselected alternative airports and given as criteria in Table 7.1, which represents the Decision-Matrix (Hwang and Yoon 1981, Janić and Reggiani 2002).

The first two attributes X_1 and X_2 are 'Population—POP' and 'Per Capita Income—PCI', respectively (EC 1997/1999). The attribute 'Population' for airports Paris (CDG) and London (H) is determined by assigning the total population of a region to the airport proportionally to its share in the total airport traffic of the region. The third attribute X_3, the 'Airport Size—AS' is expressed by the total number of passengers accommodated at a particular airport in 1998 (RBI 1995/1999). The fourth attribute X_4 is the minimum 'Generalized Access Cost—GAC' calculated by using the generalized cost function and data on travel distance, departure frequencies, charges per passenger by the airport surface public systems and the average value of passenger time. The attribute 'Quality of Service at an airport terminal' is not taken into account since its values are assumed to be very similar at select candidate airports. The 'Airline Operating Costs—AOC' are adopted as the fifth attribute X_5. They are calculated for the conditions when one hub is always kept fixed, while another is alternatively chosen from a given set of alternatives. In each case, the airline network is assumed to consist of 20 nodes representing the most famous EU

Table 7.1: Decision-Matrix for the given example—seven alternative airports with nine attributes/criteria (Janić and Reggiani 2002)

Alternative/Airport					Attributes (Criteria)				
Symbol	POP	PCI	AS	GAC	AOC	AAC	AC	MS	UAC
Notation	X_1	X_2	X_3	X_4	X_5	X_6	X_7	X_8	X_9
Sign	+	+	+	-	-	-	+	-	-
A_1-Brussels	1.1	15423	18.5	13.28	1.56	5.16	70	66	77
A_2-Paris (CDG)	6.3	16468	38.6	21.73	1.61	2.71	84	63	74
A_3-Frankfurt	3.6	18308	42.7	8.12	1.62	2.16	72	61	84
A_4-Dusseldorf	3.0	18200	15.8	9.30	2.18	6.62	34	33	79
A_5-Amsterdam	1.1	15111	34.4	8.32	1.65	2.84	90	66	68
A_6-London (H)	4.2*	13293	60.7	21.64	1.68	1.76	78	39	93
A_7-Milan (M)	4.3	15589	13.6	14.47	2.25	7.37	32	64	59

POP – Population of airport catchment area (million) [*] – the modified values according to the share of the airport traffic in the total air traffic of the region]; PCI – Per Capita Income (ECU/inhabitant); AS – Airport size (millions of pax/yr; GAC – Minimum generalized access cost (€/pax); TAC – Total airline cost of operating two-hub and spoke network (million €); AAC – The average airport cost per service (€/WLU); AC – Airport capacity (aircraft/hr); MS – Market share of the incumbent at given airport (%); UC – Utilization of airport capacity during peaks (%); € – EURO; pax – passenger.

airports among which two are the hubs and the rest the spokes. The spokes are assigned to each hub according to the minimum—great circle—distance. Then, the traffic scenario in terms of the volume of passenger inter-airport O-D flows and flight frequencies serving them is set up. The data from 1995 related to 380 main intra-European inter-city one-way passenger flows, flight frequencies, aircraft capacity (size) and the average load factor are sorted out to quantify this scenario (ICAO 1997). The average airline cost per passenger-kilometer is estimated by the cost function given in Table 7.2.

Table 7.2: Models used to determine the airline and airport unit cost per service in a given example (Janić and Reggiani 2002)

The airline unit cost (c)

$c = 6.206 \, (N^*\lambda)^{-0.397} \, L^{-0.344}$

    (3.266) (4.339) (4.733)

$R^2 = 0.896; \, F = 77.477; \, DW = 1.692; \, N = 21$

where c is expressed by €/pax-km; N is seat capacity of an aircraft; λ is the load factor; L is the route length (the adopted average values are: $N = 146$ and $\lambda = 0.65$). The values in the parenthesis below the particular coefficients are t – statistics, which illustrate the relative importance of the particular coefficients for the regression model (Janić 1997).

The cost of airport service (C)

$C = 72.366 \, W^{-0.882}$
$R^2 = 0.561; \, N = 30$

where C is expressed by €/WLU; W is the annual volume of Workload Units (WLU) accommodated at an airport; WLU is an equivalent for one passenger or 100 kg of freight (ACI 1997, Doganis 1992, RBI 1995,1999).

The fixed cost of setting up the new hub is assumed to be the same for each alternative airport, so it is not included in the values of attribute (criteria) X_5. The potential intercontinental traffic at particular airports is not taken into account, since the airline is assumed to first start its business at the EU scale. The values of the 'cost of airport service' attribute/criterion are estimated depending on the annual volume of services accommodated at a given (preselected) airport. This is carried out in two steps—first, the regression model is calibrated by using the appropriate cross-sectional data for 30 European airports (this model is given in Table 7.2); second, the 'Average Airport Cost – AAC' per service is computed by inserting the annual volume of services accommodated at each candidate airport into the regression model as the sixth attribute X_6. The 'Airport Capacity—AC' is the seventh attribute X_7 (EEC 1998). The 'Incumbent's Market Share—MS' is the eighth attribute X_8 determined as the ratio between the number of the incumbent's weekly flights and the number of weekly flights carried out by all other airlines at a given airport (ABC 1998). The 'Utilization of Airport Capacity—UAC' is compiled from various sources and given in Table 7.1 as the ninth attribute X_9 together with the above-mentioned attributes (EEC 1998, RBI 1995/1999, Urbatzka and Wilken 1997). The attributes 'Airport-induced Delay—AD' and the 'Environmental Constraints—EC' are not taken into account due to lack of precise data in the former and similarity of impacts in the latter case.

The attributes/criteria X_1 – POP, X_2 – PCI, X_3 – AS and X_7 – AC in Table 7.1 are considered by the airline as the DM as 'benefit' and the others as the 'cost' attributes/criteria. The 'benefit' criteria are marked by sign "+" and the 'cost' by sign "–".

For the sensitivity analysis, three scenarios are used for assigning the importance (weights) to attributes/criteria: scenario (a) assumes equal weights for particular attributes/criteria, which implies their relative equal importance for the DM; scenario (b) uses the weights generated from the uniform distribution [0, 1] by simulation. The set of random numbers equivalent to the number of attributes/criteria is generated and then the weights calculated by normalization, i.e. by dividing each simulated value by the sum of all generated values in order to provide the sum of weights to be equal to one. As such, this scenario may reflect the preferences of an indecisive DM, as the hypothetical (EU) airline may be at this stage; and scenario (c) the SAW and TOPSIS method use the above-mentioned entropy method and the AHP method uses its own weighting procedure for assigning weights to attributes/ criteria.

7.3.4.3 Results

(a) The SAW and TOPSIS method

Step 1: Calculation of the normalized-decision matrix $R[r_{ij}]$ given below, based upon the Decision-Matrix $A[a_{ij}]$ in Table 7.1 are given in Tables 7.3 and 7.4.

Table 7.3: The SAW method – $R[r_{ij}]$

Alt./ Crit	X_1 +	X_2 +	X_3 +	X_4 -	X_5 -	X_6 -	X_7 +	X_8 -	X_9 -
A_1	0.175	0.842	0.305	0.611	1.000	0.341	0.778	0.500	0.766
A_2	1.000	0.899	0.636	0.374	0.969	0.649	0.933	0.524	0.797
A_3	0.571	1.000	0.703	1.000	0.963	0.815	0.800	0.541	0.702
A_4	0.476	1.000	0.255	0.873	0.716	0.266	0.378	1.000	0.747
A_5	0.175	0.825	0.567	0.976	0.945	0.620	1.000	0.500	0.868
A_6	0.667	0.726	1.000	0.375	0.929	1.000	0.867	0.846	0.634
A_7	0.683	0.851	0.224	0.561	0.693	0.239	0.356	0.516	1.000

Table 7.4: The TOPSIS method – $R[r_{ij}]$

Alt./ Crit.	X_1 +	X_2 +	X_3 +	X_4 -	X_5 -	X_6 -	X_7 +	X_8 -	X_9 -
A_1	0.110	0.361	0.195	0.338	0.325	0.424	0.382	0.434	0.385
A_2	0.624	0.385	0.408	0.552	0.336	0.223	0.459	0.414	0.37
A_3	0.356	0.428	0.451	0.206	0.338	0.178	0.393	0.401	0.42
A_4	0.297	0.427	0.167	0.236	0.454	0.544	0.186	0217	0.395
A_5	0.109	0.353	0.363	0.211	0.344	0.223	0.492	0.434	0.34
A_6	0.416	0.311	0.641	0.55	0.35	0.145	0.426	0.257	0.425
A_7	0.426	0.365	0.144	0.368	0.469	0.606	0.175	0.421	0.295

Step 2: Determination of the relative importance, i.e. weight of particular criteria for the SAW and TOPSIS method given below in Table 7.5 according to scenarios (a), (b) and (c):

Table 7.5: Weights for criteria for the SAW and TOPSIS methods
(Janić and Reggiani 2002)

Weight – w	Attributes (Criteria)								
	X_1 +	X_2 +	X_3 +	X_4 -	X_5 -	X_6 -	X_7 +	X_8 -	X_9 -
Scenario (a)	0.111	0.111	0.111	0.111	0.111	0.111	0.111	0.111	0.111
Scenario (b)	0.066	0.148	0.131	0.087	0.110	0.108	0.089	0.115	0.148
Scenario (c)	0.238	0.010	0.212	0.129	0.020	0.225	0.099	0.050	0.017

As is seen, in scenario (a) the weights are equal, in scenario (b) they are generated by the uniform distribution [0, 1] and in scenario (c) they are calculated by the above-mentioned entropy method. The third group of values indicates that the criteria 'Population' and 'Average cost per airport service' are the most and criteria 'Per Capita Income', 'Incumbent's market share' and 'Utilization of airport capacity' are the least important criteria. This is due to the nature of the entropy method itself, which tends to assign the greatest importance to criteria with the greatest difference in their values.

Step 3: Calculation of the weighted-decision matrix $V[v_{ij}]$:

• SAW – $V[v_{ij}]$

For scenarios (a), (b) and (c), the normalized-weighted matrix V is calculated straightforwardly and the row values corresponding to the particular alternatives summed up. Thus, the overall score for each alternative S_i is obtained.

• .TOPSIS – $V[v_{ij}]$, v^+ *and* v^-

Scenario (a)

The normalized-weighted matrix V is calculated by using the normalized matrix $R[r_{ij}]$ and corresponding weights of criteria for scenario (a) and is given in Table 7.6a.

Table 7.6a: TOPSIS – Scenario (a): The normalized-weighted matrix V

Alt./ Crit.	X_1 +	X_2 +	X_3 +	X_4 -	X_5 -	X_6 -	X_7 +	X_8 -	X_9 -
A_1	0.012	0.040	0.022	0.038	0.036	0.047	0.042	0.048	0.043
A_2	0.069	0.043	0.045	0.061	0.037	0.025	0.051	0.046	0.041
A_3	0.040	0.048	0.050	0.023	0.038	0.020	0.044	0.045	0.047
A_4	0.033	0.047	0.019	0.026	0.050	0.060	0.021	0.024	0.044
A_5	0.012	0.039	0.040	0.023	0.038	0.025	0.055	0.048	0.038
A_6	0.046	0.035	0.071	0.061	0.039	0.016	0.047	0.029	0.047
A_7	0.046	0.041	0.016	0.041	0.052	0.067	0.019	0.047	0.033

The ideal and negative ideal solutions (v^+) and (v^-) are obtained from the matrix (V) by using Equation 7.2 (a, b) and given in Table 7.6b as follows:

Table 7.6b: TOPSIS – Scenario (a): The ideal and negative ideal solutions

Ids/ Crit	X_1 +	X_2 +	X_3 +	X_4 -	X_5 -	X_6 -	X_7 +	X_8 -	X_9 -
v^+	0.069	0.048	0.071	0.023	0.036	0.016	0.055	0.024	0.033
v^-	0.012	0.035	0.016	0.061	0.052	0.067	0.019	0.048	0.047

Then, the Euclidean distance of each alternative to the ideal and negative ideal solution (S_i^*) and (S_i^-), respectively and its closeness to the ideal solution (C_i^*) are calculated by using Equation 7.3.

Scenario (b)

The normalized-weighted matrix V is calculated similarly as scenario (a) by using the corresponding weights of criteria for scenario (b). It is given in Table 7.7a.

The ideal and negative ideal solutions (v^+) and (v^-) are obtained from the matrix (V) by using Equation 7.2 (a, b) and given in Table 7.7b.

Then, the Euclidean distance of each alternative to the ideal and negative ideal solution (S_i^*) and (S_i^-), respectively and its closeness to the ideal solution (C_i^*) are calculated by using Equation 7.3.

Table 7.7a: TOPSIS – Scenario (b): The normalized-weighted matrix V

Alt./Crit.	X_1 +	X_2 +	X_3 +	X_4 -	X_5 -	X_6 -	X_7 +	X_8 -	X_9 -
A_1	0.0073	0.0531	0.0255	0.0291	0.0358	0.0458	0.0340	0.0499	0.0570
A_2	0.0412	0.0566	0.0534	0.0449	0.0370	0.0241	0.0409	0.0476	0.0548
A_3	0.0235	0.0629	0.0591	0.0177	0.0372	0.0192	0.0350	0.0461	0.0622
A_4	0.0196	0.0628	0.0219	0.0203	0.0499	0.0588	0.0166	0.0250	0.0581
A_5	0.0072	0.0519	0.0476	0.0181	0.0378	0.0241	0.0438	0.0499	0.0503
A_6	0.0275	0.0457	0.0840	0.0473	0.0385	0.0157	0.0379	0.0296	0.0629
A_7	0.0281	0.0537	0.0189	0.0316	0.0516	0.0654	0.0156	0.0484	0.0437

Table 7.7b: TOPSIS – Scenario (b): The ideal and negative ideal solutions

Ids/Crit	X_1 +	X_2 +	X_3 +	X_4 -	X_5 -	X_6 -	X_7 +	X_8 -	X_9 -
v^+	0.0412	0.0629	0.0591	0.0177	0.0370	0.0157	0.0438	0.0250	0.0437
v^-	0.0072	0.0457	0.0189	0.0473	0.0516	0.0654	0.0156	0.0499	0.0629

Scenario (c)

The normalized-weighted matrix V given below is calculated similarly as in scenarios (a) and (b) by using the weights of criteria for scenario (c) determined by the entropy method. This is given in Table 7.8a.

The ideal and negative ideal solutions (v^+) and (v^-) are obtained from the matrix V by using Equation 7.2 (a, b) and given in Table 7.8b.

As in scenario cases (a) and (b), the Euclidean distance of each alternative to the ideal and negative ideal solution (S_i^*) and (S_i^-), respectively and its closeness to the ideal solution (C_i^*) is calculated by Equation 7.3.

Step 4: The selection of the best alternative obtained by SAW and TOPSIS in scenarios (a), (b) and (c) is given in Table 7.9.

As can be seen, both methods produce the same results for the given scenario of assigning the weights to criteria. The results are also the same for scenarios (a) and (b), in which both methods rank Frankfurt main airport as the best alternative. In addition, both methods produce the same results in scenario (c), where they rank London Heathrow airport as the preferable (best) alternative. In addition, while ranking other alternatives, the SAW method produces more similar ranks across different scenarios than the TOPSIS method, which may indicate its lesser sensitivity to the changes of procedures (methods) for assigning the weights to criteria. This may be the reason why this method, apart from its simplicity, is frequently used as a benchmarking method.

(b) The AHP method

Inputs: In the scope of AHP, the problem of selection of a new hub is approached according to the diagram shown in Fig. 7.1.

As can be seen, there are three levels. At the first level, the overall objective is established. At the second level, the attributes/criteria are established. At the last level, the airports to be evaluated as alternatives are established. The number of criteria is reduced from nine (at the SAW and TOPSIS) to four. Thus, the criterion 'Market—MAR' includes the sub-criteria 'Population—POP', 'Per Capita Income—PCI' and 'Airport size—AS'. The criterion 'Accessibility—ACC' coincides with the sub-criteria 'Generalized Access Cost—GAC'. The criterion 'Cost—COS' embraces the sub-criteria 'total airline operating costs' and 'average airport cost of service'. Finally, the criterion 'Capacity—CAP' takes into account the sub-criteria 'airport capacity', 'incumbent's market share' and 'Utilization of Airport Capacity - UAC'. The alternatives (i.e. candidate airports) (A_i) (i = 1, 2, . . .,7) are put at the lowest-third level in Fig. 7.1.

Table 7.8a: TOPSIS – Scenario (c): The normalized-weighted matrix V

Alt./Crit.	X_1 +	X_2 +	X_3 +	X_4 -	X_5 -	X_6 -	X_7 +	X_8 -	X_9 -
A_1	0.0260	0.0036	0.0410	0.0440	0.0070	0.0950	0.0380	0.0220	0.0070
A_2	0.1490	0.0039	0.0860	0.0710	0.0070	0.0500	0.0450	0.0210	0.0060
A_3	0.0850	0.0043	0.0960	0.0270	0.0070	0.0400	0.0390	0.0200	0.0070
A_4	0.0710	0.0043	0.0350	0.0300	0.0091	0.1220	0.0180	0.0110	0.0070
A_5	0.0260	0.0045	0.0770	0.0270	0.0070	0.0500	0.0490	0.0220	0.0060
A_6	0.0990	0.0031	0.1360	0.0710	0.0070	0.0330	0.0420	0.0130	0.0070
A_7	0.1014	0.0037	0.0305	0.0475	0.0094	0.1364	0.0173	0.0211	0.0050

Table 7.8b: TOPSIS – Scenario (c): The ideal and negative ideal solutions

Ids/Crit	X_1 +	X_2 +	X_3 +	X_4 -	X_5 -	X_6 -	X_7 +	X_8 -	X_9 -
v^+	0.1490	0.0043	0.1360	0.0270	0.0070	0.0330	0.0490	0.0110	0.0050
v^-	0.0260	0.0031	0.0305	0.0710	0.0094	0.1364	0.0173	0.0220	0.0070

Table 7.9: The SAW and TOPSIS ranking of alternatives in the given example (Janić and Reggiani 2002)

Alternative	The MCDM method												
	SAW						TOPSIS						
	S_i	S_i	S_i	Rank			C_i^*	C_i^*	C_i^*	Rank		
Overall score Scenario	(a)	(b)	(c)	(a)	(b)	(c)	(a)	(b)	(c)	(a)	(b)	(c)
A_1 – Brussels	0.590	0.615	0.405	7	6	7	0.330	0.666	0.245	5	6	7
A_2 – Paris (CDG)	0.783	0.752	0.728	2	3	3	0.616	0.675	0.700	2	2	2
A_3 – Frankfurt	0.788	0.794	0.745	1	1	2	0.643	0.689	0.645	1	1	3
A_4 – Düsseldorf	0.634	0.654	0.464	5	5	5	0.195	0.380	0.286	7	5	6
A_5 – Amsterdam	0.719	0.737	0.593	4	4	4	0.537	0.530	0.516	3	4	4
A_6 – London (H)	0.782	0.792	0.809	3	2	1	0.486	0.649	0.714	4	3	1
A_7 – Milan (M)	0.569	0.589	0.437	6	7	6	0.324	0.230	0.330	6	7	5

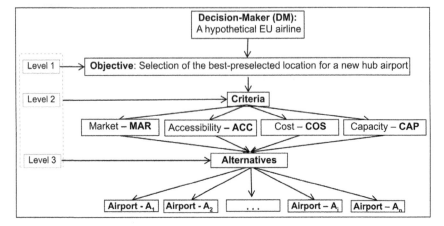

Fig. 7.1: The AHP hierarchical structuring of the MCDM problem in the given example (Janić and Reggiani 2002).

The comparative judgement includes pairwise comparison of the alternatives and criteria at two levels as previously discussed. Since seven alternative airports are evaluated with respect to four criteria, five pairwise comparison matrices of dimension 7×7 containing the judgements on each alternative with respect to each criterion (the first four) are designed. In addition, a pairwise comparison matrix, containing the judgements on each criterion with respect to the overall objective, is designed as the fifth one. The AHP original scale is used to determine the values of these matrices being the authors' choices. The importance, i.e. weights for particular criteria, CI (Consistency Index), RI (Random Index) and checking of consistency of evaluation are also calculated as mentioned above (Saaty 1980, Winston 1994). The two-level evaluation is given below:

Table 7.10: Pairwise comparison of seven alternative airports with respect to four criteria - Level 1

(a) Market – MAR

	A_1	A_2	A_3	A_4	A_5	A_6	A_7	Priority $- v_{i1}$	
A_1	1	1/7	1/3	1/3	1	1/7	½	0.038	
A_2	7	1	5	5	5	3	5	0.381	
A_3	3	1/5	1	2	2	1/5	3	0.109	
A_4	3	1/5	1/2	1	3	1/5	1	0.085	
A_5	1	1/5	1/2	1/3	1	1/5	2	0.058	
A_6	7	1/3	5	5	5	1	5	0.282	
A_7	1	1/5	1/3	1	1/5	1/5	1	0.047	$CI/RI = 0.088/1.32 = 0.067$

(b) Accessibility – ACC

(i)	A₁	A₂	A₃	A₄	A₅	A₆	A₇	Priority - v_{i2}	
A₁	1	3	1/3	1/5	1/3	3	2	0.087	
A₂	1/3	1	1/5	1/5	1/5	1/2	1/3	0.039	
A₃	3	5	1	2	2	5	3	0.284	
A₄	5	5	1/2	1	½	5	3	0.207	
A₅	5	5	1/2	2	1	5	3	0.248	
A₆	1/3	2	1/5	1/5	1/5	1	1/3	0.046	
A₇	½	3	1/3	1/3	1/3	3	1	0.089	$CI/RI = 0.089/1.32 = 0.068$

(c) Cost – COS

(i)	A₁	A₂	A₃	A₄	A₅	A₆	A₇	Priority - v_{i3}	
A₁	1	1/6	1/7	5	1/7	1/9	7	0.065	
A₂	6	1	1/3	7	3	1/3	7	0.161	
A₃	7	3	1	7	5	1/2	8	0.259	
A₄	1/5	1/7	1/7	1	1/8	1/6	3	0.034	
A₅	7	1/3	1/5	8	1	1/5	9	0.132	
A₆	9	3	2	6	5	1	7	0.327	
A₇	1/7	1/7	1/8	1/3	1/9	1/7	1	0.022	$CI/RI = 0.093/1.32 = 0.070$

(d) Capacity – CAP

(i)	A₁	A₂	A₃	A₄	A₅	A₆	A₇	Priority - v_{i4}	
A₁	1	1/2	2	2	1/3	5	1/5	0.095	
A₂	2	1	3	2	1/3	5	4	0.125	
A₃	½	1/3	1	2	1/5	5	1/7	0.072	
A₄	½	1/2	1/2	1	1/3	5	5	0.067	
A₅	3	3	5	3	1	5	1/3	0.214	
A₆	1/5	1/5	1/5	1/5	1/6	1	1/9	0.025	
A₇	5	4	7	6	3	9	1	0.402	$CI/RI = 0.075/1.32 = 0.057$

For example, the element $a_{15} = 1$ the matrix (a) indicates that the criterion 'Market' is equally important at Brussels-International and Amsterdam Schiphol airport, i.e. these two markets are considered approximately equivalent for the DM. The element: $a_{21} = 7$ indicates that the criteria 'Market' is about (approximately) seven times more important at Paris Charles de

Gaulle than at Brussels International airport. The element: $a_{35} = 2$ indicates that Frankfurt Main Airport as a 'Market' is considered about twice more important than Amsterdam Schiphol Airport. Similar explanations apply to other candidate airports and criteria in other Decision Matrices given above.

The matrix of the criteria comparison is composed in Table 7.11.

Table 7.11: Pairwise comparison of four criteria with respect to the overall objective – Level 2

	MAR	ACC	COS	CAP	Priority Weights - w_j	
MAR	1	4	½	2	0.275	
ACC	1/4	1	¼	1/4	0.076	
COS	2	4	1	4	0.473	
CAP	½	4	¼	1	0.176	$CI/RI = 0.074/0.90 = 0.082$

As can be seen, the criterion 'market' is considered to be about four times more important than the criterion 'accessibility' and twice more important than the criterion 'capacity'. The criterion 'cost' is considered about three times more important than the criterion 'market' and approximately four times more important than the criterion 'accessibility'. The criterion 'cost' is considered to be about twice more important than the criterion 'capacity'. Finally, the criterion 'capacity' is assumed to be about four times more important than the criterion 'accessibility'. Consequently, it can be seen that the proposed weighting by using the AHP scale may look like a judgement of the LCC.

The vectors of priorities for particular alternatives with respect to particular criteria (v_{ij}) ($i = 1$-7; $j = 1$-4) and the weights of particular criteria (w_j) ($j = 1$-4) for scenarios (a), (b) and (c) are synthesized and given in Table 7.12.

Results: The synthesis of priorities is carried out by calculating the overall score (S_i) for each alternative (airport) by using the last two above-mentioned synthetic matrices and Equation 7.5. The ranking of alternatives is shown in Table 7.13.

As is seen, the AHP, similarly as the SAW and TOPSIS, produces different results when it uses different methods for assigning the weights to criteria. This illustrates the sensitivity of the method to such types of changes. However, for the corresponding scenarios, the same preferable alternative is chosen as in the case of using the SAW and TOPSIS method. It is Frankfurt main airport in scenarios (a) and (b) and London's Heathrow airport in scenario (c). The results may illustrate an inherent consistency of this with the other two methods (SAW and TOPSIS) and vice versa.

Table 7.12: The vectors of priorities for particular alternatives with respect to particular criteria

$(i)/(j)$	A_1	A_2	A_3	A_4	A_5	A_6	A_7	Priority – Weights – w_j		
								Scenario (a)	Scenario (b)	Scenario (c)
MAR	0.038	0.381	0.109	0.085	0.058	0.282	0.047	0.250	0.220	0.275
ACC	0.087	0.039	0.284	0.207	0.248	0.046	0.089	0.250	0.226	0.076
COS	0.065	0.161	0.259	0.034	0.132	0.327	0.022	0.250	0.280	0.473
CAP	0.095	0.125	0.072	0.067	0.214	0.025	0.402	0.250	0.274	0.176

Table 7.13: The AHP ranking of alternatives in the given example (Janić and Reggiani, 2002)

Alternative/Airport	Scenario (a)		Scenario (b)		Scenario (c)	
	Score – S_i	Rank	Score – S_i	Rank	Score – S_i	Rank
A_1 – Brussels	0.071	7	0.072	7	0.065	7
A_2 - Paris (CDG)	0.177	2	0.172	2	0.206	2
A_3 – Frankfurt	0.181	1	0.180	1	0.187	3
A_4 – Dusseldorf	0.098	6	0.093	6	0.067	6
A_5 – Amsterdam	0.163	4	0.164	4	0.135	4
A_6 – London (H)	0.170	3	0.171	3	0.240	1
A_7 – Milan (M)	0.140	5	0.147	5	0.101	5

(c) Comparison of the results from different studies

The summary of outcomes from different studies related to the problem of selection of a new hub airport for a hypothetical European airline is summarized in Table 7.14.

As can be seen, the outputs are different when different single or multiple criteria methods are applied. The results from different single criterion methods are different, depending on the objective function used for evaluation. The selected multi-criteria methods produce the same results if the same procedure (method) for assigning weights to criteria is used. For example, according to scenario (a) in which equal weights are assigned to criteria and scenario (b) in which the weights are generated from uniform distribution [0, 1] by simulation, all three methods rank the same alternative as the preferable, i.e. Frankfurt Main Airport (A_3). In scenario (c), when the entropy method is used to assign weights to criteria, all three methods rank the same alternative as the preferable one, which, however, is different than in scenarios (a) and (b). This alternative is London Heathrow airport (A_6). In each scenario, three methods give the same results despite the fact that the number of criteria at the SAW and TOPSIS method on the one hand and the AHP method, on the other, is different. This indicates consistency and integrity of the selected methods for such applications. It also indicates that the methods for assigning the weights to criteria rather than the MCDM method are of the crucial importance for the results, which points out the importance of choosing the proper method.

The preferable airport in scenarios (a) and (b) is Frankfurt main airport. This airport appears to be most attractive due to its relatively high potential 'strength' in generating air transport demand, modest generalized airport access cost, modest total airline costs, relatively low airport cost per service, relatively high airport capacity and a reasonably high level of utilization of this capacity.

The preferable airport in scenario (c) is London's Heathrow airport. It appears most attractive due to its specificity in comparison to other airports, which is highlighted by use of the entropy method for assigning the weights to criteria. This specificity is visible through the size of the potential market and the size of the airport itself, reasonable airline cost to incorporate the airport into its 'renovated' hub-and-spoke network, relatively low costs of airport service and the relatively modest domination of the incumbent. The disadvantages in terms of higher generalized access costs and the relatively high utilization of the airport capacity are shown to be less relevant.

7.3.5 Interim Summary

This section deals with developing the methodology for evaluation of the candidate/alternative airports as the potential locations for the new hub of

Table 7.14: Comparison of the results from different studies (Janić and Reggiani 2002)

Alternative/Airport	Single criterion				Multi-criteria[1]								
		Methodology											
	Minimum cost[1]	Maximum profits[2]			SAW			TOPSIS (Scenario)			AHP		
		BR	PT1	PT2	(a)	(b)	(c)	(a)	(b)	(c)	(a)	(b)	(c)
A₁ - Brussels	1	4	4	4	7	6	7	5	6	7	7	7	7
A₂ - Paris (CDG)	2	5	5	–	2	3	3	2	2	2	2	3	2
A₃ - Frankfurt (M)	3	3	–	–	**1**	**1**	2	**1**	**1**	3	**1**	**1**	3
A₄ – Düsseldorf	6	2	–	–	5	5	5	7	5	5	6	6	6
A₅ – Amsterdam	4	**1**	**1**	**1**	4	4	4	3	4	4	4	4	4
A₆ - London (H)	5	–	–	–	3	2	**1**	4	3	**1**	3	2	**1**
A₇ - Milan (M)	7	–	–	–	6	7	6	6	7	6	5	5	5

[1] The author's calculations; [2] Berechman and de Wit 1996; Simulation runs: BR – Base Run; PT1 – Policy Test 1; PT2 – Policy Test 2.

a European (EU) airline. This emerged as an opportunity after liberalization of the EU air transport market. The methodology consists of models of the relevant indicators and measures of performances of the candidate/alternative airports considered by the airline, the MCDM (Multi-Criteria Decision-Making)—SAW (Simple Additive Weighting), TOPSIS (Technique for Order Preference by Similarity to the Ideal Solution) and AHP (Analytic Hierarchy Process) methods and scenarios of assigning the weights to the evaluation attributes of the airport performances used as evaluation criteria in the selection process. The methodology was applied to seven European airports as alternatives with nine relevant attributes of their performances. The results revealed the following: (i) three MCDM methods produced the same results under conditions when the same method of estimating the weights for criteria was used; (ii) at the same MCDM method, the weights for criteria obtained by different methods produced both the same and different results. This indicates that the weights for criteria rather than the MCDM method should be considered more carefully while dealing with this and similar MCDM problems; and (iii) consequently, due to such inherent sensitivity, the chosen discrete MCDM methods could be recommended for some ultimately initial decision(s) and not for the final.

7.4 New Runway for an Airport System

7.4.1 Background

An airport system consists of a few airports serving large volumes of commercial air-transport demand generated and attracted by a large metropolitan area.[4] Thanks to the airlines they host, particular airports of the system can compete and/or cooperate with each other for air passenger and cargo demand and other (more or less related) services. This usually leads to a relatively high concentration of large volumes of relatively stable air transport demand at one airport, which becomes the primary and rather lower volumes of inherently volatile demand at other airports, which become secondary ones. Consequently, these airports can be distinguished by the dominant airline network(s), the number and types of destinations/routes, flight frequencies per destination/route, etc. In addition, the accessibility of these airports to/from the metropolitan area can be different regarding the distance, number and type of ground transport modes deployed (De Neufville and Odoni 2003). One typical example in Europe and the largest in the world in terms of the

---

[4] A metropolitan area is considered as a relatively wide region consisting of a densely populated urban core and its less-populated surrounding territories sharing industry, infrastructure and housing (http://en.wikipedia.org/wiki/Metropolitan_area).

annual number of passengers is the London airport system (United Kingdom), which includes Heathrow (as the primary) and Gatwick, Stansted, London-City, Luton and Southend (as secondary) airports (handled about $135*10^6$ passengers in 2012). The second in the world's top 50 is the US (United States) New York airport system including John F. Kennedy, LaGuardia and Newark Liberty International (as the primary) and Westchester County, Long Island and Stewart (as the secondary) airports (handled about $112*10^6$ passengers in 2012).

One of the most important and persistent problems facing the airport systems worldwide and particularly the first one mentioned above have for a long time been matching their capacity to the generally growing air transport (passenger and freight/cargo) demand.

This section describes an application of the MCDM (Multi Criteria Decision Making) methods to the evaluation of solutions and alternatives for matching the airside (runway) system capacity to the demand at given airport system—London (UK). In such a context, 'building a new runway' is considered as the solution and candidate airports of the system as the alternatives. These alternative airports are characterized by their physical/spatial, operational, economic, environmental and social performances represented by the corresponding indicator systems, which, after being assessed according to the given operating scenario(s), are used as evaluation attributes/criteria by selected MCDM methods (Janić 2015).

7.4.2 The System and Problem—Matching Capacity to Demand at an Airport System

The available options for matching the airport system capacity to demand include courses of actions generally consisting of (i) increasing capacity, (ii) managing air transport demand, (iii) switching part of the current and prospective demand to the other transport modes and (iv) combinations of all the previous options.

7.4.2.1 Increasing Capacity

This includes solutions such as (1) building a new airport(s); (2) building new airside and landside infrastructure at one or several existing airports of the system; (3) upgrading the operating modes of the existing airside and landside infrastructure; (4) introducing innovative technologies and operational procedures and thus increasing the airside and landside capacity; and (5) combinations thereof.

7.4.2.2 Managing Air Transport Demand

This usually implies solutions in terms of imposing different operational, economic, environmental and social constraints on the demand access exclusively and/or in various combinations.

7.4.2.3 Switching Part of the Current and Prospective Demand to Other Transport Modes

This implies solutions directed at other transport modes to take over demand from air transport through complementarity and/or competition at the particular airport(s) of the airport system. In Europe, this mainly occurs at the primary airport(s) included/connected to the conventional and HSR (High Speed Rail) network (Janić 2010).

7.4.2.4 Combination of Previous Options

This implies combining the above-mentioned options and their solutions at the level of the airport system and its particular airports. The alternatives could be airports of the given airport system where particular above-mentioned options are applied and their solutions implemented individually and/or simultaneously in different combinations. Identifying the preferred solution(s) within the particular options and alternatives in the given context is inherently, at least for researchers, a MCDM (Multi-Criteria Decision Making) problem. This is because, once an option is specified, its solutions and alternatives characterized by the indicator systems of their performances are usually considered by the DM (Decision Maker) as the evaluation attributes/criteria. Most often, the DM is a single high-level body balancing the interests and preferences of particular actors/stakeholders involved. In the given case, they can be providers of air transport services (airports and airlines), local, regional and central authorities/governments, local community members, users of air transport services (air passengers and freight/cargo shippers), etc. During the process, the indicator systems of performances of the considered alternative/candidate airports considering the specified solution(s) are estimated and assigned weights reflecting their relative importance for the DM. As such, they become evaluation attributes/criteria, enabling application of the MCDM methods—from expert judgment to analytical methods.

7.4.3 Evaluation of Solutions for Matching Capacity to Demand at an Airport System

7.4.3.1 Some Related Research

Different single- and multiple-objective evaluation methods can be applied to selecting the preferred solution(s) and alternative(s) for matching capacity to demand at the given airport system. This mainly depends on the way of expressing the evaluation attributes/criteria of their performances, the number of actors/stakeholders involved and the relative importance of the former over the latter. In general, the most commonly used methods include the above-mentioned EAT (Economic Analysis Technique) or BAU (Business As Usual) and different MCDM (Multi Criteria Decision Making) or MCA (Multi Criteria Analysis) methods (EC 1998a[5], Giuliano 1985, Tabucanon and Mo-Lee 1995).

Some academic-research and professional and practical successful applications of the MCDM methods include the above-mentioned SAW, TOPSIS and AHP method (Hwang and Yoon 1981, Janić 2014, Sauian 2010). Specifically, academic efforts have been made in applying MCDM methods to evaluation of transport infrastructure projects, including, airport expansion plans, individually or in combination with the EAT and BAU methods (Brucker et al. 2011, Schutte and Brits 2012, Vreeker et al. 2002). In addition, MCDM methods have been applied to evaluate airline service quality (Tsaura et al. 2002), airline competitiveness (Lee et al. 2003) and to select an aircraft for a given airline (Ozdemir et al. 2011). In addition, the SAW, TOPSIS and AHP are used to evaluate innovative freight bundling networks in Europe (Janić et al. 1999), the just elaborated selection of the new hub airport for a European airline (Janić and Reggiani 2002) and the HS (High Speed) transport technologies (Janić 2014). This research, and in particular the case of locating new additional runway within a given airport system, represents an additional illustrative example of the potential use of the MCDM methods (Janić 2015).

7.4.3.2 Objectives and Assumptions

The main objective of this section is to elaborate the methodology based on the use of selected MCDM methods for choosing the preferable alternative solution for matching the airport system capacity to demand under the given conditions. In the given context, this is an airport in a given airport system where 'building a new runway' represents a solution for the medium- to long-term matching capacity to demand. The methodology is based on the following assumptions:

- The airport system consists of a few airports serving the air transport demand of the large metropolitan area.
- The option for matching the airside (runway) capacity to demand is *a priori* considered to be 'increasing capacity' by the solution 'building a new runway' at one of the airports of the system as alternatives.

---

[5] In particular, the EC COST 328 Action proposed the MCA (Multi-Criteria Analysis) as a useful and convenient method for evaluating the projects in the scope of the TENs [Trans-European Transport Network(s)] (EC 1998a).

- The solution influences the physical/spatial or infrastructural, operational, economic, environmental and social performances of particular alternative airports of the given airport system (the policy performances are implicitly contained in the proposed methodology).
- The performances of particular alternative/candidate airports can be expressed by the indicator systems, which, after being assigned the weights expressing their relative importance for the DM (Decision Maker), become the evaluation attributes/criteria.
- The indicator systems of performances containing indicators and measures, which taken as evaluation attributes/criteria express the 'benefits' or 'costs' of the particular alternatives.
- Balancing the airside and landside capacity at the airport system and particular airports after implementation of the preferred alternative is implied.

7.4.3.3 The Evaluation Methodology

(a) The indicator systems for performances as evaluation attributes/criteria: The indicator systems for performances as the evaluation attributes/criteria of particular alternatives of the solution—'building a new runway'—for matching capacity to demand at the given airport system are assumed to reflect the aspects of the above-mentioned particular actors/stakeholders involved in the DM (Decision Making) process. Therefore, they are categorized into physical/spatial or infrastructural, operational, economic, environmental and social indicator systems of performances as given in Table 7.15 (Janić 2015).

Table 7.15: Indicator systems of performances as the evaluation attributes/criteria for the solution 'building the new runway' at the candidate/alternative airport(s) (Janić 2015)

Performances	Indicator/Measure	Dimension	Preferred sign
Physical/Spatial or Infrastructural			
Existing infrastructure	• Number of runways	Counts	+
	• Number of apron/gate stands	Counts	+
		Counts	+
	• Number of passenger terminals		
Convenience of location	• Distance by rail/road	km	-
	• Travel time (by rail)	min	-
	• Access time to/from the catchment area	min	-

(Contd.)

Table 7.15: (*Contd.*)

Operational			
Increase in the total capacity	• Relative contribution		
	○ Atms[1]	%	+
	○ Passengers	%	+
Utilization of the total capacity	• Demand/capacity ratio		
	○ Atms	%	+
	○ Passengers	%	+
Attractiveness for airlines	• Number of currently operating airlines	Counts	+
Attractiveness for air passengers	• Total number of destinations	Counts	+
	• Number of international destinations	Counts	+
Economic			
Efficiency and effectiveness of investments	• Cost	£	-
	• Profits	£	+
	• Effectiveness (profits/ investments ratio)	%	+
Contribution to the welfare	• Employment	Counts	+
	• Economy	£	+
Effects/impacts from (non)-accommodated demand	• Losses from non-accommodated demand	£	-
	• Gains from accommodated (switched) demand	£	+
Environmental			
Fuel consumption and emissions of GHG by LTO[2] cycles on the new runways (CO_{2e})	• Cumulative amounts on the new runway	ton	-
Land use	• Total area of the occupied land	ha	-
	• Relative increase in the total occupied land	(%)	-
	• Land use intensity	Atm/ha/year	+
Social			
Noise	• Factor of additional noise	Count Atm/ affected persons	-
	• Noise efficiency		-
Safety	• Third party risk	Exposed persons/Atm	-

[1]Atm – air transport movement (1 Atm is equivalent to one landing or one take-off);
[2]LTO – Landing-Take-Off

As can be seen, the particular indicators and measures express the existing performances of the alternative/candidate airport(s) for the new runway, while some others reflect the potential performances of these during and/or after implementation of the new runway. Since the above indicators and measures are expressed in quantitative terms, the sign (+) indicates their positive (i.e. 'benefit') and the sign (-) negative (i.e. 'cost') preference for the DM. Consequently, the former sign implies the highest possible and the latter sign just the opposite, i.e. the lowest possible preferred value for the given indicator and/or measure for the DM while using it as evaluation attribute/criterion.[6]

Physical/spatial or infrastructural performances

The indicator system of physical/spatial or infrastructural performances includes the indicators and measures as follows:

'Existing infrastructure' is measured by the number of runways, aircraft apron/gate parking stands and passenger terminals at the alternative/candidate airports, thus reflecting their current size. It is assumed that a larger airport(s) is inherently under higher pressure for increasing capacity [thus, the following three measures are preferred to be as high as possible—sign (+)]:

- N_R is the number of existing runways;
- N_A is the number of existing apron/gate aircraft parking stands; and
- N_T is the number of existing passenger terminals.

'Convenience of location' has three measures: travel distance (road/rail), travel time (by rail) between the alternative/candidate airport(s) and the ultimate center of the metropolitan area and the average actual airport access time from/to the catchment area. It is assumed that it is more convenient if the airport is generally closer and thus more efficiently and effectively accessible from its catchment area [thus these are preferred to be as low/short as possible—sign (–)].

The travel time by rail/bus between the airport and the ultimate center of the metropolitan area can be estimated as follows:

$$t(D) = 1/2T/F(T) + D/v(D) \tag{7.9}$$

where

D is the travel distance between the airport and the ultimate center of the metropolitan area (by road/rail) (km).

T is the time interval in which rail/bus services are offered (hr of d);

---

[6] An indicator or measure becomes attribute/ criterion when the DM emphasizes its relative importance, i.e. assigns the relative weight to it in the given context.

$F(T)$ is the transport service frequency during time interval (T); and
$v(D)$ is the average speed of the rail/bus services on the distance (D) (km/h).

The first term of Equation 7.9 indicates the schedule delay and the second term, the in-vehicle time of the given rail/bus service.

Operational performances

The indicator system of operational performances consists of the following indicators and measures:

'*Increase in the total capacity*' is measured by the relative contribution of capacity of the new runway to the existing airside (runway) and associated landside (passenger terminal) capacity at the alternative/candidate airport(s). It is assumed that the new runway should increase the corresponding and enable increase in the passenger terminal capacity [thus both measures are preferred to be as high as possible—sign (+)].

Relative increase in the runway system capacity can be estimated as follows:

$$\Delta C_{RW}(\tau_3) = 1 + c_{rw}(\tau_3)/C_{RW}(\tau_1) \tag{7.10a}$$

Relative increase in the passenger terminal capacity can be estimated as follows:

$$\Delta C_{PS}(\tau_3) = 1 + c_{ps}(\tau_3)/C_{PS}(\tau_1) \tag{7.10b}$$

where

τ_i is the sub-period of the observed period of time ($i = 1$ before, $i = 2$ during and $i = 3$, after implementation of the new runway) (TU; TU – Time Unit);

$c_{rw}(\tau_3), c_{ps}(\tau_3)$ is the additional runway system and passenger terminal capacity, respectively, by implementing the new runway (Atm/hr and pass/hr); and

$C_{RW}(\tau_1), C_{PS}(\tau_2)$ is the existing runway system and passenger terminal capacity, respectively, before implementation of the new runway (Atm/hr and pass/hr).

'*Utilization of the total capacity*' is measured by the demand/capacity ratio for both Atm and passengers after implementation of the new runway given the scenarios of developing corresponding demand at the alternative/candidate airports. It is assumed that despite the increase, the upgraded capacity of the new runway should be used as much as possible [thus, this is preferably as high as possible—sign (+)].

Utilization of the runway capacity can be estimated as follows:

$$u_{RW}(\tau_3) = d_{rw}(\tau_3)/[c_{rw}(\tau_3) + C_{RW}(\tau_1)] \tag{7.11a}$$

Utilization of the passenger terminal capacity can be estimated as follows:

$$u_{PS}(\tau_3) = d_{ps}(\tau_3)/[c_{ps}(\tau_3) + C_{PS}(\tau_1)] \qquad (7.11b)$$

where

$d_{rw}(\tau_3), d_{ps}(\tau_3)$ is the demand, respectively, after implementation of the new runway (Atm/TU, pass/TU, respectively).

The other symbols are analogous to those in Equation 7.10 (a, b).

'Attractiveness for airlines' is measured by the number of airlines currently operating at the alternative/candidate airport(s). It is assumed that higher number of airlines is more beneficial to the airport(s) due to guaranteeing higher and inherently more stable demand. This can be achieved by providing sufficient (runway) capacity [thus, this is preferably as high as possible—sign (+)].

n_a is the number of airlines currently operating at the candidate/alternative airport(s).

'Attractiveness for air passengers' is measured by the number and diversity of destinations (domestic, international, continental, intercontinental) currently offered at the alternative/candidate airport(s). It is assumed that the additional capacity can contribute to further increase in number and diversity of destinations and flight frequencies, thus offering better choice [accordingly, both measures are preferred to be as high as possible—sign (+)].

n_d, n_{dd} is the current number of destinations and the number of intercontinental destinations at the candidate/alternative airport(s), respectively.

Economic performances

The indicator system of economic performances includes indicators and measures as follows:

'Efficiency and effectiveness of investments' is measured by the cost of investments, profits [i.e. for the alternative/candidate airport(s)] from the investments and the effectiveness (profitability) of the investments (the ratio between the candidate airport profits and the cost of investments) during the sub-period after implementation of the new runway. It is assumed that the total investment costs of building the new runway should be as low as possible and the profits and effectiveness of the investments as high [thus, the former measure is preferably as low as possible sign (–), while the latter two measures are preferably as high as possible—sign (+)].

- $I(\tau_2)$ are the investments in the new runway at the candidate airport(s) during sub-period (τ_2) (£); and

- $P(\tau_3)$ are the profits from the investments for the candidate airport during the sub-period (τ_3) after implementation of the new runway (£/TU).

Effectiveness of the investments at the candidate airport(s) during/after implementation of the new runway can be estimated as:

$$e_i(\tau_3) = P(\tau_3)/I(\tau_2) \tag{7.12}$$

'Contribution to welfare' is measured by direct, indirect and induced employment and the contribution to economy; the latter in terms of the total earnings by the aviation and non-aviation activities locally, regionally and nationally at the end and during the sub-period after implementation of the new runway, respectively. It is assumed that the increase in capacity enables traffic growth and consequently generates new employment, which in addition to the other effects contributes to the economy and consequently overall welfare [thus, both measures are preferably as high as possible—sign (+)].
Employment can be estimated as follows:

$$E[\Delta d_{ps}(\tau_3)] = a_0 + a_1[d_{ps}(\tau_2) + \Delta d_{ps}(\tau_3)] \tag{7.13a}$$

where
a_0, a_1 are the coefficients of the regression equation; and
$\Delta d_{ps}(\tau_3)$ is the passenger demand accommodated at the candidate airport(s) after implementation of the new runway (pass/TU).

Economy (contribution to the welfare) can be estimated as follows:

$$w(\tau_3) = d(\tau_3) * p(\tau_3) \tag{7.13b}$$

where
$d(\tau_3)$ is the number of units of demand accommodated at the candidate airport after implementation of the new runway (Atm/TU or pass/TU);
$p(\tau_3)$ is the average profit for society by accommodating an additional unit of demand at the candidate airport after implementation of the new runway (£/TU).

'Effects/impacts from non-accommodated demand' are measured by the cumulative loss of profits of the alternative/candidate airport(s) due to their lack of capacity to accommodate prospective demand on the one hand and the cumulative gains of profits of those airports able to accommodate their own demand as well as that of other airports thanks to their free capacity, on the other—both during and after implementation of the new runway. It is assumed that the existing and prospective demand can spill over from the airport system, which is negative, or switch to the airport(s) with spare capacity and

thus remain within the system, which is positive. Still excessive demand is assumed to spill over from the airport system which is again negative [thus, in the former and the last case the losses are preferably as low as possible—sign (−); in the second case, the gains are preferably as high as possible—sign (+)].

Losses from non-accommodated demand at the candidate airport(s) during and after implementation of the new runway (period $\tau_2 + \tau_3$) (due to the switch to the other airports of the system with free capacity) can be estimated as follows:

$$L_{km} = \begin{bmatrix} (d_{km} - c_{km}) * p_{km}, & \text{if } \quad d_{km} > c_{km} \\ 0, & \text{otherwise} \end{bmatrix} \tag{7.14a}$$

Gains for the airport(s) with free capacity from accommodated demand switched from the candidate airport(s) of the same airport system with shortage of capacity during and after implementation of the new runway (period $\tau_{2+\tau_3}$) can be estimated as follows:

$$G_{lm} = \begin{bmatrix} \sum_{k=1}^{K} q_{klm}(d_{km} - c_{km}) * p_{lkm}, \text{if } d_{lm} + \sum_{k=1}^{K} q_{klm}(d_{km} - c_{km}) \le c_{lm} \\ (c_{lm} - d_{lm}) * \sum_{k=1}^{K} q_{klm} p_{lkm} - \left\{ d_{lm} + \sum_{k=1}^{K} q_{klm}[d_{km} - c_{km}]p_{lkm} \right\}, \text{if } d_{lm} \\ + \sum_{k=1}^{K} q_{klm}(d_{km} - c_{km}) > c_{lm} \end{bmatrix}$$

$$\tag{7.14b}$$

where

k is the candidate airport(s) from which demand switches to other (free capacity) airports ($k = 1, 2,..., K$);

l is the airport with free capacity to which demand switches from the candidate airport(s) ($l = 1, 2,..., L$);

m is the time unit of the observed period in which the demand switches between airports ($m = 1, 2,..., M$);

d_{km} is the demand at the candidate airport (k) during the time unit (m) of the observed period (Atm/yr or pass/yr);

c_{km} is the capacity of the candidate airport (k) during the time unit (m) of the observed period (Atm/hr or pass/hr);

p_{km} is the profit from the unit of demand to be accommodated at the candidate airport (k) during the time unit (m) of the observed period (£/Atm-yr or £/pass-yr);

q_{klm} is the proportion of demand switching from the candidate airport *(k)* to the airport (*l*) during the time unit (*m*) of the observed period;

p_{lkm} is the profit of the airport (l) from accommodating the unit of demand switched from the candidate airport (k) during the time unit (m) of the observed period (£/yr);

c_{lm} is the capacity of the airport (l) during the time unit (m) of the observed period (Atm/yr or pass/yr); and

d_{lm} is the original demand at the airport (l) during the time unit (m) of the observed period (Atm/yr or pass/yr).

Environmental performances

The indicator system of environmental performances consists of the following indicators and measures:

'Fuel consumption and emissions of GHG' is measured by the cumulative amount of consumed fuel and corresponding emissions of GHG (CO_{2e} or Carbon Dioxide Equivalents), respectively, during the LTO (Landing-and-Take-Off) cycles carried out on the new runway during the sub-period after its implementation. It is assumed that despite the additional capacity by the new runway to accommodate more demand, the corresponding above-mentioned impacts are preferably as low as possible [thus, this is preferably as low as possible—sign (–)].

Fuel consumption and emissions of GHG during LTO cycles after implementation of the new runway at the candidate airport(s) can be estimated, respectively, as follows:

$$F_m = \sum_{j=1}^{J} p_{mj} * \frac{N_m}{2} * f_{mj} \qquad (7.15a)$$

$$E_m = \sum_{j=1}^{J} p_{mj} * \frac{N_m}{2} * f_{mj}\, e_{mj} \qquad (7.15b)$$

where

m is the time unit of the period (τ_3) after implementation of the new runway at the candidate airport (m = 1, 2,..., M) (yr);

j is the aircraft category operating on the new runway (l = 1, 2,..., J);

p_{mj} is proportion of the aircraft category (j) operating on the new runway during the time unit (m) of the period (τ_3) after its implementation;

N_m is the number of Atm carried out on the new runway during the time unit (m) of the period (τ_3) after its implementation (Atm/yr);

f_{mj} is the average fuel consumption of the aircraft category (j) during an LTO cycle carried out on the new runway in the time unit (m) of the period (τ_3) after its implementation (ton/yr); and

e_{mj} is the average emission rate of GHG per unit of fuel consumed by the aircraft category (j) during a LTO cycle carried out on the new runway in the time unit (m) of the period (τ_3) after its implementation (ton/yr).

'*Land use*' is measured by the area of currently occupied land, the relative increase in this currently occupied land and the intensity of land use after implementation of the new runway at the alternative/candidate airport. It is assumed that the area of currently and additionally taken land should be as small as possible while the intensity of its use should be as high as possible [thus the former two measures are preferably as low/less as possible—sign (–); the last measure should be as high as possible—sign (+)].

The relative increase in the occupied land after implementation of the new runway can be estimated as follows:

$$\Delta A(\tau_3) = 1 + a_r(\tau_3) / A_C(\tau_1) \tag{7.16a}$$

where

$A_C(\tau_1)$ is the area of currently occupied land by the candidate alternative(s)/ airport(s) before implementation of the new runway (ha); and

$a_r(\tau_3)$ is the land occupied by the new runway (ha).

The intensity of land use after implementation of the new runway at the candidate airport(s) can be estimated as follows:

$$U_l(\tau_3) = d_R(\tau_3) / [A_C(\tau_1) + a_r(\tau_3)] \tag{7.16b}$$

where

$d_R(\tau_3)$ is the demand accommodated at the candidate airport(s) after implementation of the new runway (Atm/TU).

Social performances

The indicator system of social performances includes the following indicators and measures:

'*Noise*' is measured by the additional noise and noise efficiency, both from Atm (Air transport movement(s)) carried out on the new runway during the sub-period after its implementation at the alternative/candidate airport. The former is expressed as the relative contribution to the total noise, while the latter is expressed as the ratio of the number of Atm carried out and the number of affected persons within the given noise contour (usually $57L_{eq}$) under given conditions during the observed period of time. It is assumed that despite generation of noise by Atm around the new runway, the increase in the total noise burden and the number of additionally affected local people should be as low as possible {thus, both measures are preferably as low as possible—sign (–)}:

The factor of increasing noise due to daily operations on the new runway after its implementation can be estimated as follows:

$$\Delta dB(\tau_3) = 10 \log_{10} n(\tau_3) \tag{7.17a}$$

where

$n(\tau_3)$ is the number Atm carried out on the new runway after its implementation at the alternative/candidate airport(s) (Atm/d).

The noise efficiency can be estimated as follows:

$$N_{eff/m} = n_m / POP_m(57L_{eq}) \tag{7.17b}$$

where

n_m is the number of Atm carried out on the new runway during the time unit (m) of the observed period (τ_3) after its implementation at the alternative/candidate airport(s) (Atm/yr); and

$POP_m(57L_{eq})$ is the number of population within the given noise contour in the time unit (m) of the observed period (τ_3), after implementation of the new runway at the alternative/candidate airport(s) (pop/yr).

'*Safety*' is measured by the third party risk as the number of incidents/accidents per Atm and an exposed person living closeby to the new runway. It is assumed that the number of these incidents/accidents should not differ (or should be even lower) than that of the existing ones, including the number of potentially exposed local population [thus, this is preferred to be as low as possible—(sign (–)].

Third party risk can be estimated as follows:

$$r_{am} = n_{am} / POP_m(57L_{eq}) \tag{7.18}$$

where

n_{am} is the number of perceived incidents/accidents on the new runway during the time unit (m) of the observed period (τ_3), after its implementation at the candidate alternative(s)/airport(s) (events/Atm/yr).

The other symbols are analogous to those in the previous equations.

(b) The MCDM methods

Two above-mentioned MCDM methods—SAW and TOPSIS—are chosen to deal with the problem of selection of the alternative airport in the given airport system where a new runway would be built as the solution for medium- to long-term matching capacity to demand (Janić 2015).

7.4.4 Application of the Evaluation Methodology

7.4.4.1 *Case—London Airport System*

(a) Background

The most recent initiative in Europe for matching capacity to demand is currently taking place at the London airport system. The considered options include: (1) 'doing nothing' and (2) 'increasing the airside (runway) capacity' by alternative solutions as follows: (a) 'building a new runway' at one of three alternative/candidate airports—Heathrow, Gatwick or Stansted, and (b) building a new airport near the Thames Estuary or converting Stansted into a four-parallel runway airport. These solutions and particularly the one of 'building a new (third parallel) runway' at Heathrow airport have been under consideration for a long time mainly due to the inherent complexity, sensitivity and controversy of the DM process, i.e. the necessity to articulate often the quite opposite interests and preferences of particular stakeholders involved—airport operators, airlines, users-air passengers, local community members and local, regional and central government(s). The main reasons are as follows:

- An inherent uncertainty as to whether sufficient demand will use the new infrastructure (the new runway and associated infrastructure) at the airport(s) where built.
- Building new or developing one of the existing smaller airports (Stansted) into a large hub implies closing the currently primary one—Heathrow. Such a decision could be economically and politically difficult, complex and risky since there are no guarantees for sufficiency of the switched demand and related services to justify the overall investments in the medium- to long-term. In addition, developing a new airport is inherently a time-consuming process during which substantial existing and particularly new long-haul air transport demand could shift from already saturated/congested London to other neighboring airport systems, such as Paris and Amsterdam.
- The chosen solution should simultaneously (i) provide sufficient runway capacity to match the current and future demand of the London metropolitan area and the entire southeast of England over the period 2025/26-2055/65 and later, efficiently, effectively and safely; and (ii) be sustainable, i.e. have the least possible impact on the environment and society while at the same time contributing to the overall (local and global) social-economic welfare.

(b) Characteristics of the options and solutions for matching capacity to demand

The main characteristics of the above-mentioned options for matching the airside (runway) capacity to demand at the London airport system are summarized in Table 7.16. The corresponding capacity of passenger terminals is also given therein.

As can be seen, option (1) has one and option (2) offers two types of alternative solutions for implementing at one of the four alternative/candidate

Table 7.16: Characteristics of particular options for the medium- to long-term matching of the capacity to demand at the London airport system (Janić 2015)

Option	Solution	Alternative	Total number of runways	Annual capacity (106) Atm passengers	
(1) Do nothing	-	-	4	1.00 - 1.04[1]	160
(2) Increase capacity	1 new runway	(1) Heathrow	5	1.26 - 1.30[1]	200
		(2) Gatwick			
	3/4 new runways	(3) Stansted	-	-	
		(4) "New" airport	6	1.56[2]	240

[1] Changing the runway system operating mode at Heathrow airport; [2] Closure of Heathrow airport

airports. In addition, the solutions of options (1) and (2) imply that the runway operating mode at Heathrow airport could be changed from the current segregated into the mixed mode (although this will require extension of one of two existing parallel runways and this is not considered as a separate alternative).

Option (1) implies leaving the capacity as it is at present since the demand has seemingly stagnated also due to exhausting its main driving forces within the London metropolitan area, such as GDP, population and pricing. If this is correct, these forces, in addition to the current noise cap at Heathrow, will be able to manage the demand at the present level, i.e. at or just below saturation of the available capacity at both Heathrow and Gatwick airport.

Option (2) implies that the stagnation in current demand occurred due to the runway capacity shortage both at Heathrow and Gatwick airports. At Heathrow airport, this shortage was caused by the imposed noise cap. Therefore, the future demand could continue to grow driven by external recovered driving forces and matched by the capacity of the new runway at either of the three airports of the system [alternative (a)]. This runway would be parallel to- and widely-spaced [at least 1,312 m (4,300 ft)] from the existing runway(s), thus enabling independent operations and accommodation of all existing and future aircraft types/categories. Fig. 7.2 (a, b, c) shows the simplified layouts of the three airports.

Without operational, economic, environmental and/or social/policy constraints, its annual capacity would be $260*10^3$ Atm and $40*10^6$ passengers. The alternative solution (b) of option (2) with 'three or four new runways' implies that demand is to be driven again by its main external driving forces. It would be matched by the existing capacity of Gatwick airport and the capacity of either a completely new four-parallel runway airport built near the Thames Estuary, or by the capacity of Stansted airport converted into a large four-parallel runway hub by building three additional runways. In such a case, Heathrow airport would be closed.

The solution (a) of option (2) with 'building a new runway' at one of the three candidate/alternative London airports appears most realistic under the given conditions and therefore it is further elaborated by analyzing its effects on matching capacity to the future medium- to long-term Atm and passenger demand developing according to the specified scenarios and continuing to the past and present developments of demand and capacity as shown in Figs. 7.3 (a, b), 7.4 (a, b) and 7.5 (a, b) (CAA 2012, 2013).

Figure 7.3 (a, b) shows the long-term development of the annual number of Atm and passenger demand, respectively, at Heathrow airport.

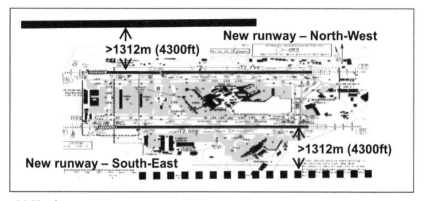

(a) Heathrow

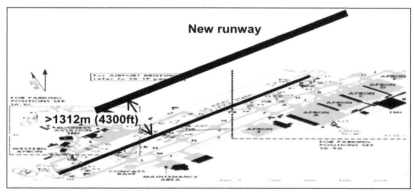

(b) Gatwick

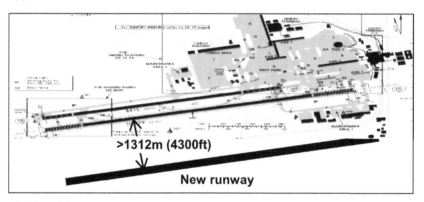

(c) Stansted

Fig. 7.2: Scheme of alternatives of Solution 1 'building a new runway'
at one of the three London airports (Janić 2015).

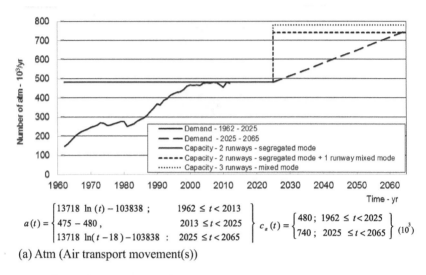

$$a(t) = \begin{cases} 13718 \ \ln{(t)} - 103838 \ ; & 1962 \le t < 2013 \\ 475 - 480 \ , & 2013 \le t < 2025 \\ 13718 \ \ln(t - 18) - 103838 : & 2025 \le t < 2065 \end{cases} \quad c_a(t) = \begin{cases} 480 \ ; & 1962 \le t < 2025 \\ 740 \ ; & 2025 \le t < 2065 \end{cases} (10^3)$$

(a) Atm (Air transport movement(s))

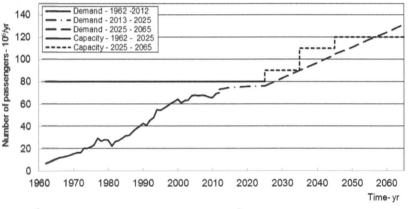

$$p(t) = \begin{cases} 2801 \ \ln{(t)} - 21232 \ ; & 1962 \le t < 2013 \\ 76 - 80 ; & 2013 \le t < 2025 \\ 2801 \ \ln(t - 12) - 21232 \ ; & 2025 \le t < 2065 \end{cases} \quad c_p(t) = \begin{cases} 80 ; & 1962 \le t < 2025 \\ 120 ; & 2025 \le t < 2065 \end{cases} (10^6)$$

(b) Air passengers

Fig. 7.3: Development of demand and capacity at London Heathrow airport (t – time) (years), $a(t)$, $c_a(t)$—demand and capacity of air transport movements, respectively; $p(t)$, $c_p(t)$—demand and capacity for air passengers, respectively (Janić 2015).

As can be seen, the past Atm and passenger demand have grown continuously (1962-2010/2013 sub-period). One of the main driving forces for such growth has been the role of the airport as the main hub of British Airways and also of Virgin Atlantic, both with a relatively high proportion of long-haul flights (about 60 per cent) and transit/transfer/connecting passengers. Consequently, over time, such relatively stable demand growth has almost saturated both the airport airside (runway system) and landside (passenger terminal) capacity. In addition to growth of demand, the main additional reason for saturated airside (runway system) capacity at about 99 per cent in 2012-13 has been the noise cap dictating use of the two parallel runways in the segregated mode (one exclusively for landings and the other exclusively for taking-offs) during a limited time of the day (due to the night ban). Furthermore, despite the new (fifth) passenger terminal recently being opened, the available passenger terminal capacity will be fully saturated by the year 2020, assuming that the trend of growing passenger demand will continue similarly as in the past, i.e. by simultaneous growth in the number of Atm within the existing constraints and the number of passengers per Atm, i.e. by increase in the average aircraft size. If a new runway was built over the 2015-2025 period operating in the mixed mode (simultaneously for both landings and take-offs), the capacity of the three-parallel runway system would be increased by an additional 260 thousand Atm/yr, thus bringing the total runway system capacity in 2025 from the current 480 to 740 thousand Atm/yr. Such capacity would again open the opportunity for growth of Atm, at least until 2055-60, when saturation would again occur according to the scenario of growing demand as shown in Fig. 7.3a. In parallel, the passenger terminal capacity would be gradually updated starting from 2025 as shown in Fig. 7.3b.

Figure 7.4 (a, b) shows the long-term development of the annual number of Atm and passengers, respectively, at Gatwick airport.

As can be seen, as at Heathrow, the single runway system capacity came very close to saturation (95 per cent) around 2010 due to the continuously growing demand as shown in Fig. 7.3a. At the same time, the passenger demand has also been growing, but at present, there is still some spare capacity for accommodating its future growth by 2025 as shown in Fig. 7.4b.

Such a development has mainly been driven by the nature of the airport traffic; namely, Gatwick hosts a relatively large proportion (about 55 per cent) of seasonal (charter) short- to medium-haul flights, including an increased proportion of LCC (Low Cot Carrier) flights. Demand stability has been provided by the presence of scheduled flights of BA (British Airways) and other partners from the One World alliance. A new runway implemented during 2015-2025 would double the current runway capacity of 260 thousand Atm/yr since a two-parallel runway system operating in the mixed mode would provide a capacity of 520 thousand Atm/yr as shown in Fig. 7.4a. This would be sufficient to accommodate the growing Atm demand until about 2055-60. At the same time, starting from 2025, the passenger terminal capacity should be gradually updated from the existing 40 to 80 million passengers/yr, until 2055-60, as shown in Fig. 7.4b.

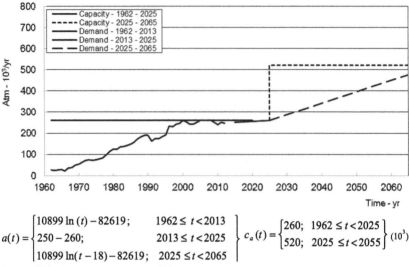

$$a(t) = \begin{cases} 10899 \ln(t) - 82619; & 1962 \le t < 2013 \\ 250 - 260; & 2013 \le t < 2025 \\ 10899 \ln(t - 18) - 82619; & 2025 \le t < 2065 \end{cases} \quad c_a(t) = \begin{cases} 260; & 1962 \le t < 2025 \\ 520; & 2025 \le t < 2055 \end{cases} (10^3)$$

(a) Atm (Air transport movement(s))

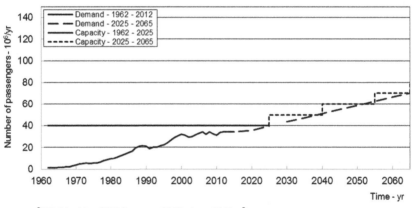

$$p(t) = \begin{cases} 1556 \ln(t) - 11799; & 1962 \le t < 2013 \\ 34 - 40; & 2013 \le t < 2025 \\ 1556 \ln(t - 10) - 11799; & 2025 \le t < 2065 \end{cases} \quad c_p(t) = \begin{cases} 40; & 1962 \le t < 2025 \\ 80; & 2025 \le t < 2055 \end{cases} (10^6)$$

(b) Air passengers

Fig. 7.4: Development of demand and capacity at London Gatwick airport (t – time) (years), $a(t)$, $c_a(t)$—demand and capacity of air transport movements, respectively; $p(t)$, $c_p(t)$—demand and capacity for air passengers, respectively (Janić 2015).

Figure 7.5(a, b) shows development of the annual numbers of Atm and passengers, respectively, at Stansted airport.

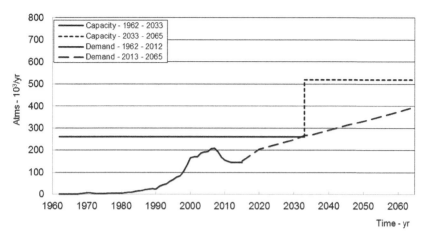

$$a(t) = \left\{ 8695 \ln (t) - 65972 \; ; \; 1962 \leq t < 2065 \right\} \quad c_a(t) = \begin{cases} 260 ; & 1962 \leq t < 2035 \\ 520 ; & 2035 \leq t < 2055 \end{cases} (10^3)$$

(a) Air transport movements

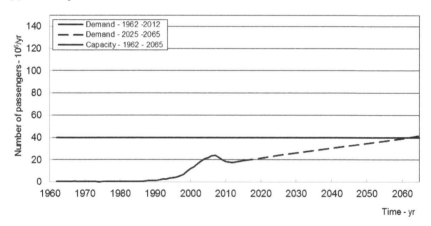

$$p(t) = \left\{ 898 \ln (t) - 6813 \; ; \; 1962 \leq t < 2065 \right\} \quad c_p(t) = \left\{ 40 ; \; 1962 \leq t < 2055 \right\} (10^6)$$

(b) Air passengers

Fig. 7.5: Development of demand and capacity at London Stansted airport (t – time) (years), $a(t)$, $c_a(t)$—demand and capacity of air transport movements, respectively; $p(t)$, $c_p(t)$—demand and capacity for air passengers, respectively (Janić 2015).

As can be seen, both Atm and passenger demand grew almost exponentially until about the year 2005. Since that time, the demand has become highly volatile mainly due to its inherent vulnerability originating mainly from European LCCs (Low Cost Carriers). These strengthened their short- to medium-haul flights over the UK and the rest of Europe during the period(s) of the overall economic prosperity and curbed them sharply during the latest economic crisis (2008-2012 period). Consequently, the available runway capacity was saturated at the level of about 55 per cent in 2012. However, if the future growth of demand continues again similarly as in the past, i.e. before the most recent crisis, the single runway capacity of about 260 thousand Atm/yr will reach saturation around 2035. Thus, a new runway implemented during the 2025-2035 period would double the existing capacity and be sufficient to accommodate the future growing Atm demand far beyond 2055-60 as shown in Fig. 7.5a. At the same time, the potential passenger terminal capacity of about 40 million passengers/year (supported by the single runway) would be sufficient to accommodate the corresponding (growing) demand until 2055-60 as shown in Fig. 7.5b. In this case, the prospective development of demand thanks to its switching from the other two airports—Heathrow and Gatwick—is not shown but it is taken into account in the further evaluation process.

7.4.4.2 Inputs

Inputs for the application of the SAW and TOPSIS method include introducing case-specific assumptions and estimating/quantifying particular indicators of performances of the three alternatives, i.e. London's Heathrow, Gatwick and Stansted airports—of the solution 1: 'building a new runway' using the above-mentioned scenarios for matching the capacity to demand over the future long-term period of time (2015-2055/60). The sub-periods of this period are (1) before deciding on Solution 1—'building a new runway' τ_1 = 1962-2015; (2) during implementation of Solution 1—τ_2 = 2015-2025; and (3) after implementation of Solution 1 τ_3 = 2026-2055-60).

(a) Assumptions

The case-specific assumptions are as follows:

- The three candidate airports for Solution 1—'building a new runway' in Table 7.16 are considered as independent alternatives implying neglecting the effects of Heathrow airport and managing Stansted airport.
- The new runway at either airport will operate in the mixed mode, thus providing capacity of $260*10^3$ Atms/yr and supporting an annual passenger terminal capacity of $40*10^6$/yr.
- Each airport will play a similar role in the airport system during and after implementation of the new runway to the one it played before. This implies that the character and structure of passenger demand and related aircraft fleet will not change substantially, while the volumes will generally continue to grow under the given conditions (*see* Figs. 7.3, 7.4 and 7.5).

- The potential new demand during the sub-period of implementation of the new runway (2015-2025) will be able to switch between airports. This implies that due to capacity constraints, the potential demand which cannot be accommodated at Heathrow or Gatwick airports will switch to Stansted airport until fully saturating its capacity. The remaining potential demand will spill over from the London airport system.
- The existing and new demand will not be compromised by potentially including either of the three airports into the HSR network.
- The aircraft fleet will continue to improve efficiency regarding fuel consumption (exclusively Jet-A fuel—kerosene) and related emissions of GHG, noise and safety, as the main driving forces for continuing diminishing impacts on the environment and society.
- The attribute/criterion 'safety' and its measure is assumed to be the same at all candidate/alternative airports; thus, it is not explicitly taken into account.
- Freight transport demand and related capacity are not taken into consideration.

(b) Estimation of the indicator systems of performances

The indicator systems of particular performances of the above-mentioned three alternatives of solution 1 and their measures are estimated and given in the self-explanatory Tables 7.17, 7.18, 7.19, 7.20 and 7.21. They are summarized again in Table 7.22.

Physical/spatial or infrastructural performances

Table 7.17: Indicator system of the *physical/spatial* or *infrastructural* performances in the given example (sub-period τ_2 = 1962-2015) (Janić 2015)

Indicator/Measure	Airport alternative Value of indicator/Measure		
	Heathrow	**Gatwick**	**Stansted**
Existing infrastructure			
• Number of runways	2	1	1
• Number of apron/gate stands (total)	186	112	110
• Number of passenger terminals	5	2	2/1
Convenience of location			
• Distance by rail/road[1] (km)			
• Travel time (by rail)[2] (min)	23	47.5	48
• Access time to/from the catchment area[3] (min.)	15-21	30	53
	105	90-105	90

[1]The rail/road distance to Central London; [2]Scheduled travel time; [3]Average door-to-door time from/to the catchment area (CAA 2011, 2012).

Operational performances

Table 7.18: Indicator system of the *operational performances* in the given example (sub-period: τ_1 = 1962-2015; τ_2 = 2026-2055/60) (Janić 2015)

Indicator/Measure	Airport alternative Value of indicator/Measure		
	Heathrow	Gatwick	Stansted
Increase in the total capacity			
• Relative contribution			
o Atm (%)	35	50	50
o Passengers (%)	33	50	50
Utilization of the total capacity[1]			
• Demand/capacity ratio			
o Atm (%)	92.2	86.5	68.0
o Passengers (%)	98.1	78.3	92.4
Attractiveness for airlines[2]			
• Number of currently operating airlines	84	45	12
Attractiveness for air passengers[2]			
• Total number of destinations	184	200	150
• Total number of intercontinental destinations	27	4	0

[1]At the end of the sub-period 2026-2055/60 (Figures 2, 3, 4); [2]Current values for the period 2012-2013 (CAA 2013)

Economic performances

Table 7.19: Indicator system of the *economic performances* in the given example (sub-period: τ_2 = 2015-2025; τ_3 = 2026-2055/60) (Janić 2015)

Indicator/Measure	Airport alternative Value of indicator/Measure		
	Heathrow	Gatwick	Stansted
Efficiency and effectiveness of investments			
• Cost (10^9£)[1]	15	10	10
• Profits (10^9£)[2]	5.11	2.29	0.44
• Effectiveness (profits/investments ratio)[3]	3.04	2.19	0.88
Contribution to the welfare			
• Employment (10^3)[4]	713	278	126
• Economy (10^9£)[5]	63.5	19.63	8.32

(Contd.)

Table 7.19: (*Contd.*)

Effects/impacts from (non)-accommodated demand			
• Losses $(10^9£)^6$	-1.8	- 0.737	0
• Gains $(10^9£)^6$	- 0.636	- 0.218	160/100

[1]Estimates (AC 2013a); [2]the 2026-2055/60 sub-period (this is the product of the cumulative number of Atm during the entire 2026-2055/60 sub-period and the average operating profit per Atm in 2012); [3]At the end of the period—2055/60; [3]the 2026-2055/60 sub-period—this is calculated by the total profits (the airport operating and contributions to the economy) during the entire 2026-2055/60 sub-period and the investment costs, i.e. as the ratio of the total earnings per unit of the investments; [4]at the end of the sub-period—2055-60 (ACI 1998, EC 2010); [5]the 2026-2055/60 sub-period (this is based on the contribution in 2012-13 of 9.7 bilion£ and 2.0 billion£ for Heathrow and Gatwick, respectively, divided by 480 and 260 thousand Atm, respectively and then multiplied by the cumulative number of Atm on the new runway over the given sub-period (GAL 2013, HAL 2013); [6]based on the airport profits per Atm in 2012-13; demand from Heathrow or Gatwick does not 1 and does switch 2 to Stansted (2015-2025 sub-period) (The gains for Stansted airport are based on the amount of the switched Atm (sub-period 2015-2025) and its current (2012-13) profitability).

Environmental performances

Table 7.20: Indicator system of the *environmental performances* in the given example (sub-period: $\tau_3 = 2026\text{-}2055/60$) (Janić 2015)

Indicator/Measure	Airport alternative Value of indicator/Measure		
	Heathrow	Gatwick	Stansted
Fuel consumption and emissions of GHG by LTO cycles on the new runway[1]			
• Fuel (10^6ton)	4.066	2.687	0.989
• GHG (10^6ton of CO_{2e})	12.098	7.684	2.739
Land use			
• Total occupied land (ha)[2]	1477	933	1037
• Relative increase in the total occupied land (%)	20	37	8.4
• Land use intensity (Atm/ha/yr)[3]	201	557	251

[1]Cumulative amounts during the 2026-2055/60 sub-period; the fleet structure at the airports remains the same as in 2012-13 with improvements in the fuel and GHG (CO_{2e}) emissions efficiency of the aircraft fleet—1 per cent per year during the 2026-2055/60 sub-period (Horton 2010, ICAO 2010, Janić 2007a); [2]a single runway with associated taxiways occupies an area of land of about 250 ha (Janić 2016); [3]the number of Atm at the level of runway system capacity.

Social performances

Table 7.21: Indicator system of the *social performances* in the given example (sub-period: τ_3 = 2026-2055/60) (Janić 2015)

Indicator/Measure	Airport alternative Value of indicator/Measure		
	Heathrow	**Gatwick**	**Stansted**
Noise			
• Additional noise (counts)[1]	1.9	1.8	1.1
• Noise efficiency (Atm/affected person)[2]	2.23	72.48	166.46
Safety			
• Third party risk	-	-	-

[1]Compared to $57L_{eq/24h}$ noise level (DfT 2013); [2]Valid for the end of the 2025-2055 sub-period (the new runway at either airport is assumed to create an additional $57L_{eq}$ noise contour, but all contours will fall by about 15 per cent compared to those in 2012-13 thanks to the noise improvements of -3dB during the period 2026-2055/60 (i.e. –0.1dB/yr) and under the condition that the average population density within the contours will not substantially change compared to the period 2012-13 (AC 2013b, ICAO 2010, Janić 2007a).

The estimated values of particular indicators and measures of performances in Tables 7.17-7.21 are summarized in Table 7.22 as the final input for application of the above-mentioned MCDM methods—SAW and TOPSIS.

As can be seen, of a total of 27 attributes/criteria, 11 have appeared as 'cost' and the remaining 16 as 'benefit' attributes/criteria.

7.4.4.3 Results

The results from application of the SAW and TOPSIS method using the above-mentioned 27 evaluation criteria are given in Tables 7.23 (a, b) and 7.24. In particular, Table 7.23a gives the values of weights of particular attributes/criteria obtained by applying the entropy method. Table 7.23b gives the values of ideal solution A^* and negative ideal solution A^- from the normalized decision matrix of the TOPSIS method.

Table 7.22: Summary of the estimated indicators and measures of performances of the three alternatives of Solution 1 used as the evaluation attributes/criteria in the given example (Janić 2015)

Indicator system of performances indicators/measures	Airport alternative[1] Value of indicator/Measure		
	Heathrow	**Gatwick**	**Stansted**
Physical/Spatial or Infrastructural			
Existing infrastructure			
1. Number of runways (counts)	+ 2	+ 1	+ 1
2. Number of apron/gate stands (counts)	+ 186	+ 112	+ 110
3. Number of passenger terminals (counts)	+ 5	+ 1	+ 1-2
Convenience of location			
4. Distance by rail/road (km)	- 23	- 47.5	- 48
5. Travel time (by rail) (min)	-15-21	- 30	- 35
6. Access time from/to the catchment area (min.)	- 105	- 90-105	- 90
Operational			
Increase in the total capacity			
Relative contribution (%)			
7. Atm	+ 35	+ 50	+ 50
8. Passengers	+33	+50	+50
Utilization of the total capacity			
Demand/capacity ratio (%)			
9. Atm	+ 92.2	+ 86.5	+ 68.0
10. Passengers	+98.1	+78.3	+92.4
Attractiveness for airlines			
11. Number of currently operating airlines (counts)	+ 84	+ 45	+ 12
Attractiveness for air passengers			
12. Total number of destinations (counts)	+ 184	+ 200	+ 150
13. Number of intercontinental destinations (counts)	+ 27	+4	0

(Contd.)

Table 7.22: (*Contd.*)

Indicator system of performances indicators/measures	Airport alternative[1] Value of indicator/Measure		
	Heathrow	**Gatwick**	**Stansted**
Economic			
Efficiency and effectiveness of investments			
14. Costs (10^6£)	- 15	- 10	- 10
15. Profits (10^6£)	+5.11	+2.29	+0.14
16. Effectiveness (ratio profits/ investments)	+3.04	+2.19	+0.88
Contribution to welfare			
17. Employment (10^3)	+ 713	+ 278	+ 126
18. Economy (10^9£)	+ 63.5	+19.6	+8.3
Effects/impacts from (non)- accommodated demand			
19. Losses (10^9£)	- 1.8	- .737	0
20. Gains (10^9£)	-0.036	-0.218	+160/100
Environmental			
Fuel consumption and emissions of GHG by LTO cycles on the new runway			
21. Fuel consumption (10^6 ton)	- 4.066	- 2.687	- 0.989
22. Emissions of GHG (10^6 ton of CO_{2e})	- 12.098	- 7.684	- 2.739
Land use			
23. Total occupied land (ha)	- 1447	- 933	- 1037
24. Relative increase in the total occupied land (%)	- 20	- 37	- 8.4
25. Land use intensity (Atm/ha/ year)	+ 201	+ 557	+ 251
Social			
26. Factor of additional noise (counts)	- 1.9	- 1.8	- 1.1
27. Noise efficiency (Atm/affected person)	- 2.23	- 72.48	- 166.46

[1]Sign "+" for the 'benefit' and sign "–" for the 'cost' attribute/criterion are given in front of the estimated values

Table 7.23: The inputs and some intermediate outputs from the selected
MCDM methods (Janić 2015)

(a) The SAW and TOSPIS—weights for particular criteria in the given example

i/j	Indicator/Measure as criterion	Weight of criteria w_j
	Physical/Spatial or Infrastructural	
1	Number of runways	0.00123
2	Number of apron/gate stands	0.00658
3	Number of passenger terminals	0.04922
4	Distance by rail/road	0.00998
5	Travel time (by rail)	0.00726
6	Access time from/to the catchment area	0.00792
	Operational	
	Relative contribution to the total capacity (%)	
7	Atm	0.00272
8	Passengers	0.00363
	Demand/capacity ratio (%)	
9	Atm	0.00166
10	Passengers	0.00091
11	Number of currently operating airlines	0.04463
12	Total number of destinations	0.00143
13	Number of intercontinental destinations	0.14726
	Economic	
14	Costs	0.00404
15	Profits	0.06170
16	Effectiveness	0.01186
17	Employment	0.04559
18	Economy	0.06147
19	Losses	0.10117
20	Gains	0.21785
	Environmental	
21	Fuel consumption	0.02699
22	Emissions of GHG	0.02915
23	Total occupied land	0.00381
24	Relative increase in the total occupied land	0.03108
25	Land use intensity	0.02132

(Contd.)

Table 7.23: (*Contd.*)

i/j	Indicator/Measure as criterion	Weight of criteria w_j
	Social	
26	Factor of additional noise	0.00544
27	Noise efficiency	0.09045

(b) The TOPSIS—ideal solution A^+ and negative ideal solution A^- from the normalized decision matrix

A^+	[(0.14567; 0.06749; 0.21184); 0.00028; 0.00026; 0.00024]
A^-	[(0.09970; 0.03833; 0.06483); 0.00057; 0.00046; 0.00054]

As can be seen in Table 7.23a, the most important indicators/measures of physical/spatial or infrastructural performances as criteria appear to be the 'number of passenger terminals' and 'airport accessibility'. The most important indicators/measures of the operational performances as criteria include the 'number of intercontinental destinations', the 'number of operating airlines' and 'relative contribution to the total passenger capacity'. As far as the economic performances are concerned, the most important indicator/measures as criteria appear to be 'gains', 'losses', 'profits' and 'economy'. The most important indicators/measures of the environmental performances as criteria are 'increase in the occupied land', 'emissions of GHG' and 'fuel consumption'. 'Noise efficiency' appears to be the most relevant indicator/measure as a criterion of the social performances.

Consequently, when all 27 criteria were taken into consideration, both SAW and TOPSIS methods produced the same results, i.e. a new runway would preferably be built first at the currently largest Heathrow airport as given in Table 7.24.

Table 7.24: Results from the MCDM ranking of the three airport alternatives of Solution 2) in the given example (Janić 2015)

Alternative/Airport	SAW		TOPSIS			
	A^*	Rank	S_{i+}	$S_{i=}$	C_{i*}	Rank
1. Heathrow	0.58735	1	0.51296	0.34307	0.40077	1
2. Gatwick	0.39882	3	0.23659	0.12479	0.34532	2
3. Stansted	0.49180	2	0.75116	0.28079	0.27210	3

The main reasons for Heathrow airport to score the best for the new runway include its current size, its proximity to the core of the metropolitan (London) area including efficient and effective ground access systems, the presence of many airlines offering a large diversity of destinations and particularly intercontinental ones, its great advantage in terms of the overall economic contribution, which prevail over the losses from the prospectively non-accommodated demand, relatively high utilization of the completely occupied land and the lowest factor of additional noise, all resulting from the new runway there.

However, the SAW method has ranked Stansted as the second and Gatwick airport as the third-last best alternative for the new runway. One strong reason seems to be the ability of Stansted airport to accommodate a large amount of the shifted demand from the other two already saturated airports in order to prevent its spillage outside the London airport system. Under such conditions, Stansted airport would create substantial social-economic benefits. Contrary, the TOPSIS method ranks Gatwick as the second and Stansted as the third best alternative for the new runway. In this case, Gatwick airport comes second due to some very similar but at the same time much weaker reasons than those mentioned above for Heathrow airport. Stansted airport remains at the last place despite the substantial gains that could be obtained, thanks to accommodating the demand switched from the other two already saturated airports.

This application has also shown that different MCDM methods, in this case SAW and TOPSIS, produced the same results, at least regarding the most preferable alternative. As such, they could be used as a support in the later stages of the decision-making process in the given and other similar cases.

7.4.5 Interim Summary

This section deals with developing the methodology for evaluating the alternative airports in the given airport system where a new runway could be built. This new runway has generally been considered as the long-term solution for increasing the runway airside (runway) system capacity of the given airport and the airport system in order to match the prospectively growing demand efficiently, effectively and safely. The methodology is based on developing the models of indicator systems consisting of indicators and measures of the physical/spatial or infrastructural, operational, economic, environmental and social performances of the candidate airports for a new runway, reflecting the interests of the particular main stakeholders involved. Then, after being quantified, these indicators and measures were used as evaluation criteria by the selected MCDM methods. These were SAW and TOPSIS method applied in combination with the entropy method for estimating the weights of particular attributes to be used as evaluation criteria.

The models were applied to select the preferred among the three airports of the London airport system—Heathrow, Gatwick and Stansted—where the new runway could be built. The results from both SAW and TOPSIS method have shown that under the given conditions, it would be most beneficial to build a new runway at Heathrow airport, as intuitively expected. Gatwick and Stansted airport appear as second and third best alternatives, respectively, using both the TOPSIS and SAW methods. In general, the proposed evaluation models have been useful in supporting the above-mentioned DM (Decision Making) process at least in its initial/preliminary stage.

7.5 Freight Transport Corridors

7.5.1 Background

The EC (European Commission) transport policy provides an institutional framework for development of the freight transport sector which is expected to serve the growing demand on the one hand while at the same time mitigating its impacts on the environment and society, on the other, in the forthcoming decades. This implies improved sustainability of the sector (CEC 2001, 2011). Such development is expected to be achieved by shifting more freight volumes from the currently dominating road to rail transport mode by developing rail and intermodal rail/road freight transport corridors throughout Europe in addition to the other transport policy measures. In European countries, such as Austria, Switzerland and Spain/France, these corridors aim to overcome topographical constraints like the Alps and Pyrenees, respectively (EC 2011).

This section deals with analysis, modelling and evaluation of the rail freight transport corridors based on their infrastructural, technical/technological, operational, economic, environmental and social performances. For such a purpose, a convenient methodology is developed including two components: (i) defining and modelling the indicator systems consisting of indicators and measures of the corridors' performances; and (ii) selecting and applying the existing (multi-criteria) evaluation methods for ranking particular corridors as competing alternatives. The main features of these multi-criteria evaluation methods are described in Section 7.2.

7.5.2 The System and Problem—European Freight Transport Policy and Rail Freight Transport Corridors

The main objectives of the EC transport policy are enumerated in white papers with some of the qualitative objectives as follows (CEC 2001, 2011):

- Shifting the balance between modes of transport by 2010, by curbing the demand for road transport and revitalizing alternative transport modes, such as railways and maritime and inland waterways; and
- Making the transport systems more efficient and safer.

The above-mentioned measures for achieving the quantitative objectives of the EC transport policy by reversing the present negative market trends in the rail freight sector have been concretized, in addition to setting up the global TEN-T network with priority axes, through the concept of rail freight transport corridors[7] as follows:

- ERTMS (European Rail Traffic Management System) corridors with the core task of deploying the European Train Control System and consequently promoting and providing interoperability throughout the European rail network.
- RNE (Rail Net Europe) corridors to deal with allocation of the rail infrastructure capacity and timetabling.
- The RFCs (Rail Freight Corridor(s)) concept aiming to provide sufficient infrastructure capacity and service performance in order to meet the requirements of current and prospective volumes of freight demand, both quantitatively and qualitatively.

In the above-mentioned concepts, parts of particular corridors overlap with each other and are usually named according to the geographical features, the EU Member States and the principal routes they pass through. The RFCs are also characterized by the time of implementation. For nine of them this will be the period 2013-2015 (EC 2011). In addition, the EU-funded FMPs [Framework Program(s)] have named particular corridors so as to reflect the main topics of research. Nevertheless, they have all been characterized by the rail lines connecting the major rail freight terminals, marshaling yards, major inland intermodal rail/road and rail/barge freight terminals and rail intermodal terminals at seaports. In all these corridors, both pure rail and intermodal (rail/road) freight transport services are carried out.

---

[7] In addition to the global TEN-T (Trans European Network-Transport) spreading throughout the EC Member States, the second pan-European Transport Conference in Crete (1994) defined 10 pan-European transport corridors as passenger and freight transport routes in Central and Eastern Europe. Some additions were made at the third conference in Helsinki (1997), which was why these corridors were referred to as 'Crete corridors' or 'Helsinki corridors', respectively. In particular, after the end of the civil war in the former Yugoslavia, the tenth corridor was defined.

7.5.3 A Methodology for Evaluation of Rail Freight Transport Corridors

7.5.3.1 Some Related Research

Some related research on analyzing, modelling and evaluating freight transport corridors in Europe can be broadly divided into qualitative and quantitative categories[8]. The qualitative research has dealt with corridors from the spatial policy and governance perspective (Priemus and Zonneveld 2003). The former has generally considered corridors as linear spatial structures with bulk of all kinds of infrastructure and policies integrating transport, infrastructure, the economy, urbanization and environmental developments (Chapman et al. 2003). The latter has focused on mega corridors as large infrastructure axes spread between major urban areas and requiring innovative governance at the local, regional, national and international level. As such, through stimulating relationships between particular actors involved, society and space, they contribute to international integration within the European area. One strong reason is the increase in the number of international entrepreneurs (Vries and Priemus 2003, Romein et al. 2003).

The quantitative research deals with (a) investigation of performances of the rail and intermodal rail-based freight transport corridors in Europe; and (b) application of the multi-criteria evaluation methods generally to transport and freight transport corridors.

(a) Performances of the rail and intermodal rail-based freight transport corridors

This research can be roughly divided into that published in scientific journals and that published as a part of EC-funded projects.

- In the academic context, the topic is constantly under review and/or overview (Bontekoning et al. 2004, Janić and Reggiani 2001, Janić 2006). In addition, research on assessing the full (internal and external) costs of the rail, rail-road intermodal and road freight transport in Europe indicates the advantages of rail and rail-road intermodal over road-truck freight transport along medium- and long-distance corridors (Janić 2007b, 2008, Janić and Vleugel 2012).
- EC-funded research and development projects are carried out in the scope of (i) topical networks, (ii) concerted actions and (iii) integrated projects. Some of these include COST Transport Actions, Framework Programs, the Marco Polo Program and research on 'monitoring' and investigating

---

[8] Although research carried out in the US and elsewhere has not been considered, this does not diminish the generosity of this overview.

liberalization of the rail freight transport markets in particular EU countries.

1. In the COST Transport Actions (European Co-operation in the Field of Scientific and Technical Research), some related projects on rail/road intermodal freight transport deal with the network design and operation and the relationship between particular modes including interconnectivity and interoperability. Some of these projects include COST 310 'Freight Transport Logistics', COST 328 'Integrated Strategic Infrastructure Networks in Europe', COST 340 'Towards an European Intermodal Transport Network: Lesson From History' and COST 356 'EST— Towards the definition of a measurable environmentally sustainable transport' (http://www.cost.esf.org).

2. FMP (Framework Program) projects deal mainly with rail freight transport, integrated Trans-European transport networks including Trans-European corridors and the assessment of their particular social and environmental impacts. Some of these include DIOMIS—Developing Infrastructure and Operating Models for Intermodal Shift; CPRC— PERCEPTION OF FREIGHT—Changing the Perception of Rail Cargo; RETRACK—Reorganisation of Transport Networks by Advanced Rail Freight Concepts; CREAM—Customer-driven Rail-Freight Services on a European Mega-Corridor Based on Advanced Business and Operating Models; LOGCHAIN MTC NRW—BALKAN—Analysing the Flow of Traffic Goods—North Sea/North Rhine-Westphalia/ Balkans; LOGCHAIN EAST-WEST CARGO FLOW—Freightchain: Re-Engineering East-West Rail Cargo Flows for Service and Speed; REORIENT—Implementing Change in the European Railway System; BRAVO—Brenner Rail Freight Action Strategy Aimed at Achieving a Sustainable Increase in Intermodal Transport Volume by Enhancing Quality, Efficiency and System Technologies; Study of road freight traffic across the Alps; Green freight corridor in Europe; RECONNECT— Reducing Congestion by Introducing New Concepts of Transport; SCANDINET—Promoting Integrated Transport in Peripheral Areas of the Union; ITESIC—Integration of technologies for European short intermodal corridors; RECORDIT—Real Cost Reduction of Door-to-Door Intermodal Transport; APRICOT—Advanced pilot tri-modal transport chains for the corridors West to South/South-East Europe for combined transport; and EUFRANET—Improving the Competitiveness of Rail Freight Services (http://www.transport-research.info/web/ projects/transport_themes.cfm).

3. The Marco Polo program started in the year 2003 aimed at reducing road congestion and improving the environmental performances of the

freight transport system within the EU by stimulating shifting freight volumes from road to non-road, short-sea shipping, rail and inland waterways transport modes under current market conditions. Some relevant projects include the Rotterdam-Istanbul shuttle train (DARIS) and Cologne-Kosekoy block train application (TRITS) (http://europa. eu.int/comm/transport/marcopolo/projects).

4. Research also focuses on 'monitoring' and investigating the achieved level of liberalization in the rail freight transport market in particular EU countries evaluated by convenient indicators (Kirchner 2011). In addition, the data sources and in-depth analysis was made as part of the REORIENT project (the EU 6 FMP) as well as in a few academic/ research papers (Ludvigsen 2009, Warren et al. 2009).

(b) Application of multi-criteria evaluation methods

Different single- and multiple-objective evaluation methods were used by both researchers and practitioners as tools in the scope of given DSS (Decision Support System) for selecting the preferred alternative(s) for supplying the transport infrastructure and services of different transport modes aimed at matching their capacity to demand under the given conditions. Specifically, applying the MCDM methods to evaluation of transport infrastructure projects in a combination with the EAT (Economic Analysis Technique) or BAU (Business As Usual) methods has been a matter of rather wide academic interest (Brucker et al. 2011, Giuliano 1985, Hwang and Yoon 1981, Janić 2014, Sauian 2010, Schutte and Brits 2012, Tabucanon and Mo-Lee, 1995; Vreeker et al. 2002). In addition, some applications of the SAW (Simple Additive Weighting), TOPSIS (Technique for Order Preference by Similarity to Ideal Solution) and AHP (Analytical Hierarchy Process) method have considered the values of particular stakeholders in the transport corridor planning processes (Bethany et al. 2011) and the evaluation of the performances of (i) transport corridors (Ding et al. 2008), (ii) general logistics systems (Sawicka et al. 2010), (iii) innovative freight bundling networks in Europe (Janić et al. 1999) and iv) HS (High Speed) transport technologies (Janić 2014). Consequently, in Europe, the above-mentioned COST 328 Action gave one of the strongest recommendations for using the MCDM instead of the 'pure monetary' EAT or BAU method (*see* footnote No. 5).

7.5.3.2 Objectives and Assumptions

In spite of the rather large above-mentioned and remaining body of academic and consultancy/professional literature dealing with estimation of the performances of the RFCs, the systematic academic research on their analysis, modelling and evaluation, the latter also by using different MCDM methods, is still relatively scarce. Therefore, the objectives are to develop and present the methodology for analyzing, modelling and evaluating the rail and intermodal rail/road freight transport corridors as competing alternatives either between themselves and/or to other non-rail transport modes such as road, inland waterways and short-sea shipping. The methodology consists of two parts. The first part—analysis and modelling—defines the indicator systems of the corridors' performances and develops their analytical models. These performances are physical/spatial or infrastructural, technical/technological, operational, economic, social and environmental. The indicator systems consist of indicators and measures of the above-mentioned performances. The second part—evaluating—includes selection among existing MCDM methods for ranking the alternative corridors, using the above-mentioned indicator systems of their performances as the evaluation attributes/criteria. These methods are used in combination with the above-mentioned entropy method for assigning weights to particular attributes of performances as the evaluation criteria. In addition, the objective is also to compare outcomes from the selected MCDM methods in order to assess their convenience in application in the given and similar contexts.

In order to fulfil the expectations, the proposed methodology should be:

- Sufficiently generous to be applied to different rail and intermodal rail-based freight transport corridors and with slight modifications to the corridors operated by other transport modes (road, inland waterways and short-sea shipping);
- Relatively simple, transparent and thus understandable for the particular actors/stakeholders involved;
- Easily applicable in terms of estimation of the particular indicator systems, i.e. indicators and measures of performances by using the available data; and
- Able to express particular indicators and measures of performances conveniently as attributes/criteria in evaluation of the above-mentioned alternative corridors.

Consequently, the methodology could particularly be useful, in addition to researchers/consultants, for the managers and/or governors of the corridor(s), providers of transport infrastructure and services and the policy makers at different institutional levels (regional, national, international). The managers and/or governors of corridors could use the methodology for estimating and monitoring competitiveness of their corridor(s) to those operated by the same rail and/or other transport modes (road, inland waterways, short-sea shipping), when serving the same or closely neighboring markets. The providers of transport (rail and intermodal rail-based) infrastructure and services could use the methodology for assessing the realized overall efficiency, effectiveness and social-economic feasibility (i.e. sustainability) of their activities/

operations under the given conditions. Transport and other policy makers could use the methodology as an initial guidance for assessing the feasibility of their decisions, mainly related to prioritizing investments in the alterative corridors' infrastructure and/or subsidizing some (usually innovative) rail and/or intermodal rail-based freight services.

The above-mentioned methodology is based on the following assumptions:

(a) General

- The indicator systems consisting of indicators and measures of performances are specified for the given period of time;
- Some indicators and measures of performances represented as parameters only and others derived from the analytical equations are used as the evaluation attributes/criteria; and
- The weights representing the relative importance of particular attributes/ criteria are model-derived and not the outcome of the subjective judgements of particular actors/stakeholders involved.

(b) Characterization of the corridors

- The rail or intermodal rail/road freight transport corridor is considered to be of a linear spatial layout, spreading longitudinally through different regions and countries. The transport infrastructure consists of bi-directional rail tracks connecting a set of the sequentially located rail/rail or intermodal rail/road freight terminals. As such, this infrastructure enables operation of intermodal freight trains to provide the corresponding transport services.
- The goods/freight flows consolidated into containers of the basic size—TEU (Twenty Foot Equivalent Unit) are loaded and unloaded (i.e. transshipped) at the rail/rail or rail/road intermodal terminals. The loading takes place after they are brought from the shippers to the origin terminal and the unloading happens before they are delivered to the receivers from the destination terminal. Delivery from/to the TEU shippers/receivers, respectively, is carried out either by regional rail (industrial tracks) or by trucks. This implies that the terminals are access locations for TEU flows where they enter and leave the corridor from/to the gravitational areas of these terminal(s), respectively. The gravitational area of particular terminals can be of a different size and shape in the horizontal plane (circle, square, rectangle, trapezoid, etc.) (Larson and Odoni 2007). In addition, the TEU flows can transit through the terminal(s), which usually happens on the same incoming and outgoing train(s), i.e. without transshipments between different trains.
- The set(s) of O-D (Origin-Destination) terminals of TEU flow(s) and the rail tracks connecting them define the route(s) along the corridor.

- The above-mentioned corridor(s) and its routes can be considered as a line transport network with sequentially located nodes/terminals connected by links/rail tracks in both directions. This implies that each node/terminal is connected by two incoming and two outgoing links, each from a different side. Fig. 7.6 shows the simplified spatial scheme.

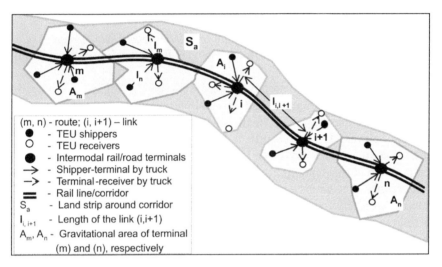

Fig. 7.6: Scheme of the intermodal rail/road freight transport corridor as a line transport network.

The intermodal rail/road terminals of the intermodal rail/road freight corridor in Fig. 7.6 are analogous to the rail/rail terminals of the rail freight corridor(s).

(c) Operations on the corridors

- The rail/rail or rail/road intermodal freight transport services carried out on the given route of the corresponding corridor during the specified period of time have the same characteristics. This implies that the collecting/distributing regional trains and road trucks, respectively and the (intermodal) freight trains serving given TEU flows have the same average size/capacity and utilization (i.e. load factor), speed, operating and external cost, energy consumption and related emissions of GHG, etc.
- Shippers and receivers of TEU flows are assumed to be uniformly distributed in the gravitational area of particular intermodal terminals.
- The volumes of TEU flows on the given route of the corridor are constant during the specified period of time.

- The TEU flows generated and attracted by individual shippers and receivers, respectively, in the gravitational area of the given rail/rail or intermodal rail/road terminal(s) are in balance, i.e. they are approximately the same.

7.5.3.3 Models of the Indicator Systems of Performances

(a) Physical/spatial or infrastructural performances

The indicator system of physical/spatial or infrastructural performances of a given corridor and its specified route includes the following indicators and measures:

Corridor length (km) is the distance between the begin and end rail/rail or intermodal rail/road freight terminal measured along the shortest rail line(s) connecting them. The length reflects the spatial extensiveness of the corridor relevant for the corridor's managers/governors and infrastructure and transport services providers. On the one hand, they are often faced with the problem of overcoming incompatibilities and various associated barriers; on the other, they count on a greater number of more spatially concentrated (closer) users and consequently greater volume of TEU flows. It is expressed as follows:

$$L = \sum_{i=1}^{N-1} l_{i,i+1} \tag{7.19a}$$

Accessibility (terminals/100 km) is the ratio between the number of intermodal terminals and length of a corridor. It is particularly relevant for users, i.e. shippers and receivers of TEUs, since it represents the quality of spatial access to the corridor's transport services. It is expressed as follows:

$$AS = \frac{N}{L} \tag{7.19b}$$

Area coverage (km²) is the sum of gravitational areas of the individual terminals along a corridor. As a measure of spatial availability of services within the entire area around a corridor, this is particularly relevant for users, i.e. shippers and receivers of TEUs. It is expressed as follows:

$$ID = \frac{L}{S_a} \tag{7.19c}$$

Infrastructure density (km/km²) is the ratio between the length of a corridor and the size of its coverage area. It reflects the extensiveness of a corridor respecting the area it serves. As such, it is relevant for the corridor's managers/governors and rail transport operators expecting greater TEU flows from users, i.e. shippers and receivers of TEUs either located over a wider area or concentrated in a narrower area around the corridor. It is expressed as follows:

$$ID = \frac{L}{S_a} \tag{7.19d}$$

Route length (km) is the shortest rail line distance between any two intermodal terminals as the origins and destinations of the TEU flow(s) and related intermodal train services. This reflects the spatial extensiveness of TEU flows, which can be relevant for the rail infrastructure and transport service providers—for the former due to maintaining the longer route requiring more overall resources, and for the latter due to the need for engaging generally a greater train fleet on the longer routes under given conditions and vice versa. It is expressed as follows:

$$L_{nm} = \sum_{i=m}^{n-1} l_{i,i+1} \tag{7.19e}$$

Maximum axle load (ton/axis) is the allowed load per axis of a rail wagon on particular link(s) of a corridor. It influences the types of wagons used, their utilization and consequently composition and utilization of the entire train(s). As such, it is mainly relevant for rail operators responding to the requirements of their users—shippers and receivers—of TEUs. It is expressed as follows:

$$AXL_{mn} \tag{7.19f}$$

where

N	is the number of nodes/terminals in the network/along a corridor;
$N\text{-}1$	is the number of links in the network/along a corridor;
m, n, i, j	is the index of terminals ($m, n, i, j \in N$);
$l_{i,\,i+1}$	is the length of a link connecting the terminals (i) and ($i+1$) (km);
S_a	is the gravitational area (strip of land) spread on both sides and along a corridor (km^2);
A_j	is the size of gravitational area of the terminal (j), (km^2); and
l_m, l_n	is the average (road) distance for collecting and distributing TEU flows within the gravitational area of the terminal (m) and (n), respectively (km).

(b) Technical/technological performances

The indicator system of technical/technological performances of a given corridor and its specified route includes the following indicators and measures:

Propulsion systems (counts) is the number of differently powered engines used for running intermodal trains along the corridor. Generally, these can be diesel and electric. This appears important if the corridor is not electrified along the entire length. The necessary consequence is changing the engine usually at the different countries' border(s), which takes time and consequently decreases the trains' operating speed and increases the transit and delivery time of TEUs. This is relevant for rail operators and users, i.e. shippers and receivers of TEUs. It is expressed as follows:

$$k \qquad (7.20a)$$

Electric propulsion systems (count; kV) is the number of different electric propulsion systems and their corresponding power (kV – Kilovolts) available along the corridor, again mainly being country-specific. In case of a lack of multi-system engines, the impacts for the actors/stakeholders involved can be similar as in cases of changes between diesel and electric engines. It is expressed as follows:

$$m, VP_{mn} \qquad (7.20b)$$

Length/weight, payload capacity and technical speed of trains (m/ton; TEU/train; km/h) are conditioned by the signaling system and characteristics of rail tracks, respectively, along the particular links of the corridor. The former three are relevant for rail operators and the last one for both rail operators and users of their services. It is expressed as follows:

$$Z_{mn}/W_{mn}, PW_{mn}, V_{mn} \qquad (7.20c)$$

Length/weight and payload capacity of train's wagons (m/ton; TEU/wagon) characterize their length, gross weight and payload capacity. These are primarily relevant for the rail operators and generally influence the train length, its gross weight and payload capacity. It is expressed as follows:

$$z_{mn}/w_{mn}, pw_{mn} \qquad (7.20d)$$

Number of wagons per train (counts/train) depends on the train's length/weight influenced by the volumes of TEU flows, length/weight of individual wagons and train control, i.e. signaling system along a corridor and its particular route(s). This is again mainly relevant for the rail transport service providers. It is expressed as follows:

$$N_{mn} \qquad (7.20e)$$

Payload capacity of a road truck (TEU/truck) is the maximum number of TEUs per truck operating from shippers to the intermodal terminal at one and from the intermodal terminal to the receivers of TEUs at the other end of the given route of a corridor. This is relevant for the road transport operators in planning the type, size and utilization of their fleets under given conditions. It is expressed as follows:

$$w_{r/mn} \qquad (7.20f)$$

Length of rail tracks in terminal(s) (m) reflects the capacity of a terminal expressed by the number of simultaneously accommodated rail wagons. This number is additionally influenced by the above-mentioned wagon length. As such, this is mainly relevant for the terminal and partially rail operators. It is expressed as follows:

$$L_j \tag{7.20g}$$

Number, service rate and utilization of the terminal transshipment facilities and equipment (counts, TEU/TU, %) reflect the transshipment capacity of terminals under given operational regime. As such, these are mainly relevant for terminal operators while offering their services to both (particularly new) road and rail transport operators. It is expressed as follows:

$$n_j, \ \theta_j, \ \delta_j \tag{7.20h}$$

Interoperability (counts) is the number of different propulsion and electric propulsion systems per the country's border crossings. As such, it is mainly relevant for rail operators while planning deployment of the multi-system engines. It is expressed as follows:

$$IOP = (k+m)/NBC_{nm} \tag{7.20i}$$

where

TU	is time unit (hr, d, wk, mon, or yr);
k	is the number of different propulsion systems along a corridor (i.e. diesel and/or electric);
m, VP_{mn}	is the number of different electric propulsion systems and their power along the route (m, n) of a corridor [counts; voltage – kV(kilovolt)];
Z_{mn}, W_{mn}, PW_{mn}	is the maximum length, gross weight and payload capacity, respectively, of a train on the route (m, n) (m; ton; ton);
V_{mn}	is the maximum technical speed of a train on the route (m, n) (km/h);
z_{mn}, pw_{mn}	is the length and payload capacity of a wagon of trains operating along the route (m, n) of a corridor (m; TEU/wagon);
N_{mn}	is the number of wagons per train operating on the route (m, n) (counts); $(z_{mn}*N_{mn} \le Z_{mn})$;
$w_{r/mn}$	is the payload capacity of a truck operating in the gravitational area of the terminal (m) and (n) (TEU/truck);
L_j	is the average length of rail tracks at the terminal (j) $[j \in (m, n)]$ (m) $(Z_{mn} \le L_j)$;
n_j, θ_j, δ_j	is the number, service rate and utilization of the transshipment facilities and equipment at the terminal (j) $(j \in (m, n)$ (counts; TEU/TU; %) $(\delta_j \le 100$ per cent); and
NBC_{mn}	is the number of border crossings along the route (m, n) of a corridor (counts).

The other symbols are analogous to those in Equation 7.19.

(c) Operational performances

The indicator system of operational performances of a given corridor and its specified route(s) includes the following indicators and measures:

Demand density (TEU/km²) is the ratio between the volumes of TEU flows generated and attracted by the terminals along a corridor and the size of gravitational area of these terminals. This can be useful particularly for the intermodal rail transport operators, including new entrants in assessing attractiveness of the corridor. It is expressed as follows:

$$DD = \sum_{j=1}^{N} \frac{P_j(q_{jg} + q_{ja})}{A_j} \qquad (7.21a)$$

Traffic and transport capacity[9] is the maximum number of intermodal trains and volumes of TEUs, respectively, that can be transported and pass a fixed point of the given route of a corridor during a specified period of time under conditions of constant demand for service. Traffic capacity is represented by the maximum frequency of trains, which can be dispatched along the route in the same direction as the reciprocal of the minimum time interval between them. This minimum time interval depends on the block signaling system along particular links of the route. Transport capacity is the product of the train's payload capacity and the traffic capacity. These capacities relevant for the rail infrastructure and transport service providers indicate the overall potential capacity to handle current and prospective volumes of TEUs. As such they could be an indicator of attractiveness of a corridor/route for the new (rail service) entrants as well as for prospective users to locate their businesses nearby. It is expressed as follows:

• Traffic (trains/TU)

$$TC_{mn} = \min\left\{ f_{max/mn}; \left[\min_{i \in (m,n-1)} [\mu_{i,i+1}]; \ \min_{j \in (m,n)} [\mu_j] \right] \right\} \qquad (7.21b)$$

where $\mu_{i,i+1} = f_{max/i,i+1}$ and $f_{max/i,i+1} = 1/h_{min/i,i+1}$ and

$f_{max/mn} = 1/h_{min/mn}$ or $f_{max/mn} \equiv Q_{max/mn} / \Lambda_{mn} M_{mn} pw_{mn}$ and

---

[9] The traffic and transport capacity of the route L_{mn} and consisting links are based on the known maximum flow—minimum cut theorem (Ford and Fulkerson 1962). Thus, the route capacity is determined as the minimum among the corresponding capacities of the nodes and links included. The maximum frequency is inversely proportional to the minimum time interval between dispatching successive trains on the link (route). The capacities of nodes/terminals depend on the number, service rate and utilization of transshipment facilities.

$$\mu_j = \frac{N_j L_j n_j \theta_j \delta_j}{\Lambda_{mn} M_{mn}^2 pw_{mn}}$$

- Transport (TEU/TU)

$$TRC_{mn} = \min\left\{\eta_{max/mn}\left[\min_{i\in(m,n-1)}[\eta_{i,i+1}]; \ \min_{j\in(m,n)}[\eta_j]\right]\right\} \qquad (7.21c)$$

where $\eta_{i,i+1} = f_{i\,max/,i+1}\Lambda_{mn} M_{mn} pw_{mn}$

and $\eta_{max/mn} = TC_{mn}\Lambda_{mn} M_{mn} pw_{mn}$ and $\eta_j = \mu_j \Lambda_{mn} M_{mn} pw_{mn}$

Traffic and transport concentration are the maximum number of intermodal trains and volumes of TEUs, respectively, simultaneously operating on a given route of a corridor at the above-mentioned corresponding capacities. As such, they could be primarily relevant for traffic managers monitoring current traffic on the route and rail transport operators considering the size of train fleet to be engaged. It is expressed as follows:

- Traffic (trains/route)

$$TI_{mn} = TC_{mn} * \bar{t}_{mn} \qquad (7.21d)$$

- Transport (TEU/route)

$$TRI_{mn} = TRC_{mn} * \bar{t}_{mn} \qquad (7.21e)$$

Traffic and transport intensity are the maximum number of intermodal trains and volumes of TEUs, respectively, per unit of length of a given route operating at the corresponding (traffic and transport) capacities. Their relevance is similar as that of 'traffic and transport concentration'. It is expressed as follows:

- Traffic (trains/km)

$$TD_{mn} = \frac{TI_{mn}}{L_{mn}} \qquad (7.21f)$$

- Transport (TEU/km)

$$TRD_{mn} = \frac{TRI_{mn}}{L_{mn}} \qquad (7.21g)$$

Transport work (TEU-km) is defined as the maximum volume of TEUs transported on the given route operating at the transport capacity. As the basic measure of output, this is relevant for the rail infrastructure and transport service providers. It is expressed as follows:

$$TW_{mn} = TRC_{mn} * L_{mn} \tag{7.21h}$$

Productive capacity (TEU-km/TU) is the product of transport capacity and the train's operating speed. It incorporates the transport capacity relevant for transport service providers and the transit speed relevant for users, i.e. shippers and receivers of TEUs. It is expressed as follows:

$$TP_{mn} = TRC_{mn} * \overline{V}_{nm} \tag{7.21i}$$

TEU transit speed and time are the average operating speed and ratio between length and corresponding average operating speed, including the anticipated delays of the train's services, respectively. They are of high relevance for both users and (rail/road) transport service providers – the former interested in as high as possible speed and as short as possible time of delivering TEUs and the latter interested in providing them both just as expected (planned). It is expressed as follows:

- Speed (km/h)

$$\overline{V}_{mn} = \frac{\sum_{i=m}^{n-1} v_{i,i+1}(l_{i,i+1})}{(n-1-m)} \tag{7.21j}$$

- Time (TU)

$$\overline{t}_{mn} = \sum_{i=m}^{n-1} [\frac{l_{i,i+1}}{v_{i,i+1}(l_{i,i+1})} + D_{i,i+1}] \tag{7.21k}$$

TEU delivery speed and time are the corresponding averages between the doors of shippers and receivers of TEUs located in the gravitational area of the start and the end terminal(s) of the given route of a corridor. They are also preferred to be as high as possible and as short as possible, respectively, for both users and (rail/road) transport service providers under given conditions. It is expressed as follows:

- Speed (km/h)

$$V_{mn} = \frac{l_m + L_{mn} + l_n}{\left[\frac{l_m}{v_m} + \frac{1}{2}\tau_m + D_m + \sum_{i=m}^{n-1}\left(\frac{l_{i,i+1}}{v_{i,i+1}(l_{i,i+1})} + D_{i,i+1}\right) + \frac{1}{2}\tau_n + D_n + \frac{l_n}{v_n}\right]} \tag{7.21l}$$

where $\tau_m = \dfrac{Q_{max/mn}}{n_m \theta_m \delta_m}$; $\tau_n = \dfrac{Q_{max/mn}}{n_n \theta_n \delta_n}$; $l_m \approx c_m \sqrt{A_m}$; $l_n \approx c_n \sqrt{A_n}$

- Time (TU)

$$t_{mn} = \left[\frac{l_m}{v_m} + \frac{1}{2}\tau_m + D_m + \sum_{i=m}^{n-1}\left(\frac{l_{i,i+1}}{v_{i,i+1}(l_{i,i+1})} + D_{i,i+1} \right) + \frac{1}{2}\tau_n + D_n + \frac{l_n}{v_n} \right] \quad (7.21m)$$

Reliability of services (%) is the ratio between the actually realized and planned/scheduled rail transport services during a specified period of time. This is relevant for both users, i.e. shippers and receivers of TEUs and providers of transport services along the given route of a corridor. It is expressed as follows:

$$R_{mn} \quad (7.21n)$$

Punctuality of services (%) is the ratio between the number of delayed and on-time rail transport services during a specified period of time. This is also relevant for both users, i.e. shippers and receivers of TEUs and transport service providers. It is expressed as follows:

$$P_{nm} \quad (7.21o)$$

Train fleet size (sets) is the number of train sets of given (the same) composition operating along the route of a corridor. This is relevant for providers of rail transport services while planning the train fleet to be engaged under given conditions. It is expressed as follows:

$$N_{1/mn} = TC_{mn} * 2 * \left(\frac{1}{2}\tau_m + \overline{t_{mn}} + \frac{1}{2}\tau_n \right) \quad (7.21p)$$

Utilization of train fleet (i.e. load factor) (%) is the ratio between the volumes of transported TEUs and the offered payload capacity of a train fleet on the given route of a corridor during a specified period of time. This is particularly relevant for the providers of rail transport services. It is expressed as follows:

$$U_{1/mn} = \frac{Q_{max/mn}}{TRC_{nm}} * 100 \quad (7.21q)$$

where

P_j, A_j is the population of users (i.e. shippers and receivers of TEUs) and the size of gravitational area, respectively, of the terminal (j), (counts, km²);

q_{jg}, q_{ja} is the volume of TEUs generated and/or attracted, respectively, by users (shippers and/or receivers) in the gravitational area of terminal (j) ($j \in (m, n)$ (TEUs);

$Q_{max/mn}$ is the maximum volume of TEUs on the route (m, n) (TEUs);

Λ_{mn}	is the average load factor of a train operating on the route (m, n) $(\Lambda_{i, i+1} \leq 1)$;
λ_{mn}	is the average load factor of a truck operating in the gravitational area of terminal (m) and (n) $(\lambda_{mn} \leq 1)$;
$f_{i,i+1}, f_{mn}$	is the train service frequency on the link $(i, i+1)$ and the route (m, n), respectively (trains/TU);
$h_{i,i+1}, h_{mn}$	is the minimum time between successive departures of trains on the link $(i, i + 1)$ and the route (m, n), respectively (TU/train);
$v_{i,i+1}, t_{i,i+1}$	is the average train's operating speed and transit time, respectively, on the link $(i, i +1)$ (km/h; TU);
$D_{i, i+1}$	is the average train anticipated delay on the link $(i, i + 1)$(TU);
R_{mn}, P_{mn}	is the reliability and punctuality, respectively, of the delivery services along the given route (m, n) of a corridor (%; %);
τ_m, τ_n	is the average train's service time at the terminals (m) and (n), respectively (TU);
D_m, D_n	is the average train's anticipated delay while being served in the terminals (m) and (n), respectively (TU); and
v_m, v_n	is the average speed of truck(s) in the gravitational area of terminals (m) and (n) along the distance l_m and l_n, respectively (km/h).

The other symbols are analogous to those in Equations 7.19-7.20.

(d) Economic performances

The indicator system of economic performances of a given corridor and its specified route consists of the following indicators and measures:

Operational cost (€/train)[10] is the monetary expense of delivering the given volume(s) of TEUs from shippers to receivers at both ends of the given route of a corridor. This indicates efficiency of the transport service providers, i.e. intermodal (rail, road, terminal) operators, competing internally between themselves and externally with other transport mode counterparts. For users, i.e. shippers and receivers of TEUs, it is relevant due to reflecting charges (prices) of the offered services. It is expressed as follows:

$$C_{o/mn} = Q_{max/mn} * (c_{2o/m} l_m + c_{o/m} + L_{mn} c_{1o/mn} + c_{o/n} + c_{2o/n} l_n) \qquad (7.22a)$$

---

[10]Due to simplicity, the unit operational and external costs of particular phases are shown constant, but actually they decrease more than proportionally with the increase in both transport distance and volume of TEUs at both road and rail transport mode, and intermodal terminals as well. The external cost includes the cost of emissions of GHG, land use, noise and traffic incidents/accidents (Janić 2007b).

External cost (€/train) is the monetary expense of damages to society and environment by delivering TEUs from shippers to receivers at both ends of the given route of a corridor. As internalized, this cost is again relevant for the overall efficiency of transport service providers after being included in charges (prices) of their services. It is expressed as follows:

$$C_{e/mn} = Q_{max/mn} * (c_{2e/m}l_m + c_{e/m} + L_{mn}c_{1e/mn} + c_{e/n} + c_{2e/n}l_n) \qquad (7.22b)$$

Time cost (€/train) is the (monetary) value of time of TEUs during their delivery from shippers to receivers at both ends of the given route of a corridor. This can be particularly relevant for shippers and receivers dealing with time sensitive (decaying) and/or high value goods. It is expressed as follows:

$$C_{t/mn} = Q_{max/mn} \left\{ \left[\frac{l_m}{v_m} + \frac{1}{2}\frac{Q_{max/mn}}{\beta_m} \right] \alpha_m + \bar{t}_{mn}\alpha_{mn} + \left[\frac{1}{2}\frac{Q_{max/mn}}{\beta_n} + \frac{l_n}{v_n} \right] \alpha_n \right\}$$

$$(7.22c)$$

Total cost (€/train) is the sum of operational, external and time cost of delivering TEUs along the given route of a corridor under given conditions. As such, they are relevant for the managers/governors of a corridor while assessing its competitiveness. It is expressed as follows:

$$C_{mn} = C_{o/mn} + C_{e/mn} + C_{t/mn} + C_{i/mn} + C_{w/mn} \qquad (7.22d)$$

Investments and subsidies (€/train) include expenses for maintenance of the infrastructure and supporting facilities and equipment along the given route of a corridor, acquisition of rail and road rolling stock and eventually subsidizing new intermodal (rail/road) transport services. On the one hand, this is relevant for the above-mentioned receivers of funds and on the other, for investors and subsidizing institutions/authorities interested in the effects of their placed funds. It is expressed as follows:

$$C_{i/mn} = L_{mn} * c_{mn} + TW_{mn} * c_{s/mn} \qquad (7.22e)$$

Contribution to the welfare (€/train) is the monetary contribution of the intermodal rail/road services carried out along the given route of a corridor to the regional and national social welfare, usually expressed as contribution to GDP (Gross Domestic Product). This can be particularly relevant for the local population around the route and a corridor. It is expressed as follows:

$$C_{w/mn} = PR_{mn} * TW_{mn} \qquad (7.22f)$$

where

$c_{1o/mn}, c_{1e/mn}$ is the average unit operational and external cost, respectively, of a train operating on the route (m, n) (€/TEU-km);

$c_{2o/m}, c_{2e/n}$	is the average unit operational and external cost, respectively, of a truck operating in the gravitational area of terminals (m) and (n), respectively (€/TEU-km);
$c_{o/m}, c_{o/n}$	is the average unit operational cost of the terminals (m) and (n), respectively (€/TEU);
PR_{nm}	is the average unit social-economic benefits from the rail freight transport services on the route (m, n) (€/t-km);
$c_{e/m}, c_{e/n}$	is the average unit external cost of the terminals (m) and (n), respectively (€/TEU);
$\alpha_m, \alpha_{mn}, \alpha_n$	is the average unit cost of TEU's time while at the terminal (m), route (m, n) and the terminal (n) respectively (€/TEU-TU);
β_m, β_n	is the average rate of collecting and distributing TEU flows to/from the terminals (m) and (n), respectively (TEU/TU); and
$c_{mn}, c_{s/mn}$	is the average unit investment cost and subsidies, respectively, for the rail freight transport services, respectively, on the route (m, n) (€/km; €/TEU-km).

The other symbols are analogous to those in Equations 7.19 - 7.21.

(e) Environmental performances

The indicator system of environmental performances of a given corridor and its specified route consists of the following indicators and measures:

Energy/fuel consumption and *related emissions of GHG*[11] are the quantities consumed and emitted, respectively, by transporting TEUs between shippers and receivers at both ends of the given route of a corridor. They are relevant for all actors/stakeholders directly and/or indirectly involved or affected by operations on the route. Transport service providers intend to maximize the energy/fuel and GHG emissions efficiency by deploying innovative technical/technological and operational measures in order to minimize the local and global impacts of their operations. It is expressed as follows:

- Energy/fuel consumption (kWh)

$$EC_{mn} = TW_{mn} * e_{1/mn} + Q_{max/mn} * (l_m e_{2m} + e_m + e_n + l_n e_{2n}) \qquad (7.23a)$$

- Emissions of GHG[11] (ton)

$$EM_{mn} = TW_{mn} * e_{1/mn} * r_{1/mn} + Q_{max/mn} * (l_m e_{2m} r_{2m} + e_m r_m + e_n r_n + l_n e_{2n} r_{2n})$$
$$(7.23b)$$

---

[11] The energy/fuel consumption is directly proportional to the product of the volumes of TEUs and the unit energy/fuel consumption; the latter is assumed to be constant under given conditions. The emissions of GHG expressed in CO_{2e} (CO, CO_2, NO_X, H_2O, particles, etc.) are proportional to the total energy/fuel consumption and the corresponding emission rates of GHG.

Land use (km^2) is the total area of land used for setting down the corridor's infrastructure, including the intermodal terminals, road network(s) in their gravitational areas and the rail links connecting them. It is expressed as follows:

$$LU = \sum_{j=1}^{N} A_j + \sum_{i=1}^{N-1} [l_{i,i+1} - d_i - d_{i+1}] * s_{i,i+1} \qquad (7.23c)$$

where

$e_{2m}, e_{2n}, r_{2m}, r_{2n}$	is the average rate of energy/fuel consumption and related emissions of GHG, respectively, of a truck or regional train operating in the gravitational area of terminal (m) and (n), respectively (kWh/TEU-km; kgCO$_{2e}$/kWh; alternatively, the energy/fuel consumption can be expressed in l of fuel/100 km);
$e_{1/mn}, r_{1/mn}$	is the average rate of energy/fuel consumption and related emissions of GHG, respectively, of a train operating on the route (m, n) (kWh/TEU-km), (kgCO$_{2e}$/kWh);
$e_m, e_n; r_m, r_n$	is the average rate of energy consumption and related emissions of GHG, respectively, of the terminals (m) and (n), respectively (kWh/TEU; kgCO$_{2e}$/kWh),
d_i, d_{i+1}	is the approximate radius of the gravitational area of the terminals constraining the link (i) and $(i+1)$, respectively (km); and
$s_{i,i+1}$	is the width of the land strip around the link $(i, i+1)$ not overlapping with the gravitational area of terminals on its both ends (km).

The other symbols are analogous to those in Equations 7.19-7.22.

(f) Social performances

The indicator system of social performances of a given corridor and its specified route include the following indicators and measures:

Noise—cumulative level and spatial intensity[12] is the noise and the noise per unit length of the route, respectively, generated by trains or by trucks and trains while delivering TEUs between their shippers and receivers at both ends of the given route of a corridor. This appears relevant for the local population, which is already exposed to noise by other sources. It is expressed as follows:

---

[12]The noise intensity is assumed to be uniformly distributed along the door-to-door distance in line with distribution of the primary sources—vehicles—trucks and trains. The noise due to transshipment of TEUs in the terminals is not taken into account.

- Cumulative level [dB(A)]

$$L_{eq/mn} = 10\log \sum_{k=1}^{TC_{mn}} 10^{\frac{L_{1/eq}(k)}{10}} + 10\log \sum_{k=1}^{N_{2/mn}} 10^{\frac{L_{2m/eq}(k)+L_{2n/eq}(k)}{10}} \qquad (7.24a)$$

where $N_{2/mn} = \dfrac{Q_{max/mn}}{\lambda_{mn} * w_{mn}}$ (trucks/train)

- Spatial intensity (dB(A)/km)

$$\overline{L}_{eq/mn} = \frac{L_{eq/mn}}{l_m + L_{mn} + l_n} \qquad (7.24b)$$

Congestion (TU) is the time loss of cars due to interfering with trucks or of the passenger trains due to interfering with regional trains around the intermodal rail/road or rail/rail freight terminals, respectively, while transporting TEUs on the given route of a corridor. This can be relevant for the above-mentioned affected parties. It is expressed as follows:

$$D_{mn} \qquad (7.24c)$$

Safety (counts/TU) is the risk of traffic incidents and accidents, which could happen due to transporting TEUs between the users, i.e. shippers and receivers at both ends of the given route of a corridor. This is relevant for all actors/stakeholders involved in operations on the route (and a corridor). Also, it is relevant for the third party, i.e. the local population, exposed to the risk of such incidents and accidents and their consequences—injuries, loss of life and property damage. It is expressed as follows:

$$TAC_{mn} = TW_{mn} * a_{1/mn} + Q_{max/mn} * (l_m + l_n) * a_{2/mn} \qquad (7.24d)$$

where

$L_{1/eq}(k)$ is the noise generated by (k)-th train operating on the route (m, n) [dB(A)];

$L_{2m/eq}(k), L_{2n/eq}(k)$ is the noise generated by (k)-th truck or regional freight train operating in the gravitational area of terminal (m) and (n), respectively [dB(A)]; and

D_{mn} is the average delay, i.e. the time loss, as the difference between the actual and planned delivery of TEUs along the route (m, n) (TU); and

$a_{1/mn}, a_{2/mn}$ is the average rate of traffic incidents/accidents/fatalities of the freight (intermodal) trains and regional trains or road trucks, respectively, operating on the route (m, n) and in the gravitational area of terminals (m) and (n), respectively (counts/TEU-km).

The other symbols are analogous to those in Equations 7.19-7. 23.

7.5.3.4 The Indicators of Performances as the Evaluation Attributes/Criteria for the MCDM Methods

The multi-criteria evaluation of the alternative rail freight transport corridors can be carried out by using some of the above-mentioned MCDM methods, in this case SAW and TOPSIS and the indicators and measures of corridors' performances as the evaluation attributes/criteria, which are summarized in Table 7.25.

7.5.5 Application of the Evaluation Methodology

7.5.5.1 Inputs

(a) Geography of the case corridors

The proposed methodology has been applied to evaluation of two intermodal rail/road trans-European freight transport corridors named in the EU-funded research as RETRACK (REorganization of Transport Networks by Advanced RAil Freight Concepts) and CREAM (Customer-driven Rail-freight services on a European mega-corridor based on Advanced business and operating Models) (EC 2008, 2012b).

The RETRACK corridor spreads between the North Sea and the Black Sea gateways, from Rotterdam (Netherlands) and Antwerp (Belgium) to Constanta (Romania). This corridor passes through The Netherlands, Germany, Austria, Hungary, Romania, Bulgaria and finally Turkey. As such, its overall length between the farthest origin(s) and destination(s) of freight/goods (TEU) flow exceeds 1,500 km, of which the main route between Cologne (Germany) and Gyor (Hungary) is 1,220 km. The RETRACK corridor partially overlaps with the RNE Corridor C02 in its northern part: Rotterdam/Antwerp-Cologne and RNE Corridor C09, TNT-T Priority axis 22 in its middle and south-eastern part: Budapest – Gyor – Bucureşti – Constanţa/Kulata/Svilengrad/Varna/Burgas and partially with the ERTMS E corridor Dresden – Constanţa. The CREAM corridor passes through the Benelux countries (The Netherlands, Belgium), Germany, Austria, Hungary (the main route of 908km), Romania, Bulgaria, Serbia-Montenegro and Turkey/Greece. The CREAM corridor of the total length of 3,150 km partially overlaps with the RNE Corridor C11: Munich – Salzburg – Ljubljana – Zagreb – Beograd – Sofia – Istanbul (EC 2008, http://www.rne.eu). Figure 7.7 shows a simplified layout of both the corridors. The intermodal transport services in the RETRACK corridor are provided mainly by private rail operators with the focus mainly on containerized and marginally on non-containerized goods/freight shipments. In the CREAM corridor, these are mainly national rail operators focusing exclusively on containerized goods/freight shipments. In order to make both corridors comparable for the

Table 7.25: Indicators and measures of performances as the MCDM
attributes/criteria

Indicator/Measure	Notation	Preference
Physical/spatial or infrastructural		
Corridor length (km)	L	+
Accessibility (terminals/km)	A	+
Area coverage (km²)	S_a	+
Infrastructure density (km/km²)	ID	+
Route length (m, n) (km)	L_{mn}	+
Maximum axle load (ton/axis)	AXL_{mn}	+
Technical/technological		
Propulsion systems (counts)	k	-
Electric propulsion systems (counts, power) (-; kV)	$m, VP_{i,i+1}$	-, +
Length/weight, payload capacity and technical speed of a train (m/t; TEU/train; km/TU)	$Z_{mn}/W_{mn}, PW_{mn}, V_{mn}$	+,+,+,+
Length/weight and payload capacity of a train wagon (m/t; TEU/wagon)	$z_{mn}/w_{mn}/pw_{mn}$	+,+ +
Number of wagons per train (counts/train)	N_{mn}	+
Payload capacity of road truck (TEU/truck)	$W_{r/mn}$	+
Length of rail tracks in terminal(s) (counts, m)	L_j	+
Number, service rate and utilization of terminal transshipment facilities and equipment (counts, TEU/TU, %)	n_j, θ_j, δ_j	-,+,+
Interoperability (%)	IOP	+
Operational		
Demand density (TEU/km²)	DD	+
Capacity • Traffic (trains/TU) • Transport (TEU/TU)	TC_{mn} TRC_{mn}	+ +
Concentration • Traffic (trains/route) • Transport (TEU/route)	TI_{mn} TRI_{mn}	+ +

(Contd.)

Table 7.25: (*Contd.*)

Intensity		
• Traffic (trains/km)	TD_{mn}	+
• Transport (TEU/km)	TRD_{mn}	+
Transport work (TEU-km)	TW_{mn}	+
Productive capacity (TEU-km/TU)	TP_{mn}	+
TEU transit		
• Speed (km/TU)	$\overline{V}_{mn}$	+
• Time (TU)		-
	$\overline{t}_{mn}$	
TEU delivery		
• Speed (km/TU)	V_{mn}	+
• Time (TU)	t_{mn}	-
Reliability of services (%)	R_{mn}	+
Punctuality of services (%)	P_{mn}	+
Train fleet size (sets)	$N_{1/mn}$	-
Utilization of train fleet (%)	$U_{1/mn}$	+
Economic		
Operational + time cost (€/TU)	$C_{o/mn}$	-
External costs (€/TU)	$C_{e/mn}$	-
Total cost (€/TU)	C_{mn}	-
Investments and subsidies (€)	$C_{i/mn}$	-
Contribution to the welfare (€)	$C_{w/mn}$	-
Environmental		
Energy/fuel consumption (MWh/TU)	EC_{mn}	-
Emissions of GHG (t/TU)	EM_{mn}	-
Land use (km²)	LU	-
Social		
Noise		
Cumulative level (dBA)	$L_{eq/mn}$	-
Spatial intensity (dBA/km)		-
	$\overline{L}_{eq/mn}$	
Congestion (TU)	D_{mn}	-
Safety (counts/TU)	TAC_{mn}	-

The sign (+) or (–) indicates preference for as high as or as low as possible values of particular indicators and measures of performances, respectively.

purpose of evaluation, the particular indicators and measures of performances have been expressed in tons (t), where necessary.

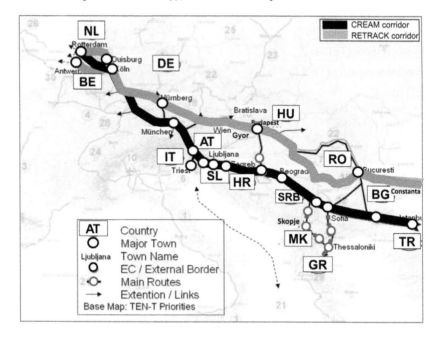

Fig. 7.7: Simplified layout of the RETRACK and CREAM corridor (EC 2008, 2012b).

(b) The evaluation attributes/criteria and their weights

The inputs for applying the models for estimating the indicators and measures of performances were obtained from the demonstration phase of the above-mentioned two EU projects. As such, these inputs strongly influenced the values of particular indicators and measures and consequently the multi-criteria evaluation score, without affecting the generosity of the proposed methodology. The values of these 48 indicators and measures of performances estimated by the models are given in the self-explanatory Table 7.26.

The weights for particular attributes of the corridors' performances as criteria, also given in Table 7.26, have been estimated by using the above-mentioned entropy method. As such, they could be considered to reflect the relative preferences of particular criteria from the research perspective.

7.5.5.2 Results

The indicators and measures of performances and their weights in Table 7.26 are used as the input for the SAW and TOPSIS methods. Due to the

Table 7.26: The estimated indicators and measures of performances as the attributes/criteria and their weights for evaluation of two alternative intermodal (rail/road) freight transport corridors

Indicator/Measure – Attribute/ Criteria	Alternative		Weight
	1 **RETRACK**	**2** **CREAM**	
Physical/spatial or infrastructural			
Corridor length (km)	1800	3150	0.025
Accessibility (terminals/100km)[1]	0.5	0.4	0.056
Area coverage (km²)[2]	70650	94200	0.007
Infrastructure density (km/km²)	0.025	0.033	0.006
Route length (km)	1220	908	0.083
Maximum axle load (ton/axis)	20	20	-
Technical/technological			
Propulsion systems (number)	1	2	0.0380
Electric propulsion systems (counts, power) (-; kV)	3	4	0.0034
Length/weight, payload capacity and technical speed of a train (m/ton; ton/train; km/h)	450/969/100	475/1026/100	0.0005
Length/weight and payload capacity of a train wagon (m/ton; ton/wagon)	25/77/57	25/77/57	0.0005
(Maximum) number of wagons per train (counts/train)	17	18	0.0005
Payload capacity of a road truck (ton/truck)	26	26	-
Length of rail tracks in terminal(s) (-, m)[4]	525	-	-
Number, service rate and utilization of terminal transshipment facilities and equipment (counts, ton/hr, %)[4]	3/35/80	-	-
Interoperability (-)	1.70	1.25	0.0035
Operational			
Demand density (t/km²)[3]	2.924	18.072	0.1869
Capacity • Traffic (trains/wk) • Transport (t/wk)	4 3876	5 5130	0.0040 0.0060

(Contd.)

Table 7.26: (*Contd.*)

Indicator/Measure – Attribute/ Criteria	Alternative		
	1 RETRACK	2 CREAM	Weight
Concentration			
• Traffic (trains/route/wk)	0.806	0.711	0.01084
• Transport (t/route/wk)	781.0	729.6	0.00415
Intensity			
• Traffic (trains/100 km)	0.066	0.00078	0.0020
• Transport (ton/km)	0.640	0.803	0.0005
Transport work (million t-km/wk)	4728.7	4658.0	0.0005
Productive capacity (t-km/h)[5]	23256	37962	0.0194
TEU transit			
• Speed (km/h)	36.0	38.0	0.0022
• Time (hr)	33.8	23.9	0.0097
TEU delivery			
• Speed (km/h)	24.0	37.0	0.0152
• Time (hr)	50.8	24.3	0.0420
Reliability of services (%)	83	86	0.00045
Punctuality of services (%)	90	95	0.00046
Train fleet size (sets)	2	2	-
Utilization of train fleet capacity (%)	70	75	0.00046
Economic			
Operational + time cost (thousand €/wk)[6]	125.098	158.288	0.29124
External costs (thousand €/wk)[6]	35.678	41.143	014667
Total cost (thousand €/wk)[6]	160.176	199.924	0.00415
Investments and subsidies (€/wk)[4]	-	-	-
Contribution to the welfare (thousand €/wk)[7]	699.4	688.9	0.00415
Social			
Noise			
• Cumulative level (dBA)	180.94	182.14	0.00005
• Spatial intensity (dBA/km)	0.137	0.185	0.00738
Congestion (hr)[4]	-	-	-
Safety (counts)[8]	$8.078 * 10^{-9}$	$2.510 * 10^{-9}$	0.00533

(*Contd.*)

Table 7.26: (*Contd.*)

Environmental			
Energy/fuel consumption (MWh/wk)[9]	116.12	110.40	0.00023
Emissions of GHG (t/wk)[9]	53.455	50.828	0.00622
Land use (km²) (thousand)[10]	180	315	0.02516

[1]Only those located nearby main cities are considered; [2]The radius of the gravitational area of each intermodal terminal is considered to be 50km; [3]Ratio of the total quantity of goods/freight and the gravitational area of all terminals (RETRACK – 206590 t; CREAM – 1702416 t); [4]Not taken into evaluation due to lack of data; [5]Based on the goods/freight delivery speed; [6]Based on Janić 2007b; [7]Estimated from EC 2012; [8]An estimate based on the past data; [9]Based on Janić and Vleugel 2012; [10]Estimated as the product of length of the corridor and radius of terminals of 50 km.

similarity (equality) and/or the lack of relevant data, some attributes/criteria (8) such as maximum axle load, characteristics of train wagons, truck payload capacity, characteristics of intermodal terminals, train fleet size, investments and subsidies and congestion have been dropped off from the evaluation procedure, thus considering the remaining 36. The results are shown in Tables 7.27 and 7.28.

Table 7.27: Results from the SAW method in the given example

Performances	Alternative 1: RETRACK A*	Alternative 2: CREAM A*	Rank/ Alternative
Physical/spatial or infrastructural	0.89019	0.82282	1/1
Technical/technological	0.99849	0.54794	1/1
Operational	0.42351	1.00000	1/2
Economic	1.00000	0.94659	1/1
Social	1.00000	0.01172	1/1
Environmental	0.99905	0.57860	1/1
All	**0.796**	**0.436**	1/1

Table 7.28: Results from TOPSIS method in the given example

Alternative/Scores	S_i^+	S_i^-	C_i	Rank
Alternative 1: RETRACK	0.61555	1.57639	0.71918	1
Alternative 2: CREAM	0.77229	1.51136	0.66182	2

Note: The symbols S_i^+, S_i^- and C_i are explained in sub-section 7.2.

As can be seen, according to both SAW and TOPSIS methods, the RETRACK corridor proved a better alternative considering all criteria, thus indicating in some sense the feasibility of using either of the MCDM methods in the given context. In addition, as given in Table 7.4, both corridors were evaluated using the SAW method with respect to particular categories of performances. In this case, the RETRACK corridor proved a better alternative with respect to all, except the operational performances. In particular, regarding the physical/spatial performances, the RETRACK corridor is a better alternative mainly due to the longer main route and higher (spatial) accessibility of the intermodal (rail/road) services. These two performances fully compensate its weaknesses reflected through the overall length of the corridor, the total area coverage and the infrastructure density.

Regarding the technical/technological performances, the RETRACK corridor has again shown to be a better alternative. This is mainly due to the higher interoperability, i.e. the smaller number of different (country specific) propulsion (and electrical propulsion) systems used and despite operating shorter and lighter trains. In both the corridors, these trains composed of similar types of wagons and carrying similar types of loading units, such as ISO and non-ISO containers, swap-bodies and semi-trailers, are pulled by multisystem electric engines, thus diminishing the impacts of differences in the electric propulsion systems along the route(s). In addition, in the CREAM corridor, diesel engines are also deployed along non-electrified links of the main route, thus additionally diminishing the already lower interoperability of the CREAM trains. Furthermore, the monitoring and tracking/tracing of trains/wagons/goods is carried out by different systems. In the RETRACK corridor, an innovative IT system comprising four components is used, whereas in the CREAM corridor, an IT system—Software Train Monitor and GPS devices + wireless personal area network(s)—were used.

The CREAM corridor was the better alternative regarding its operational performances. This is mainly due to operating transport services on a much larger scale during the demonstration phase carried out by trains with a fixed configuration as compared to those of the RETRACK counterpart of a rather flexible configuration. Consequently, almost all directly related indicators and measures of these performances, such as demand concentration, capacity concentration, intensity, transport work (frequency of the 'shuttle' train services operating on the shorter routes) and utilization of the train fleet (fixed train composition along the route with sufficient demand secured at the beginning) have been higher at CREAM than at its RETRACK counterpart. In addition, the transit and delivery speed of goods/freight shipments by the CREAM trains was higher and the corresponding times substantively shorter. These have been influenced mainly by differences in the length of route(s), very low but similar commercial speeds and the long time taken by the RETRACK wagons and wagon groups at two hubs (about one-third of the total train-service time). In addition, the reliability and punctuality of CREAM services was higher despite being carried out by longer, heavier and less interoperable trains on a shorter route.

The RETRACK corridor was a better alterative with respect to its economic performance mainly due to the lower operating time, external and consequently total cost, including also a slightly higher contribution to the overall welfare. The former three have mainly been due to carrying out a much lower scale of operations under the given conditions—lower service frequency by trains with lower payload capacity. The latter is due to carrying out operations on a substantially lower scale but on a longer route, which has compensated for the lower scale of operations. As in the case of the economic performances, the RETRACK corridor was the better alternative due to the social performances. This is mainly thanks to the lower cumulative noise and their spatial densities and the higher level of safety, i.e. lower risk of traffic incidents and accidents under the given conditions. The RETRACK corridor scored higher regarding the environmental performances thanks to the lower energy consumption, emissions of GHG and land use. However, all these three categories of performances were due to the substantially lower absolute scale of operations in the RETRACK when compared to the CREAM corridor.

Consequently, as mentioned above, thanks to performing better in most categories of performances, the RETRACK corridor proved a better alternative overall, i.e. regarding the above-mentioned 36 attributes/criteria taken into account in the multi-criteria evaluation.

7.5.6 Interim Summary

This section deals with synthesizing and application of the methodology for evaluating rail freight transport corridors. This includes defining and developing (analytical) models of the indicator systems of corridors' physical/spatial or infrastructural, technical/technological, operational, economic, environmental and social performances. These indicators and measures of performances are used as criteria in application of two existing MCDM methods—the SAW and TOPSIS—to the evaluation of two trans-European intermodal (rail/road) freight transport corridors—RETRACK and CREAM. The results show that the selected MCDM methods could work reasonably well, implying at the same time their generosity to be applied to other similar cases. Both selected MCDM methods produced the same ranking of just two alternatives characterized by a relatively large number of evaluation criteria (44 in total, of which 36 were considered).

Regarding the specificity of two alternative corridors, the RETRACK corridor seemed preferable regarding all except the operational performances, as a result of which it was identified as the preferable alternative overall, i.e. regarding all the performances considered. In both evaluation cases, the indicators and measures of performances as evaluation criteria were expressed in absolute terms, thus highlighting the high influence of the scale of operations along the corridors on the final evaluation score(s). In addition, it would be very easy to apply the above-mentioned evaluation procedure to conditions when all criteria are expressed exclusively in relative rather than in the above-mentioned mixed relative/absolute terms. Furthermore, the above-mentioned score(s) implies that the particular actors/stakeholders involved, such as the users—goods/freight shippers and receivers, providers of transport infrastructure and services, local, regional and national corridor-governing authorities and policy makers—could carefully consider prospective scenarios, conditions and related performances while guiding and managing development and improvements in particular existing rail transport corridors in Europe and elsewhere.

7.6 Concluding Remarks

This chapter deals with evaluation of transport systems by using existing MCDM (Multi Criteria Decision Making) methods. These are the SAW (Simple Additive Weighting), TOPSIS (Technique for Order Preference by Similarity to the Ideal Solution) and the AHP (Analytic Hierarchy Process) methods. They have been applied to three cases of transport systems. The first has been the case of selection of a new (second) hub airport by an airline; the second is the selection of the airport within an airport system where a new runway is to be built to increase the capacity of the given airport system; and the last case evaluates two rail freight transport corridors. In all these cases, the indicators and measures of infrastructural, technical/technological, operational, economic, environmental and social performances have been modelled and estimated using the case-specific inputs and then used as evaluation attributes/criteria by the particular MCDM method(s). The relative importance, i.e. weights of particular attributes/criteria has been estimated by the available analytical (entropy) method or simulation. This, however, should not prevent obtaining these weights from the particular actors/stakeholders involved in the evaluation processes. In some cases, the above-mentioned methods produced different results regarding the preferable alternative, which underscores their careful selection and use.

In the first above-mentioned case, seven European airports were preselected as alternatives with their relevant performance as attributes/ criteria. The results of the evaluation indicate that an airline would prefer a large, already well-developed airport to enable the space for new entry and relatively modest costs of including it in the existing airline network. In the second case, three airports of the given airport system were considered as potential alternatives for building a new runway. The evaluation results show that the already largest airport within the system (in this case London Heathrow airport) is the preferred alternative. In the last case, two alternative intermodal (rail/road) freight transport corridors were evaluated. The results show that if the particular criteria of performances are taken in absolute terms, the scale of operations significantly influences the preferred alternative corridor.

In general, the above-mentioned cases of evaluation of transport systems show that the proposed MCDM methods could be a useful support to the DM processes, at least regarding the consistency of the approach, which seems to be increasingly needed in these but also in other similar though rather complex, sensitive and controversial cases.

References

ABC. 1998. *Executive Flight Planner*. REED Travel Group, ABC International Division, UK.

AC. 2013a. *Airports Commission: Proposals for Providing Additional Runway Capacity in the Longer Term—Gatwick Airport Limited Response*. Airports Commission, London Gatwick, London, UK.

AC. 2013b. Discussion Paper 05: *Aviation Noise*. Airports Commission, London, UK.

ACI. 1997. The World's Airports in 1996: Airport Ranking by Aircraft Movements, Total Passengers and Cargo. Airport Council International, http:// www.airports. org./move96.html.

ACI Europe. 1998. *Creating Employment and Prosperity in Europe*. A study by ACI EUROPE of the Social and Economic Impact of Airports. Airport Council International, Europe, Brussels, Belgium.

Adler N., Berechman J. 2001. Evaluating Optimal Multi-hub Networks in a Deregulated Aviation Market with an Application to Western Europe. *Transportation Research* A 35: 373-390.

Ashford N. 1988. Level of Service Design Concept for Airport Passenger Terminals: A European View. *Transportation Research Record* 1199: 19-32.

AW. 2000. Think Local – Act Global. *Airliner World*, January: 34-37.

Aykin T. 1995. Networking Policies for Hub-and-Spoke Systems with Application to the Air Transportation System. *Transportation Science* 29: 201-221.

Bailey E., Graham R.D., Kaplan P.D. 1985. *Deregulating the Airlines*. The MIT Press, Cambridge, MA, USA.

Berechman J., De Wit J. 1996. An Analysis of the Effects of European Aviation Deregulation on an Airline's Network Structure and Choice of a Primary West European Hub Airport. *Journal of Transport Economics and Policy* 30: 251-274.

Bethany S., Holland H.J., Noberga A.A.R., O'Hara G.C. 2011. Using Multi-criteria Making to Highlight Stakeholders' Values in the Corridor Planning Process. *The Journal of Transport and Land Use* 4: 105-118.

Bontekoning Y., Macharis C., Trip J.J. 2004. Is a New Applied Transportation Field Emerging?—A Review of Intermodal Rail-Truck Freight Transport Literature. *Transportation Research* A 38: 1-34.

Bowen B.D., Headly D.E. 2002. *The Airline Quality Rating*. Report, The University of Nebraska at Omaha, Wichita State University, Wichita, USA.

Brucker D.K., Macharis, C., Verbeke A. 2011. Multi-Criteria Analysis in Transport Project Evaluation: An Institutional Approach. *European Transport\Trasporti Europei* 7: 3-24.

Burghouwt G., Hekfoort J., Van Eck J.R. 2002. *Airline Network Configuration in the Deregulated European Aviation Market*. European Transport Conference 2002. CD-Proceedings. 9/11 September. Cambridge, Oxford, UK.

Button K., Haynes K., Stough R. 1998. *Flying into the Future: Air Transport Policy in the European Union*. Edward Elgar Publishing Limited, Cheltenham, UK.

Button K., Swan D. 1991. Aviation Policy in Europe. pp. 85-123. *In:* Button K.J. (ed.) *Airline Deregulation: National Experiences*. David Fulton Publishers, London, UK.

Button J.K., Stough R.R. 2000. *Air Transport Networks*. Edward Elgar Publishing Limited, Cheltenham, UK.

CAA. 2000. *Quality of Service Issues*. Consultation paper. UK Civil Aviation Authority. London, UK.

CAA. 2011. *Catchment Area Analysis—Airport Market Power Assessments*. Working Paper. Civil Aviation Authority. London, UK.

CAA. 2012. *Airport Market Power Assessment—Annex*. Civil Aviation Authority. London, UK.

CAA. 2013. *UK Airport Statistics*. Civil Aviation Authority. London, UK.

CEC. 2001 *European Transport Policy for 2010: Time to Decline—White Paper*. Commission of the European Communities COM (2001) 370 final. European Communities. Brussels, Belgium.

CEC. 2011. *Roadmap to a Single European Transport Area—Towards a Competitive and Resource Efficient Transport System—White Paper*. Commission of the European Communities COM (2011) 144 final. European Communities. Brussels, Belgium.

Chapman D., Pratt D., Larkham P., Dickins, J. 2003. Concepts and Definitions of Corridors: Evidence from England's Midlands. *Journal of Transport Geography* 11: 179-191.

DeNeufville R., Odoni A. 2003. *Airport Systems: Planning, Design and Management*. The McGraw Hill Companies. New York, USA.

DfT. 2013. *Noise Exposure Contours around London Airports: Noise Exposure Reports and Contours for London Airports (Heathrow, Gatwick and Stansted)*. Department for Transport. London, UK (https://www.gov.uk/government/publications/noise-exposure-contours-around-london-airports/).

Ding Y., Yuan, Z., Li Y. 2008. Performance Evaluation Model for Transportation Corridor Based on Fuzzy AHP Approach. *Fuzzy Systems and Knowledge Discovery* (FSKD '08.), Fifth International Conference on Fuzzy Systems and Knowledge Discovery. IEEE Computer Society, Los Alamitos, California, CA, USA 3: 608-612.

Doganis R. 1992. *The Airport Business*. Routledge. London, UK.

EC. 1999. *Transport in Figures: Statistical Pocket Book—EU Transport*. European Commission. Directorate General DG VII. Brussels, Belgium.

EC. 1998a. Interactions between High Speed Rail and Air Passenger Transports in Europe. *Final Report on the Action COST 318*. European Commission. Luxembourg.

EC. 2008. *Customer-driven Rail-Freight Services on a European Mega-Corridor Based on Advanced Business and Operating Models (CREAM), (different deliverables)*. European Commission. Brussels, Belgium.

EC. 2010. *Impact Assessment of the Single Aviation Market on Employment and Working Conditions for the Period 1997-2007*. Commission Staff Working Document. European Commission. Brussels, Belgium.

EC. 2011. Handbook on Regulation Concerning a European Rail Network for Competitive Freight (Regulation EC 913/2010). European Commission. Brussels, Belgium.

EC. 2012b. *REorganization of Transport Networks by Advanced RAil Freight Concepts (RETRACK), (different deliverables)*. European Commission. Brussels, Belgium.

EEC. 1998. *Capacity Shortfall in Europe*. EEC Report No. 324, PCL 4-E. EUROCONTROL Experimental Centre. Brussels, Belgium.

EEC. 2002. *ATMF Delays to Air Transport in Europe*. EUROCONTROL Experimental Centre/ECAC, CODA Database. Brussels, Belgium.

Ford R.L., Fulkerson R.D. 1962. *Flows in Networks*. Princeton University Press, Princeton, New Jersey, USA.

GAL. 2013. *Directors' Report and Financial Statements for the Year Ended 31 March 2012*. Gatwick Airport Limited, London, UK.

Giuliano, G. 1985. A Multicriteria Method for Transportation Investment Planning. *Transportation Research* A 19A: 29-41.

HAL. 2013. *Heathrow Financial Accounts*. Heathrow Airport Limited, Hounslow, Middlesex, England, UK.

Horton G. 2010. *Future Aircraft Fuel Efficiencies—Final Report*. QinetiQ. Cody Technology Park, Farnborough, UK.

Hwang L.C., Yoon K. 1981. *Multi Attribute Decision-Making: Methods and Applications*. Lecture Series in Economics and Mathematical Systems. Springer-Verlag, Berlin, Germany.

ICAO. 1988. *Digest of Bilateral Air Transport Agreements*. Doc. 9511. International Civil Aviation Organization, Montreal, Canada.

ICAO. 1997. *Traffic by Flight Stage—1995*. Digest of Statistics No. 416, Series TF, No. 108, 3/95. International Civil Aviation Organization, Montreal, Canada.

ICAO. 2010. *Aircraft Technology Improvements*. ICAO Environmental Report 2010. International Civil Aviation Organization, Montreal, Canada.

IFAPA. 1988. *European Airline Mergers: Implications for Passengers and Policy Options*. International Foundation for Airline Passenger Associations, Zurich, Switzerland.

Janić M. 1997. Liberalization of the European Aviation: Analysis and Modelling of the Airline Behavior. *Journal of Air Transport Management* 3: 167-180.

Janić M. 2001. *Air Transport System Analysis and Modelling*. Gordon and Breach Science Publishes, The Netherlands.

Janić M., Reggiani A., Nijkamp P. 1999. Sustainability of the European Freight Transport System: Evaluation of Innovative Bundling Networks. *Transportation Planning and Technology* 23: 129-156.

Janić M., Reggiani A. 2002: An Application of the Multiple Criteria Decision Making (MCDM) Analysis to the Selection of a New Hub Airport. *The European Journal of Transport and Infrastructure Research* 2: 113-142.

Janić M. 2006. Sustainable Transport in the European Union: Review of the Past Research and Future Ideas. *Transport Reviews* 26: 81-104.

Janić M. 2007a. *The Sustainability of Air Transportation: A Quantitative Analysis and Assessment*. Asgate Publishing Limited, Aldershot, UK.

Janić M. 2007b. Modelling the Full Costs of Intermodal and Road Freight Transport Networks. *Transportation Research* D12: 33-44.

Janić M. 2008. An Assessment of Performance of European Long Intermodal Freight Trains (LIFTs). *Transportation Research* A42: 1326-1339.

Janić M. 2010. True Multimodalism for Mitigating the Airport Congestion: Substitution of Air Passenger Transport by High-Speed Rail. *Transportation Research Record* 2177: 78-87.

Janić M., Vleugel J. 2012. Estimating Potential Savings in Externalities from Rail-Road Substitution in Trans-European Freight Transport Corridors. *Transportation Research* D17: 154-160.

Janić M. 2014. *Advanced Transport Systems: Analysis: Modelling and Evaluation of Performances*. Springer, UK.

Janić M. 2015. A Multi-Criteria Evaluation of Solutions and Alternatives for Matching Capacity to Demand in an Airport System: The Case of London. *Transportation Planning and Technology* 38: 709-737.

Janić M. (2016). Analyzing, Modelling and Assessing the Performances of Land Use by Airports. *International Journal of Sustainable Transportation*. DOI:10.1080/15568318.2015. 1104566.

Kirchner C. 2011. *Rail Liberalisation Index 2011*. Market Opening: Rail Markets of the Member States of the European Union. IBM Business Consulting Services, Switzerland and Norway.

Larson C.R., Odoni R.A. 2007. *Urban Operations Research. Dynamic Ideas*. Belmont, Massachusetts, USA.

Lee S.H., Chu C.W, Chen K.K., Chou M.T. 2003. *A Fuzzy Multiple Criteria Decision Making Model for Airline Competitiveness Evaluation*. Proceedings of the Eastern Asia Society for Transportation Studies 5: 507-519.

Ludvigsen J. 2009. Liberalisation of Rail Freight Markets in Central and South-Eastern Europe: What the European Commission can do to Facilitate Rail Market Opening. *European Journal of Transport & Infrastructure Research (EJTIR)* 9: 46-62.

Morrison S., Winston C. 1994. *The Evolution of the Airline Industry*. The Brooking Institute, Washington, D.C., USA.

O'Kelly E.M. 1986. The Location of Interacting Hub Facilities. *Transportation Science* 20: 92-106.

OECD. 1988. *Deregulation and Airline Competition.* Organization for Economic Co-operation and Development, Paris, France.

Oum, H.T., Park J.H., Zhang A. 2000. *Globalization and Strategic Alliances.* Pergamon Press, Oxford, UK.

Ozdemir Y., Basligil H., Karaca M. 2011. *Aircraft Selection Using Analytic Network Process: A Case for Turkish Airlines.* Proceedings of the World Congress on Engineering 2011, Vol. II, WCE 2011, July 6-8. London, UK.

Panamure WLB. 2000. *Light at the End of the Concourse: A Review of European Airports.* Panamure West LB: Experts in Growth. London, UK.

Priemus H., Zonneveld W. 2003. What are Corridors and What Are the Issues? Introduction to Special Issue: The Governance of Corridors. *Journal of Transport Geography* 11: 167-177.

RBI. 1995/1999. *Airline Business. Various Issues.* REED Business Information. BPA International Ltd., Sutton, Surrey, UK.

Romein A., Trip J.J., Vries de J. 2003. The Multi-Scalar Complexity of Infrastructure Planning: Evidence from the Dutch-Flemish Megacorridor. *Journal of Transport Geography* 11: 205-213.

Saaty T.L. 1980. *The Analytic Hierarchy Process: Planning Setting Priorities.* McGraw Hill Text. New York, USA.

Sauian S.M. 2010. *MCDM: A Practical Approach in Making Meaningful Decisions.* Proceedings of the Regional Conference on Statistical Sciences 2010 (RCSS, 10). June 2010: 139-146.

Sawicka H., Weglinski S., Witort P. 2010. Application of Multiple Criteria Decision Methods in Logistics Systems. *Electronic Scientific Journal of Logistics* 6: 99-109.

Schutte I.C., Brits A. 2012. Prioritizing Transport Infrastructure Projects: Towards a Multi-Criterion Analysis. *Southern African Business Review* 16: 97-112.

Stainland M. 1998. *Open Skies–Fewer Planes? Public Policy and Corporate Strategy in EU-US Aviation Relations.* European Policy Paper Series. Centre for West European Studies, University of Pittsburgh, Pittsburgh, UK.

Stasinopoulos D. 1992. The Second Aviation Package of the European Community. *Journal of Transport Economic and Policy* 26: 83-87.

Stasinopoulos D. 1993. The Third Phase of Liberalization in Community Aviation and the Need for Supplementary Measures. *Journal of Transport Economics and Policy* 27: 323-328.

Straja S.R. 2000. *Application of Multiple Attribute Decision Making to the OST Peer Review Program.* Institute for Regulatory Sciences. Office for Science and Technology, US. Department of Energy. Columbia MD, USA.

Tabucanon T.M., Mo Lee H. 1995. Multiple Criteria Evaluation of Transport System Improvements. *Journal of Advanced Transportation* 29: 127-143.

Tretheway W.M. 1990. Globalization of the Airline Industry and Implications for Canada. *Logistics and Transport Review* 26: 357-367.

Tsaura S.H., Chang T.Y., Yena C.H. 2002. The Evaluation of Airline Service Quality by Fuzzy MCDM. *Tourism Management* 23: 107-115.

Urbatzka E., Wilken D. 1997. Estimating Runway Capacities of German Airports. *Transportation Planning and Technology* 20: 103-129.

Vreeker R., Nijkamp P., Welle T.C. 2002. A Multicriteria Decision Support Methodology for Evaluation Airport Expansion Plans. *Transportation Research D7*: 27-47.

Vries de J., Priemus H. 2003. Mega Corridors in North-West Europe: Issues for Transnational Spatial Governance. *Journal of Transport Geography* 11: 225-233.

Zanakis S.H., Solomon A., Wishart N., Dublish S. 1998. Multi-Attribute Decision Making: A Simulation Comparison of Selected Methods. *European Journal of Operational Research* 107: 507-529.

Warren E.W., Gerrit B. Andre van V., Tuuli, J. 2009. Assessing the Variation in Rail Interoperability in 11 European Countries and Barriers to its Improvement. *European Journal of Transport & Infrastructure Research (EJTIR)* 9: 4-30.

Winston W.L. 1994. *Operational Research: Application and Algorithms*. International Thompson Publishing, Belmont, California, USA.

Yu G. 1998. *Operations Research in the Airline Industry*. Kluwer's International Series, London, UK.

http://www.airwise.com/

http://www.ryanair.com

http://en.wikipedia.org/wiki/Metropolitan_area

http://www.cost.esf.org

http://europa.eu.int/comm/transport/marcopolo/projects

http://www.rne.eu/

CHAPTER 8

CONCLUSIONS
Summary & Lessons Learnt

This book presents case-based modelling, planning and evaluation of transport systems operating within different transport modes. Each case has been ultimately considered as a kind of 'transport system' to be elaborated, i.e. modelled, planned and/or evaluated. Such an approach implied analyzing the case-system components and operations, identifying the problem(s) and then modelling these problems by modifying existing and developing new methodologies, each consisting of several analytical models to deal within a structural and systematic manner. These models are mainly driven by and developed in the format of indicators and measures of infrastructural, technical/technological, operational, economic and social performances of the considered cases, sometimes implicitly implying their mutual dependability. Later on, after being quantified by means of real-life inputs, these indicators and measures of performances are used as criteria in the evaluation of alternatives within the particular (selected) system cases. The planning of transport systems has been considered in a more general way, though some of the planning steps are illustrated with some real-life cases.

The cases pertaining to transport systems have been selected using the following criteria related to their relative convenience for the purpose: (i) coverage of the activities—modelling, planning and evaluation, including the relative complexity of performing such activities; (ii) diversity in cases of transport systems operated by different transport modes and their performances illustrating their representativeness and in a certain sense their generosity; and (iii) availability of sources, mainly provided by the author's research.

Consequently, the book consists of one chapter analyzing and modelling performances (one case per chapter), four chapters dealing with modelling diverse performances (two or three cases per chapter), one chapter dealing with planning, and one chapter elaborating the evaluation of transport systems (three cases).

Modelling, planning and evaluation are carried out for cases under systems operated by the rail, road, air and sea transport modes and for both passengers and goods/freight shipments. Cases under the rail transport mode include HSR (High Speed Rail) (for passengers), intermodal (rail/road) and road transport networks and corridors, and logistics networks operating under regular and irregular (disrupting) conditions (for goods/freight shipments). Under the road transport mode, the cases are similar to those for the rail freight mode. As for the air transport mode, cases relate to airports operating under different conditions and the resilience of the air transport network affected by large-scale disruptive event(s) (all for passengers). Under the sea transport mode, the case of an intercontinental supply chain operated by container ships of different sizes including the mega ones is taken up for consideration.

The level of complexity of modelling, planning and evaluation and style of presenting them aims to make the book attractive for a relatively wide audience of readers at all academic levels (BSc, MSc, PhD, post-doc), professionals from the transport industry and policy makers at different institutional levels—local, national and international. The lessons learnt after writing this book are as follows:

- The material originates primarily from the author's research carried out during the past decade-and-a-half, thus indicating its convenience and role as a sort of guidance to other prospective authors who are aiming to undertake similar projects, i.e. writing books on their past and existing research in a consistent way; consequently, the chapters and their sections are organized in a manner similar to the papers published in the scientific and professional journals;
- The term 'transport systems' has been ultimately used for particular cases concerning different transport modes. In the above-mentioned cases, the entire transport networks of infrastructure and services and their components—nodes and links/routes and services—are referred to as 'systems';
- The above-mentioned transport systems operated by the same and/or different transport modes are shown to be very complex with diverse entities requiring careful modelling, planning and evaluation. This is mainly due to increasing requirements for simultaneously considering a wide range of aspects of their implementation and operations, particularly if the objective has and is to make them 'greener', i.e. more sustainable, in the given context;
- Models of performances and those for evaluation of the above-mentioned transport systems are shown to be convenient mainly due to establishing

and understanding the effects/impacts of certain influencing factors on the selected indicators and measures of performances; in addition, such (analytical) models have enabled a sensitive analysis of particular indicators and measures of performances, depending on the changes in influencing factors;

- The MCDM (Multi-Criteria Decision-Making) methods, as either complement or an exclusive alternative to the conventional NPV (Net Present Value) method contained in the BAU (Business As Usual) evaluation approach, are shown to be useful due to advantages such as their explicit coverage of a wide range of diverse attributes of the given system's alternatives for use in evaluation, including taking into account the specific interests and preferences of particular actors/stakeholders involved in DM (Decision-Making) processes; and

- Despite being used for specific cases, the models of indicators and measures of performances and the selected MCDM methods demonstrate their generality to be applied to similar as well as different corresponding cases.

Therefore, the lessons learnt are as follows:

- Always look at the system and the problem, and then define the system's performances of interest related to the problem before defining the indicators and measures of these performances under the given conditions;

- In order to ensure transparency and understanding, always try to use simpler analytical models (already available or developed for the purpose in question) for estimating particular indicators and measures of performances under given conditions;

- In choosing the models of indicators and measures of the systems' performance, always keep the availability of the relevant data for their estimation in mind, particularly, the mutual driving of the models and the available data. In any case, the models should be sufficiently generous to be used in similar cases;

- Be careful in applying different MCDM methods since they do not always produce the same results. Therefore, choose a single method in the initial stage of selecting the preferred, i.e. the best among a few alternatives, but always choose the method after careful consideration and argumentation;

- Always bear in mind that analysis, modelling and planning of transport systems and/or their particular components (the latter of which are also considered as 'systems' in this book) are mutually interrelated activities and as such should be considered either explicitly or implicitly, the latter while performing them individually;

- Be aware that the modelling, planning and evaluation of transport systems as presented above, can only be an input for some kind of pre-decision-making and not for the final decision-making itself, which is usually an *a posterior* activity carried out by the DM, i.e. the particular actors/stakeholders involved; and
- Finally, bear in mind that the transport systems are dynamic entities continuously developing in order to satisfy the growing transport demand efficiently, effectively and safely. The necessity to make them 'greener', i.e. more sustainable, will certainly require more intensive modelling, planning and evaluation. This can be carried out by existing innovative and/or completely new approaches. This book aims to provide some ideas in that direction.

INDEX

Accessibility, 5, 13, 34, 45, 49, 66, 264, 311, 315, 316, 320, 350, 360, 374, 377, 380

Accidents, 6, 49, 59, 60, 63-65, 71, 77, 99, 101, 102, 110, 117, 118, 123, 128-132, 134, 141, 151, 158, 159, 195, 220, 223, 225, 333, 368, 372, 381

ACM (Aircraft Movement), 171, 189

ADS-B (Automatic Dependent Surveillance - Broadcast), 168

Africa, 9, 263

Agreements, 292, 293, 304, 385

AHP (Analytical Hierarchy Process), 356

Air, 1, 3, 4, 5, 7, 14, 34, 45, 49, 50, 58, 70, 71, 72, 80, 85, 99, 101, 110, 111, 134, 135, 138, 140, 141, 167, 168, 195, 196, 204, 205, 208, 213, 219-235, 237-239, 242-248, 250, 254, 262, 283, 291, 292, 294, 296, 297, 299-301, 303, 305, 318, 320-323, 325, 328, 332, 334, 338, 340, 341, 347, 383-386, 390

Air France, 78,

Air transport, 3, 7, 50, 58, 72, 84, 135, 167, 195, 196, 204, 205, 219-239, 242-245, 247, 248, 250, 283, 291, 292, 296, 297, 299, 300, 318, 320-323, 325, 332, 334, 338, 340, 341, 384-386, 390

Airbus, 79

Aircraft, 2, 5, 49, 50, 69, 70, 76-79, 82-96, 101, 132, 137, 140, 151, 152, 167, 169-173, 175-187, 189, 197, 205, 222, 223, 225-227, 235, 237, 239, 252, 253, 254, 261, 282, 299, 302, 303, 305, 306, 323, 326, 331, 336, 339, 342, 343, 345, 385

Aircraft Movement, 180

Airline, 3, 78, 84, 179, 220-222, 224-227, 230, 237, 245, 247, 248, 275, 286, 291, 292, 293, 295-307, 311, 318, 320, 323, 382

Airport, 2, 3, 5, 76-84, 87, 89-92, 94-97, 132, 137, 138, 167-170, 178-191, 223-225, 228-234, 237, 239, 241-245, 250, 259-263, 282, 286, 291-308, 311, 315-336, 338-348, 350-352, 382, 383

Airside, 77-79, 83, 168, 181, 286, 298, 299, 321, 323, 324, 327, 334, 339, 351

Airspace, 78, 84, 87, 88, 181, 182, 189, 221, 223-226

Algorithm, 232, 233, 295

Alliance, 78, 293, 294, 304, 339

Alternatives, 3, 201, 265, 280, 281, 285-291, 296, 298, 302, 304, 308, 311, 313, 314, 316, 317, 320, 321-324, 337, 342, 343, 347, 350, 352, 357, 381-383

America, 9, 263
Amsterdam, 78, 79, 83, 96, 133, 225, 262, 263, 294, 304, 305, 313, 315, 316, 317, 319, 334
Analysing, 355
Analytical, 71, 84, 96, 103, 105, 108, 114, 143, 170, 171, 189, 201, 202, 279, 280, 287, 290, 322, 356, 357, 358, 381, 382, 389, 391
Annual, 34, 37, 55, 57, 60, 64, 80, 81, 97, 155, 179, 180, 181, 186, 187, 199, 260, 262, 306, 321, 335, 336, 339, 340, 342
Antonov, 254
Antwerp, 125, 268-274, 276, 277, 373
Application, 3, 55, 88, 89, 104, 108, 124, 132, 153, 154, 156, 159, 166, 179, 188, 189, 198, 215, 226, 234, 237, 244, 262, 266, 281, 285, 286, 292, 300, 303, 321, 322, 333, 342, 346, 351, 354, 356, 357, 373, 381,
Approach, 14, 33, 61, 63, 82, 83, 86, 88-90, 133, 171, 172, 181-184, 203, 221, 227, 255, 259, 280, 282, 286, 296, 300, 383, 389, 391
Apron, 78, 83, 84, 181, 324, 326, 343, 347, 349
APT (Air Passenger Transport), 34, 49, 58
Area, 5, 12, 16, 61, 66, 70, 78, 79, 82, 94, 98, 99, 101, 103, 104, 107, 122, 141, 150, 151, 164, 166, 181, 197, 203, 205, 215, 235, 257, 260, 262, 263, 292, 293, 297, 298, 301, 304, 305, 320, 323-326, 332, 334, 336, 343, 345, 347, 349, 351, 354, 358, 359, 360, 361, 363, 364, 366, 367, 368, 370, 371, 372, 374, 377, 379, 380,
Arrivals, 87, 90, 103, 110, 111, 168, 210, 239, 240
ASDE X (Airport Surface Detection Equipment – Model X), 169
Asia, 9, 13, 154, 190, 261
Assignment, 257, 258, 275, 278, 279, 282, 291, 296

Assumption, 3, 42, 45, 51, 83, 84, 103, 105, 119, 128, 143, 170, 182, 202, 216, 228, 232, 259, 273, 296, 303, 323, 342, 356, 358
ATC (Air Traffic Control), 80, 82, 85, 291, 301
ATC (Automatic Train Control), 28
ATC controllers, 90, 94, 225
Atlantic, 338
ATM (Air Traffic Management), 227
ATRS (Air Transport Research Society),
Attack, 220, 223, 226
Attraction, 257, 258
Attractiveness, 261, 264, 297, 325, 328, 344, 347, 364
Attribute, 45, 47, 49, 258, 264, 265, 280, 281, 285-287, 290, 291, 296-308, 311, 320-322, 324, 326, 343, 346-348, 351, 357, 358, 373, 374, 376-379, 381, 382, 391
Austria, 51, 124-127, 293, 352, 373
Availability, 18, 27, 45, 82, 104, 159, 198, 201, 298, 300, 360, 389, 391
Average, 18, 21, 30, 34, 37-40, 43-48, 50-59, 68-70, 76, 80, 81, 84-89, 92-97, 103-106, 108, 110-112, 114, 115, 123, 127-132, 139, 141, 145-150, 152, 155, 157, 158, 161, 162, 166, 171, 178, 180, 183, 185, 186, 188-190, 197, 200, 202, 207, 208, 212, 215-219, 223, 234, 237, 238, 244, 251, 260-262, 266, 270, 289, 290, 299-306, 308, 311, 326, 327, 329, 331, 339, 343, 345, 346, 359, 361, 363, 366, 368-372
AVG, 68, 69
Axle, 10, 11, 127, 273, 361, 374, 377, 379

BA (British Airways), 339
Ballasted, 12
Ballastless, 12, 13, 53
Barges, 5, 101, 138, 151, 154, 353
BAU (Business As Usual), 280, 323, 356, 391

BCR (Benefit-Cost Ratio), 280
Belgium, 10, 11, 37, 50, 51, 62, 124-127, 215, 292, 293, 373
Benefits, 66, 71, 94, 138, 201, 255, 286, 294, 324, 351, 370
Beograd, 373
Bilateral, 292, 293, 304
Brussels, 37, 66, 225, 294, 304, 305, 313, 315-317, 319
Bucureşti, 373
Budapest, 125, 126, 373
Burgas, 373

CAA (Civil Aviation Authority), 183, 298, 336, 343, 344
Calibrating, 259
Cancellations, 6, 220-222, 224, 226, 239
Cancelled, 62, 201, 214, 220, 222, 226, 229, 231-234, 237, 238, 242-244
Capacity, 1, 2, 4, 5, 12, 13, 21, 22, 30, 33, 34, 38-45, 53-56, 63, 68, 70, 71, 76-78, 83-97, 103-106, 108, 110, 114, 120, 132, 137-141, 143, 146, 147, 149, 153, 154, 156, 159, 161, 163, 166-168, 170-172, 178-182, 184-190, 203, 216, 217, 220, 221, 223-225, 229-231, 234, 235, 249, 250, 253-255, 257, 258, 260-262, 266, 273, 275, 278, 279, 282, 286, 298-301, 303, 305-308, 311, 315, 316, 318, 321-331, 333-336, 338-345, 347, 349-351, 353, 356, 359, 362-367, 374, 375, 377-382
Car, 4, 5, 13, 20, 21, 29, 34, 49, 69, 70, 110, 123, 129, 152, 275-279, 298
Cargo, 5, 137, 140, 154, 159, 167, 179, 204, 205, 222, 228, 229, 253, 254, 320-322, 355
Caribbean (Islands), 235
Categories, 12, 16, 37, 46, 60, 61, 85, 89, 99, 102, 105, 127, 142, 143, 154, 171, 181, 184, 199, 201, 240, 249, 336, 354, 380, 381
Caveats, 104, 115, 116

CDTI (Cockpit Display Traffic Information), 169
Chain, 2, 102, 137, 138, 140-147, 149-156, 158-167, 189, 190, 196, 197, 201, 214, 249, 355, 390
Charles de Gaulle, 78, 294, 304
China, 8, 11, 13, 17, 18, 34-36, 38, 40, 50, 58, 64, 65, 153, 156, 190, 250-252, 256, 263
Choice, 13, 45, 71, 201, 210, 257, 258, 264-266, 270, 271, 275, 280, 281, 286, 291, 314, 328
CI(Consistency Index), 314
Closely, 142, 202, 295, 357
Closely-spaced, 2, 137, 138, 167, 168-175, 177, 179, 181, 183, 184, 186, 188-190,
Code, 30, 179, 293, 294
Cognitive, 222
Collection, 27, 99, 100-110, 142, 154-156, 158, 205, 256, 257
Commercial, 7, 13, 31, 49, 50, 70, 78, 127, 235, 239, 251-254, 256, 282, 297, 298, 320, 380
Communications, 6, 27, 30, 50, 197
Competition, 35, 58, 66, 71, 100, 116, 117, 119, 133, 139, 200, 216, 260, 292-294, 297, 298, 322
Competitive, 34, 47, 58, 102, 112, 116, 117, 132, 133, 260-262, 275, 323, 355, 357, 369
Complexity, 2, 71, 169, 280, 281, 285, 334, 389, 390
Component, 1-9, 23, 25, 27, 29, 33, 38, 42, 49, 52, 55, 66, 71, 98, 100, 101, 108,109, 113, 114, 163, 169, 195, 200, 203, 210-213, 220-223, 225, 226, 228, 229, 232, 250, 256, 258, 264, 265, 278, 292, 295, 298, 302, 303, 352, 380, 389-391
Concentration, 98, 104, 198, 294, 320, 365, 374, 378, 380
Concept, 4, 12, 14, 32, 99, 100, 198, 217, 220, 222, 227, 290, 353, 355, 373

Conditions, 1-4, 8, 14, 26, 30, 32, 34, 38, 43, 49, 55, 61, 62, 76, 77, 80, 82, 84, 85, 87-89, 93-96, 102, 103, 105, 113, 115-120, 123, 131-133, 138-141, 144, 146, 147, 155, 159, 161, 167-171, 175, 177, 183, 186, 188-190, 195-197, 199, 202, 203, 205, 212, 215, 219-222, 226-233, 237, 241-244, 249, 255, 257-259, 261, 264, 265, 270, 275, 278, 286, 290, 292, 296, 298, 302-304, 320, 323, 332, 336, 342, 351, 352, 356, 358, 361-364, 366, 367, 369, 370, 381, 382, 390, 391

Configuration, 16, 19-21, 29, 78, 80-85, 89-96, 115, 132, 141-144, 189, 198, 202-206, 215, 216, 232, 233, 266, 273, 302, 380

Congestion, 9, 59, 60, 62, 65, 71, 77, 83, 88, 93, 99, 123, 128-132, 141, 151, 152, 154, 168, 279, 355, 372, 375, 378, 379

Consistency, 113, 289, 304, 314, 316, 318, 383

Consolidation, 6, 139, 204-208, 212-214, 216

Constanţa, 373

Constraints, 1, 2, 8, 77, 80, 82-84, 96, 100, 139, 167, 168, 181, 275, 279, 295, 297, 299-301, 307, 322, 336, 339, 343, 352

Consumer, 100, 143, 156, 196, 197, 200-219, 244, 278, 301

Consumption, 5, 6, 67-71, 77, 94, 99, 100, 102, 117, 121, 122, 128, 130-132, 138, 139, 141, 143, 145-147, 149-152, 155-159,166, 197-200, 202, 203, 205, 207, 210, 212, 215, 325, 331, 343, 345, 348-350, 359, 370, 371, 375, 379, 381

Container, 6, 97, 99, 126, 137, 139, 140, 153-159, 161-163, 166, 190, 193, 194, 202, 204, 205, 216, 249, 252, 254, 255, 267, 282, 358, 380, 390

Continental, 138, 197, 225, 282, 304, 328

Control, 3, 6, 18, 23, 25, 26-30, 39, 45, 71, 80, 82, 85, 94, 102, 137, 169, 189, 199, 210, 221, 225, 291, 299, 301, 303, 353, 362

Convoy, 121- 124, 129

Corridors, 3, 97, 111, 112, 115-119, 132, 170, 172, 175, 180, 186, 352-359, 373, 376, 377, 380-383, 390

COST ((European) CoOperation in Science and Technology), 119, 280, 323, 354-356,

CREAM (Customer-driven Rail-freight services on a European mega-corridor based on Advanced business and operating Models), 129, 355, 373, 376-381

CRH (Chinese Rail High), 8, 13, 18, 35

Criteria, 142, 228, 257, 258, 280, 281, 285-291, 296, 301, 303, 305-309, 311, 314, 316-322, 324, 346, 347, 349-352, 354, 356, 358, 373, 374, 376, 377, 379-383, 389

Criterion, 7, 201, 285-291, 295, 298, 306, 311, 314, 315, 316, 318, 319, 326, 343, 348-350

Crosswind, 82, 169, 175, 176, 177, 184, 186, 189

Crude, 6, 7

Curvature, 10, 11, 13, 67

Curve, 10, 11, 126, 170, 254

Customer, 93, 138, 140-184, 150, 152, 153, 156-158, 193

Cycles, 253, 325, 331, 345, 348

Decision, 281, 285, 287, 288, 295, 307, 308, 320, 334, 346, 350, 351, 358, 392

Decision-Matrix, 287, 290, 304, 305, 307

Deconsolidation,204, 205, 207, 208, 212-214, 216

Delay, 6, 15, 34, 38, 43, 45-48, 62, 76-78, 83, 35, 37, 88, 92-96, 99, 101, 106, 110, 118, 123, 128, 129, 132, 146, 154, 158, 170, 171, 178, 177, 189, 207, 208, 211-213, 220-

227, 231-235, 237-240, 242-244, 261, 299-301, 303, 327, 366, 368, 372

Delivery, 99, 100, 103, 115, 116, 127, 133, 142, 143, 197-199, 201, 203, 204, 207, 211, 212, 214-217, 219, 267, 270, 271, 362, 366, 368, 369, 372, 375, 378-380

Demand, 1, 2, 4, 5, 16, 32-38, 43, 44, 54-59, 66, 71, 76-78, 80-85, 87, 88, 90-100, 104, 105, 108, 111-117, 133, 138, 139, 143-145, 147, 159, 167, 168, 170, 178, 186, 200, 203, 223, 257-264, 266, 276, 269, 271, 275, 278, 279, 282, 285, 293, 296-299, 318, 320-325, 327-236, 338-343, 345, 348, 351, 353, 356, 364, 380, 392

Density, 11, 37, 38, 212, 290, 346, 360, 364, 374, 377, 380

Departures,42, 45, 87, 94, 101, 103, 104, 110, 111, 146, 156, 175, 239, 240, 368

Design, 5, 8-14, 16, 18, 21-23, 27, 39, 40, 49, 62, 63, 139, 149, 154, 224, 226, 228, 251, 256, 257, 293, 314, 355

Destination, 4, 31, 45, 98, 99, 103, 104, 110, 116, 143, 158, 179, 190, 196, 205, 220, 227, 257, 258, 259, 261, 270, 294, 320, 325, 328, 344, 347, 349-351, 358, 361

Diagonal, 169, 171-173, 182-184, 186, 189

Dictionary, 219,

Diesel, 6, 101, 102, 218, 221, 227, 228, 361-363, 380

DIIM (Dynamic Interoperability Input-Output Model),227

Dis(utility) 264, 265

Disasters, 195, 220, 223, 224

Disruptive, 3, 147, 195, 197, 201-203, 205, 207, 212, 214-216, 218-223, 226-232, 233-235, 238-244, 260, 262, 390

Distance, 8, 10, 11, 14, 16, 18, 24-28, 36, 40, 42, 43, 45, 49, 58, 60-62, 77, 86, 88, 89, 97, 99, 102-106, 108, 111-118, 122, 123, 128, 129, 133, 146, 147, 149, 151, 152, 157, 158, 167, 171-173, 176, 177, 181, 182, 185, 186, 189, 198-200, 207, 208, 212, 214-216, 238, 239, 260, 261, 262, 264-266, 268, 270-273, 278, 279, 288, 301, 304, 306, 309, 311, 320, 326, 327, 343, 354, 360, 361, 368, 371

Distribution, 18, 23, 81, 99-103, 105, 106, 108-110, 138, 142, 145, 154, 155, 156, 158, 161, 197, 200, 201, 205, 212, 213, 257-259, 264, 282, 291, 298, 303, 307, 308, 318 371

Disturbing, 3, 214,

Diversion, 267, 270,

DM (Decision-Maker), 218

Door-to-door, 4, 77, 100, 102, 103, 106, 108, 111-116, 133, 198, 201, 204, 215, 262, 264, 267, 270, 271, 273, 343, 371

Dubai, 167, 179, 182, 184, 189, 190

Dummy,260

DXB (Dubai International Airport), 167, 168, 170, 179, 180-190

EAT (Economic Analysis Technique), 280, 323, 256

EC (European Commission), 27, 115, 132, 352

Economic, 1, 2, 5, 32, 33, 50, 60, 65, 67, 71, 76, 77, 102, 105, 115, 132, 133, 137, 139, 144, 147, 149, 161, 166, 190, 195, 197, 201, 203, 219, 227, 228, 242, 244, 249, 255-257, 261, 262, 278, 295, 297, 321, 322, 324, 328, 334, 336, 342, 344, 350-352, 357, 368, 370, 381, 382, 389

Economy, 66, 71, 227, 237, 329, 345, 350, 354

Effect, 2, 3, 32, 50, 59, 65, 66, 71, 76, 77, 99, 102, 114, 116, 120, 129, 131, 137, 158, 166, 168, 170, 187,

189, 227, 238, 139, 249, 255, 325, 329, 336, 342, 345, 348, 369, 391

Effective, 1, 6, 13, 49, 100, 117, 198, 213, 256, 298, 351

Effectiveness, 6, 83, 99, 100, 102, 116, 139, 198, 222, 282, 295, 298, 300, 325, 328, 329, 344, 348, 349, 357

Efficiency, 5, 6, 68-70, 98-100, 102, 116, 121, 139, 164-166, 198, 227, 285, 292, 298, 300, 325, 328, 332, 333, 343-346, 348, 350, 355, 357, 368-370

Efficient, 1, 6, 13, 23,31, 32, 49, 66-68, 72, 100, 117, 138, 140, 161, 164, 198, 213, 219, 244, 256, 294, 298, 306, 351, 353, 392

ELECTRE (ELimination Et ChoixTraduisant la REalité (ELimination and Choice Expressing Reality)), 280

Electric, 6, 18, 23, 67-70, 101, 121, 127, 361-363, 380

Electricity, 7, 67, 111, 117, 118, 121, 128

Emissions, 67-71, 77, 100, 101, 104, 117, 121, 122, 128-130, 132, 141, 149, 158, 164-166, 325, 331, 343, 345, 348-350, 359, 368, 370, 371, 375, 379 ,381

Employment, 30, 31, 60, 66, 261, 325, 329, 344, 348, 349

Energy, 4-7, 18, 51, 52, 66-71, 77, 99-102, 117, 118, 121, 128, 130-132, 139, 141, 142, 149, 164, 192, 299, 359, 370, 371, 375, 379, 381

England, 292, 344,

Entropy, 259, 286, 290, 291, 303, 307, 308, 311, 318, 346, 351, 375, 376, 382,

Envelopes, 185-187

Environment, 1, 2, 5, 13, 32, 49, 71, 99, 100, 116, 117, 120, 128, 139, 142, 195, 255-258, 334, 343, 352, 369,

Environmental, 1, 2, 13-15, 32, 33, 60, 66, 67, 71, 76, 77, 80, 84, 105, 115, 117, 119, 120, 129, 132, 133, 137, 140, 143, 144, 149, 164, 166, 168,

190, 195, 249, 255, 300, 321, 322, 324, 331, 336, 345, 350-352, 354, 355, 357, 370, 381, 382

Equilibrium, 219, 278, 279

Equipment, 4, 6, 9, 12, 18, 23, 27, 28, 32, 50, 52, 56, 60, 66, 67, 71, 97, 119, 139, 141, 142, 221, 225, 249, 363, 369, 374, 377

ERTMS (European Rail Traffic Management System),24, 27-29, 353, 373

ETCCS (European Train Control and Command System), 27

EU (European Union), 70, 96, 97, 108, 113 ,116, 117, 128, 129, 199, 202, 263, 291-294, 296, 303, 304, 306, 307, 320, 353, 355, 356, 373, 376

Euro, 305

EUROCONTROL, 191, 246, 385

Europe, 7, 9, 10, 14, 16, 18, 24, 34, 35, 37, 38, 50, 51, 58, 61, 65, 69, 97, 102, 103, 108, 112, 115, 116, 119, 121, 125, 127, 132, 153, 154, 190, 198, 215, 217, 224, 250-252, 256, 262, 263, 281, 320, 322, 323, 333, 342, 352-356, 382

European Co-operation in the Field of Scientific and Technical Research, 119, 355

European, 7, 9-11, 15, 18, 24, 27, 28, 35, 50, 51, 77, 78, 96, 97, 102, 108, 110-112, 115-119, 124, 127, 128, 132, 133, 142, 168, 170, 189, 198, 199, 202, 219, 223, 224, 227, 244, 251, 263, 280, 291-295, 304, 306, 318, 320, 323, 341, 352, 353, 354, 355, 373, 381, 382

Eurostar, 19, 37, 40, 48, 68, 69

Evaluation, 2, 3, 137, 255, 257, 258, 280-282, 285-287, 292, 296, 298, 300, 304, 314, 318, 320-324, 326, 342, 346, 347, 351, 352, 354, 356-358, 373, 376, 377, 379, 381-383, 389, 390-392

Events, 102, 130, 195, 203, 214, 219-228, 230, 244, 248, 260, 262, 333

External, 1, 2, 28, 32, 48, 63, 76, 98-106, 108, 109, 111-116, 119, 132, 133, 140, 195, 197, 200, 214, 227, 264, 299, 336, 354, 359, 368, 369, 370, 381

Externalities, 2, 32, 59, 65, 67, 70, 71, 76, 77, 110, 113-117, 119, 129-133

FAA (Federal Aviation Administration), 77, 85, 168-170, 183, 184, 223, 224,

Facilities, 4, 6, 9, 12-14, 23, 32, 50, 52, 56, 66, 67, 71, 97, 117, 119, 138, 139, 141, 142, 221, 225, 249, 363, 364, 369, 374, 377

Factor, 13, 44, 45, 51, 54-56, 69, 103-106, 108, 110, 115, 120, 127, 139, 143, 146, 155, 156, 164, 237, 259-261, 266, 275, 299, 306, 332, 351, 359, 367, 368

Failures, 15, 27, 48, 195, 220, 223, 227

FDI (Foreign Direct Investments), 262

Far East, 7, 153

Fatalities, 63, 123, 128-130, 372

Fleets, 143, 159, 164, 362

Flexibility, 138, 198, 212, 217, 230, 292, 293, 296

Flight, 76-78, 83-85, 87, 88, 91-96, 132, 220, 222, 224, 225, 229, 234, 237, 239, 240, 260, 261, 292-294, 298, 299, 302, 306, 320, 328

Flow, 6, 99, 123, 129, 152, 215, 269, 278, 279, 302, 358, 361, 364, 373,

FMP (Framework Program), 355, 356

France, 10, 11, 16, 18, 24, 25, 31, 37 ,50, 51, 62, 63, 66, 69, 78, 126, 215, 225, 292, 352

Frankfurt, 66, 78, 294, 295, 303, 305, 311, 313, 316-319

Freight, 1-6, 8, 30, 32, 76-78, 97, 98, 102, 103, 108 ,109, 114-123, 125-133, 137-150, 152, 154, 155, 158, 159, 166, 167, 190, 196, 197-205, 207-216, 218-220, 226, 244, 255-258, 264-267, 269, 270, 271, 273-275, 277, 281, 282, 285, 286, 292,

294, 295, 306, 321-323, 352-360, 370, 372, 373, 377, 379-383, 390

Frequency, 12, 23, 31, 37, 43-45, 53-57, 102, 106, 108, 111, 141 ,143, 145, 147, 150, 152, 155, 159, 160, 166, 167, 197-203, 207-217, 219, 244, 261, 266, 277, 301, 327, 364, 368, 380, 381

Friability, 220, 222, 226, 228, 232-234, 241-243

Fuel, 4-7, 69, 70, 77, 94, 101, 118, 121, 122, 128, 139, 141, 149, 157, 158, 164, 166, 195, 299, 331, 343, 345, 350, 370, 371, 375, 379

Full, 2, 13, 26, 76, 77, 88, 98-100, 103, 104, 108-115, 119, 132, 133, 145, 354,

Function, 6, 49, 68, 122, 212, 221, 264, 265, 272, 279, 281, 290, 304, 306, 318

Fuzzy, 142

FAGs [FinalApproach Gate(s)], 171, 181,

Greening, 2, 142, 143

Gains, 138, 164, 187, 329, 330, 345, 350, 351

Gate, 14, 15, 78, 83, 84, 88, 89, 181, 324, 326, 343, 347, 249

Gatwick, 294, 321, 335-337, 339, 340, 342-352

GDP (Gross Domestic Product), 34-36, 60, 65, 66, 260, 262, 263, 297, 336, 369

Generic, 16, 26, 70, 105, 143, 144, 217, 228, 279,

Gography, 153

Germany, 10, 11, 16-18, 23, 24, 31, 50, 51, 62, 64, 65, 124-126, 292, 293, 273,

GHG (Green House Gases), 5, 77, 94, 101, 117, 141, 157

Goods, 1, 4- 6, 32, 77, 97, 98, 100, 103, 106, 108, 115, 116, 120, 126, 127, 133, 137-150, 154, 155, 158, 159, 190, 196-205, 207-216, 218-220,

244, 255, 258, 264-267, 269, 270, 271, 273-275, 277, 285, 358, 369, 373, 379, 380, 382, 390
Gradient, 10, 11, 26
Greece, 124, 125, 373
GS (Glide Slope), 169, 172, 184
GSMC - R (Global System for Mobile Communications—Railways), 27
Gyor, 373

Handling, 105, 106, 110, 111, 138, 141, 147-150, 157, 158, 161, 167, 196, 201, 203, 214, 301
Haul, 9, 49, 50, 66, 69, 77, 99, 102, 103, 105-107, 109, 112, 119, 179, 180, 216, 260, 334, 338, 339, 342
Heathrow, 78, 83, 167, 224, 294, 304, 311, 316, 318, 321, 334-339, 342-348, 350-352, 383
Heavy, 80, 89, 167, 180, 181, 183, 187, 220, 223,
Helsinki, 116, 117, 294, 353,
Hierarchical, 287, 288, 314
Highways, 4, 5, 9, 122,
Horizontal, 10, 173, 358
Hour, 34, 37, 39, 45, 53, 54, 76, 80, 81, 84-87, 91, 93-95, 170, 178, 230, 258, 260, 303
HS (High Speed), 8, 281, 223, 356,
HSR (High Speed Rail), 2, 7, 249, 250, 256, 322, 390
Hub, 2, 3, 76, 78, 84, 96, 132, 138, 141, 143-148, 150, 153, 153, 156-159, 161, 163, 164, 167, 179, 190, 227, 262, 286, 292-297, 299-306, 311 , 318, 323, 334, 336, 338, 382, 383
Hub-and-Spoke, 144, 388,
Hungary, 124-126, 373
Hurricane, 196, 235, 239, 241, 243, 244
Hydrogen, 7, 69

IATA (International Air Transport Association), 224
ICAO (International Civil Aviation Organization), 85, 183, 184, 292, 306, 345, 346,

ICT (Information/Communication Technologies), 198, 199
IDACS (Integrated Departure and Arrival Coordination System), 169
Ideal, 288, 309-312, 346, 350
ILS (Instrument Landing System), 88, 168, 181, 184
Impact, 130, 223,
Importance, 47, 68, 139, 201, 222, 229-231, 259, 262, 285-287, 289-291, 296, 298, 303, 306-308, 314, 318, 322-324, 326, 358, 382
Inbound, 78, 80, 82, 90, 92, 93, 95, 96, 270,
Incidents, 6, 49, 59, 60, 63, 65, 71, 77, 102, 117, 118, 123, 128-132, 141, 151, 158, 195, 220, 223, 225, 333, 368, 372, 381
Incumbent, 305, 318
Index, 64, 153, 164, 289, 314
Indicator, 96, 232, 297, 321, 322, 324, 326-328, 331, 332, 343-355, 357, 358, 360, 361, 364, 368, 370, 371, 381,
Information, 6, 13, 14, 25-28, 45, 66, 84, 197, 201, 265, 290
Infrastructural, 2, 32, 33, 137, 14, 143, 159, 190, 324, 326, 343, 350-352, 357, 360, 374, 377, 379, 381, 382, 389
Infrastructure, 1, 3-9, 16-18, 23, 30, 32, 38, 42, 50-53, 56, 63, 64, 66, 67, 70, 71, 78, 83, 99, 100, 116, 117, 119, 138, 139, 141, 143, 146, 152, 197, 201, 220, 224, 227, 249, 250, 255-257, 275, 278, 282, 291, 320, 321, 323, 324, 326, 334, 343, 347, 353, 354, 356-361, 364, 365, 369, 371, 380, 382, 390
Inland, 4, 5, 6, 15, 97, 99, 115, 116, 141, 154, 155, 198, 199, 202, 204, 213, 228, 256, 353, 356, 357
Innovative, 2, 7, 78, 117, 119, 137-139, 168-170, 182, 189, 190, 249, 280, 321, 323, 354, 356, 358, 370, 380, 392

Instrument Flight Rules, 85, 223,
Insurance, 66, 100, 153, 164
Intensity, 29, 38, 39, 70, 84, 87, 104,
    120, 122, 123, 129, 131, 152, 197,
    201-203, 216, 223, 237-239, 243,
    244, 291, 325, 332, 345, 348, 349,
    365, 371, 372, 375, 378
Intermodal, 2, 5, 13, 76, 77, 97-116, 119,
    132, 133, 138, 141, 152, 154, 202,
    216-219, 226, 244, 256, 266, 267,
    270-277, 352-355, 357, 358, 360,
    362, 364 365, 368, 369, 371-373,
    377, 379-381, 383, 390
Internal, 1, 2, 48, 50, 76, 98-100, 102,
    103-106, 108-116, 119, 133, 139,
    195, 197, 200, 214, 280, 298, 299,
    354, 368, 369
Internalizing,103, 113, 116
Interoperability,5, 27, 227, 353, 355,
    363, 374, 377, 380
Inventory, 141, 147-149, 157, 158, 161-
    164, 166, 190, 200, 201, 208-211,
    213, 216, 219, 227, 257
Investment, 63, 65, 100, 113, 201, 249-
    251, 255, 261, 262, 325, 328, 329,
    334, 344, 345, 348, 358, 369, 375,
    378, 379
IRR (Internal Rate of Return), 280
Irregular, 3, 26, 147, 195-197, 202, 205,
    214, 215, 219, 230, 244, 390
Israel, 263
Istanbul, 125, 225, 356, 373,
Italy, 10, 11, 16, 24, 46, 50, 51, 62, 69,
    215, 292, 304
IUR (International Union of Railways),
    73, 74, 136, 284

Japan, 7, 12, 13, 20, 31, 35-39, 47-49,
    50, 51, 58, 60, 63, 66, 69, 224,
    250, 263
JIT (Just-in-Time), 198
Journey, 24, 301

Kerosene, 7, 343
Kg, 11, 12, 108, 128, 149, 306
Kilometer, 68, 260,302, 303, 306

Kilovolt, 363
Kilowatt, 19-21
KLM, 78, 294
Knot, 86, 89, 90, 146, 172, 235
Kts, 82, 88, 146, 157, 159, 161, 164, 183
Kulata, 373

Labor, 51, 99, 100, 102, 299
LaGuardia, 237, 242, 243
Land, 1, 5, 15, 50, 60, 62, 67, 70, 71,
    116, 117, 141, 149, 150, 164, 166,
    171, 173, 176, 226, 256, 257,
    299,325, 332, 345, 348-351, 362,
    368, 371, 375, 379, 381
Landings, 2, 77, 80, 82-87, 90, 93, 137,
    168-172, 176-178, 181-190, 339
Landside, 78, 298, 321, 324, 327, 339
Lane, 123, 129
Large, 2, 3, 16, 30, 68, 76-78, 84, 89,
    90, 96, 98, 132, 137, 138, 140,
    154, 167, 181, 183, 184, 195, 196,
    218-220, 222, 223, 228, 230, 233-
    235, 242, 244, 249, 250, 259, 282,
    297, 300, 320, 323, 334, 336, 339,
    351, 354, 356, 381, 382, 390
Lateral, 172, 181, 182, 293
Latin (America), 263
Layer, 12, 221, 222
Layout, 5, 10, 13, 14, 16, 17, 104, 116,
    124, 256, 258, 292, 358, 373, 376
LCC (Low Cost Carrier), 49, 294, 316,
    339
Length, 7, 10-13, 15, 18-21, 24, 26, 30,
    31, 34-46, 50, 53-56, 60, 61, 63,
    70, 79, 86-89, 97, 103-106, 110,
    111, 118, 120-122, 126-129, 141,
    146, 151, 152, 154, 156, 159, 169,
    172, 181, 182, 227, 250-252, 266,
    302, 303, 306, 360-366, 371, 373,
    374, 377, 379, 380
Level, 6, 7, 12, 14, 18,24,27-29, 60-64,
    76-78, 93, 95, 118, 122, 130, 132,
    142, 151, 187, 195, 197, 214, 216,
    221, 227, 228, 232, 235, 237-244,
    255, 256, 278, 279, 288, 289, 296,
    298, 311, 314, 316, 318, 322, 336,

342, 345, 346, 354, 356, 371, 372, 375, 378, 381, 390

LHR (London Heathrow), 78, 167, 294

Liberalization, 291-293, 320, 355, 356

Line-hauling, 99-104, 106, 108, 110, 113

Liner, 153, 156, 166

Lines, 5, 8-18, 23-27, 29, 30, 36, 38, 41, 43, 45, 46, 50, 51, 58, 59, 61-63, 66, 67, 69, 220, 249-252, 256, 353

Link, 30, 34, 144, 146, 275, 278,279, 361, 364, 368, 371

Liquid (Hydrogen), 7, 69

Ljubljana, 125, 373

Load units, 97, 99-115

Load, 10, 11, 44, 45, 51, 54-56, 97, 99-115, 120, 127, 139, 143, 146, 152, 153, 155, 156, 266, 275, 306, 359, 361, 367, 368, 374, 377, 379

Loading, 6, 7, 18, 39, 139, 141, 144-150, 155, 156, 158, 159, 161-163, 205, 214, 267, 270, 358, 380

Localizer, 88

Logic, 142

Logistics, 3, 98, 126, 137, 139, 142, 143, 195-209, 214, 215, 218, 219, 226, 244, 254, 355, 356, 390

Logit, 264, 265, 270

London, 37, 66, 78, 97, 167, 225, 268, 269, 272, 274, 276, 277, 294, 304, 305, 311, 313, 317-319, 321, 333-338, 340, 341, 343, 351, 352, 383

Long, 1, 3, 12, 18, 32, 48, 58, 59, 78, 79, 83, 88, 94, 101, 116, 118, 124, 137, 140, 178, 198, 220-222, 226, 250, 253, 254, 285, 292, 293, 321, 334, 380

Longitudinal, 11, 169, 171-173, 176, 177, 182-184, 186, 187, 189

Losses, 118, 123, 152, 195, 222, 226, 227, 325, 330, 345, 348-351

LTO (Landing-Take-Off), 325, 331, 345, 348

Luton, 321

MADM (Multi-Attribute Decision Making), 280

Maglev, 256

Maintainability, 18

Makers, 1, 2, 116, 285, 357, 358, 382, 390

Management, 3, 23, 24, 27-29, 63, 71, 77, 102, 137, 142, 143, 168, 169, 189, 201, 221, 222, 227, 353

Managing, 321, 342, 382

Manufacturer, 202, 203, 205, 207-215

Marco Polo, 119, 354, 355

Maritime, 139, 140, 204, 255, 353

Market, 1, 35, 36, 47, 48, 66, 97, 103-105, 112, 113, 115-117, 131, 133, 197-199, 202, 203, 217, 241, 260, 262, 263, 265, 270, 273-278, 291-301, 303, 305, 306, 308, 311, 314-316, 318, 320, 353, 356

Matching, 168, 278, 321-324, 333-336, 342, 356

Material, 3, 12, 51, 99, 143, 198, 201, 390

Matrix, 259, 287-290, 303-312, 314-316, 346, 350

MCA (Multi-Criteria Analysis), 280, 323

MCDM (Multi-Criteria Decision Making), 3, 280-282, 285-287, 292, 296, 300, 303, 304, 313, 314, 318, 320-323, 333, 346, 349-351, 355-357, 373, 374, 380-383, 391

Means, 6, 26, 28, 55, 63, 67, 69, 76, 90, 93, 99-101, 116, 142, 189, 195, 203, 259, 270, 282, 286, 290, 303, 389

Measure, 14, 30, 38, 50, 61, 77, 78, 100, 118, 129, 133, 142, 146, 227, 229, 232, 235, 239, 249, 278, 286, 287, 289-291, 296, 301, 320, 324, 326-329, 331-333, 343-353, 357, 358, 360, 361, 364, 365, 368, 370-382, 389, 391

Medium, 9, 32, 49, 59, 69, 78, 97, 112, 116-119, 132, 168, 179, 221, 230, 253, 260, 281, 285, 323, 333-336, 339, 342, 354

Mega, 2, 124, 137, 138, 140, 143, 159, 161, 163, 164, 166, 167, 189, 190, 249, 282, 354, 355, 373, 390

Methodology, 3, 84, 104, 143, 144, 170, 171, 202, 203, 205, 220, 228, 234, 237, 285, 296, 303, 318-320, 323, 324, 333, 351, 352, 354, 357, 358, 373, 376, 381

Metropolitan, 260, 320, 323, 326, 334, 336, 351

Middle East, 167, 263

Mile, 8, 86, 88-90, 146, 172, 181, 235, 260

Minima, 191

Minimizing, 279

Minimum, 6, 10, 11, 15, 16, 28, 30, 39-42, 85-90, 171-173, 175-178, 182-184, 198, 223, 304-306, 319, 364, 368

Mix, 8, 17, 85, 86, 89, 90, 171, 172, 177, 180, 182, 184-189, 205, 336, 342, 389

MLS (Microwave Landing System), 168

Mobile, 5, 27, 100

Mobility, 1, 4, 5, 49, 50

Modal, 100, 115, 138, 199, 257, 258, 264-266, 270, 271, 282, 355

Mode, 4-7, 77, 82-84, 87, 90, 92, 96, 97, 99, 101, 102, 104, 106, 108, 113-116, 119, 120, 140, 152, 167, 170, 199, 201-205, 216, 217, 219, 228, 244, 264-275, 279, 282, 335, 336, 339, 342, 352, 368, 390

Model, 84, 85, 88, 104, 105, 111, 115, 137, 143, 169, 170, 197, 205, 214, 215, 227-229, 232, 233, 257, 259-265, 270, 278, 279, 282, 295, 306, 358

Modelling, 2, 3,7, 52, 76, 83, 98, 102, 104, 114, 119, 132, 133, 137, 138, 142, 166, 170, 188, 189, 195-197, 201-205, 219, 226-228, 242, 244, 257, 258, 303, 352, 354, 356, 357, 389-392

Montenegro, 373

Motorways, 118, 122

MS (Member States), 108, 118, 128, 129, 199, 263, 292, 353

MTOW (Maximum Take-Off Weight), 171

Multi, 2, 3, 4, 18, 29, 127, 257, 275, 280, 285, 293, 318-323, 352, 354, 356, 362, 363, 373, 376, 381, 382, 391

Munich, 125, 256, 295, 373,

National, 1, 4, 17, 24, 27, 34, 36, 60, 96, 116, 197, 202, 222, 224, 227, 237, 238, 255, 292, 294, 295, 298, 304, 354, 357, 369, 373, 382, 390

Netherlands, 51, 78, 124, 125, 127, 153, 156, 215, 263, 292, 393, 373

Network, 2-4, 7, 9, 10, 12, 13, 16-18, 23, 29-31, 33-36, 48, 62, 63, 76-78, 84, 97-117, 119, 125, 132, 133, 137, 138, 142, 144, 179, 181, 195-209, 211, 214-216, 218,235, 237, 239, 242-244, 249, 250, 256-259, 264, 266270, 272-282, 292, 295, 298-305, 318, 320, 322, 323, 343, 353-356, 359, 361, 371, 373, 380, 382, 390

Newton, 67

NG (Next Generation), 78, 168

Nm (nautical mile), 181

Node, 34, 206, 233, 248, 251, 259, 359

Noise, 12, 13, 14, 45, 49, 50, 59, 60-62, 65, 71, 77, 82, 83, 94, 96, 99 ,100-102,110, 117, 118, 122, 128-132, 141, 151, 154, 167, 299, 325, 332, 333, 336, 339, 343, 346, 348, 350, 351, 368, 371, 372, 375, 378, 381

NPV (Net Present Value), 391

NY (New York), 235, 237, 239, 241-244

Objectives, 2, 3, 27, 83, 103, 119, 138, 143, 170, 202, 228, 249, 256, 257, 281, 296, 323, 352, 353, 356

O-D (Origin-Destination), 126, 259, 262-267, 358

Oil, 6, 7, 158, 195, 199

Operability, 227, 237-239, 241

Operating, 3-6, 9, 12, 16, 21-27, 30, 34, 37, 39, 40, 42-45, 51-68, 71, 77, 80, 82, 83, 85, 87, 90-96, 108, 110, 117-124, 127-129, 132, 138, 142, 143, 145-147, 154, 157-164, 167, 177, 179, 181, 184-186, 189, 195-197, 202, 203, 215, 219, 220, 227, 229, 234, 244, 261, 267,271, 282, 295, 297-305, 311, 321, 325, 328, 331, 335, 336, 339, 344, 345, 347, 349, 350, 355, 359, 362, 363, 365-373, 380, 381, 389, 390

Operations, 1, 6, 9, 12, 23, 27, 30, 55-57, 59,64, 68, 69, 77-84, 95, 97-101, 105, 117-119, 129, 137, 142, 150, 151, 154, 167-169, 172, 176, 177, 181, 182, 185-189, 195,200, 201, 203, 205, 214, 221, 222, 228, 232, 238, 242, 244, 252, 292, 295, 296, 299, 332, 336, 358, 359, 370, 372, 381-383,389, 390

Operators, 6, 27, 48, 49, 63, 99, 100, 102, 103, 116, 117, 139, 197, 203, 204, 209, 213, 217, 218, 220, 255, 275, 295, 334, 360-365, 368, 373

Options, 49, 100, 138, 292, 321, 322, 334-336

Orders, 141, 143, 207-210, 214, 216, 217

Outbound,78, 80, 82, 90, 92-96, 270-273

Oxford, 219

Panamax, 154, 155, 166

Pairwise, 288, 289, 314, 316

Pallet, 215-219, 244

Parallel, 2, 13, 70, 78, 79, 137, 138, 167-177, 179, 181, 183-190, 235, 285, 292, 334, 336, 339

Paris, 18, 19, 37, 66, 78, 256, 294, 304, 305, 313, 315, 317, 319, 334

Passenger, 4-8, 13, 14, 16, 29, 30, 32,34-38, 44, 45, 49, 54-59, 63-66, 69-71, 78, 116, 118, 123, 128, 167, 180, 220, 226, 237, 252, 253, 256, 257, 260-265, 282, 295, 298-306, 320, 321, 324, 326-329, 334, 336, 338-343, 347, 349, 350, 353, 372

Passenger-kilometer,

Path, 16, 28, 41, 86-89, 169, 172, 182, 227, 239, 278, 279

Payload, 5, 118, 121, 138-143, 146, 147, 149, 153, 159, 253, 254, 273, 275, 362-364, 367, 374, 377, 379, 381

PCI (Per Capita Income), 261, 297

Pendolino, 19, 68, 69

People, 1, 14, 15, 101, 102, 153, 225, 332

Performances, 2, 4, 5, 7, 27, 30, 32, 33, 39, 50, 59, 67, 71, 76, 77, 80, 115, 132, 133, 137, 138, 140-147, 149, 151, 153, 154, 156, 159-167, 188-190, 195, 197, 202, 219, 220, 227, 235, 242, 244, 255, 257, 278, 280, 281, 286, 296, 300, 301, 320-324, 326- 328, 331, 332, 342-361, 364, 368, 370-383, 389-391

Person,333, 346, 348

Physical, 4, 5, 9, 32,78, 98, 138, 141, 197, 213, 220, 221, 255-258, 321, 324, 326, 343, 347, 349-351, 357, 360, 374, 377, 379-381

Plan, 256-258

Planner, 383

Planning, 1-3, 45, 50-53, 102, 185, 230, 249, 250, 252, 255-259, 278, 280-282, 295, 356, 362, 363, 367, 389-392

Platform, 13-16

Policies, 9, 113, 115, 202, 354

Population, 8, 62, 66, 101, 113, 118, 122, 141, 256, 257, 261, 297, 300, 304, 305, 308, 311, 333, 336, 346, 367, 369, 371, 372

Ports, 5, 138, 154-159, 161, 163, 190, 204, 256

Practical, 34, 38, 43, 170, 171, 178-190, 280, 287, 304, 323

Prices, 1, 6, 32, 56, 58, 99, 100, 112, 116, 202, 203, 216, 217, 278, 299, 368, 369

Principle,120, 144, 264, 278

Probability, 27, 153, 155-159, 164, 172, 212, 213, 264, 265, 270, 290

Problem, 3, 78, 83, 98, 140, 168, 198, 221, 278, 281, 285-288, 292, 295, 296, 311, 314, 318, 321, 322, 333, 352, 360, 389, 391

Procedures, 2-6, 9, 30, 71, 78, 88, 89, 137, 138, 158, 168-170, 182, 185, 189, 190, 226, 249, 278, 287, 290, 311, 321

Process, 3, 104, 142, 205, 249, 250, 256-258, 278, 280, 281, 285-288, 291, 292, 296, 320, 322, 324, 334, 342, 351, 352, 356, 382

Production, 111, 121, 128, 138, 141, 143, 145-152, 155, 156, 159, 197, 198, 200-205, 222, 253

Productivity, 5, 30, 34, 38, 44, 45, 101, 141, 145-147, 159-160, 166

Profits, 50, 60, 295, 325, 328, 329, 344, 345, 348-350

Program, 119, 168, 353-355

PROMETEE (Preference Ranking Organization METHod for Enrichment of Evaluations), 280

Propulsion, 60, 117, 361-363, 374, 377, 380

PULL, 200, 217

Punctuality, 34, 45-48, 62, 200, 203, 264, 265, 367, 368, 375, 378, 381,

PUSH, 115, 200, 217

Qualitative, 227, 352, 354

Quality, 8, 14, 23, 33, 34, 45, 49, 71, 76, 97, 98, 142, 154, 188, 198, 200, 226, 264, 292, 297-301, 304, 323, 355, 360

Quantitative, 222, 227, 326, 353, 354

Queue, 88, 123, 129

Queuing, 88, 93, 123, 152, 170, 178, 298

Radius, 10, 11, 371, 379

Rail, 2-7, 9-18, 23-30, 35, 45, 46, 49, 58, 59, 65, 66, 72, 76, 77, 97-104, 108, 110, 113-133, 138, 140, 141, 152, 154, 199, 204, 205, 213, 216-220, 226, 244, 249-252, 256, 266, 268,

270-278, 282, 298, 322, 324, 326, 327, 343, 347, 349, 352-374, 377, 380-383, 390

Railways, 6, 8, 29, 31, 34, 48, 65, 97, 101, 199, 202, 204, 256,353

RAMS (Reliability, Availability, Maintainability and Safety), 18

Rank, 311, 313, 317, 318, 350, 379

Ranking, 280, 281, 286, 311, 313, 316, 317, 350, 352, 357, 381

Rate, 10, 24-26, 37, 40, 42, 43, 48, 52, 53, 56, 58, 63, 64, 69, 70, 87, 93, 96, 97, 100, 106, 108, 112, 115, 118, 121-129, 139, 146-150, 156-163, 180, 197, 200-210, 215, 222, 239-242, 254, 270, 279, 280, 299, 331, 363, 364, 370-374, 377

Ratio, 44, 46, 87, 88, 93, 95, 130, 131, 229-239, 280, 288, 303, 306, 325-328, 332, 344-349, 360, 364, 366, 367

Receivers, 1, 32, 98-104, 140, 201, 220, 222, 255, 265, 275, 358-362, 366-372, 382

Regional, 1, 4, 8, 49, 60, 65, 66,201, 222, 249, 250, 260-262, 292, 298, 299, 322, 334, 354, 357-359, 369, 371, 372, 382

Regression, 93, 110, 179, 239, 240, 250, 258-265, 270, 295, 303, 306, 329

Regular, 3, 26, 27, 49, 53, 103, 118, 123, 125, 138, 147, 195-197, 202, 205, 215, 219, 221, 226, 229, 230, 244, 390

Relationship, 10, 11, 21, 22, 25, 34-38, 42, 46, 48-52, 55, 58-64, 68, 71, 84, 91-96, 112, 127, 142, 171, 179, 188, 224, 238, 239, 250-252, 271, 273, 303, 355

Reliability, 18, 27, 34, 45-48, 62, 142, 159, 197, 198, 264, 265, 298, 301, 367, 368, 375, 378, 381

Research, 2, 3, 14, 66, 68, 78, 83, 102, 103, 117, 119, 142, 143, 168, 170, 201, 202, 226-228, 280, 295, 296, 322, 323, 353, 354-356, 373, 376, 389, 390

Resilience, 3, 195, 196, 198, 200-202, 205, 219-222, 226-234, 238-244, 390

Retailer, 205-215

RETRACK (REorganization of Transport Networks by Advanced RAil Freight Concepts), 355, 373, 376-382

Revenue, 195, 226, 237, 251, 260

RFCs (Rail Freight Corridor(s)), 353, 356

RI (Random Index), 289, 314-316

Richter, 224

Risk, 14, 63, 141, 151, 152, 157, 158, 164, 166, 167, 212, 215, 294, 298, 325, 333, 346, 372, 381

RNE (Rail Net Europe), 353, 373

Road, 65, 66, 70, 76, 77, 97-133, 138-140, 152, 154, 198, 199, 202, 204, 205, 212, 216-219, 226, 244, 256, 266-278, 282, 324, 326, 343, 347, 349, 352-377, 380, 381, 383, 390

Rolling stock, 3-9, 12, 18, 23, 29, 34, 38, 43, 48, 49-57, 63-67, 71, 100, 102, 119, 200, 249-252, 282, 369

Rotterdam, 125, 153, 156, 158, 225, 268, 269, 272, 274, 276, 277, 356, 373

Route, 7, 33-48, 62-67, 70, 88, 121, 125, 126, 141, 144, 146, 153-159, 212, 227, 231, 242, 262, 270, 272, 275, 277, 278, 292, 295, 300-303, 306, 320, 359-381

RPK (Revenue Passenger Kilometer), 260

RPM (Revenue Passenger Mile), 260

Rule, 86, 234

Runway, 2, 3, 76-98, 132, 137, 167-189, 223-225, 229-231, 249, 250, 286, 320-352, 382, 383

Safe, 1, 6, 8, 9, 13, 23, 26-28, 40, 42, 49, 152, 158, 172, 198, 238

Safety, 6, 12-14 16, 27, 39, 45, 50, 59, 60, 63, 64, 71, 77, 117-119, 131, 139, 141, 151, 164, 221, 228, 232, 235, 238, 298, 343, 381

SAW (Simple Additive Weighting), 287

Salzburg, 373

Sandy, 235, 236, 238,242, 243

Saturation, 167, 187, 188, 336, 339, 342

Savings, 2, 45, 46, 68, 77, 117, 119, 123, 131-133, 142, 237

Scale, 3, 55, 59, 63-66, 115, 120, 131, 133, 155, 161, 196, 197, 201, 118, 220-224, 228, 233, 234, 242, 244, 249-251, 258, 286, 287, 289, 291, 297, 298, 306, 314, 316, 380-383, 390

Scenario, 51, 55-58, 189, 203, 214, 218, 221, 302, 306-309, 311, 316, 318, 321, 339

Schedule, 34, 45, 103, 155, 211, 213, 261, 327

Schiphol, 78-82, 89, 91, 92, 94-96, 225, 262, 263, 294, 304, 315, 316

Sea, 1, 4, 5, 7, 97, 138, 140, 141, 152-154, 158, 164, 166, 220, 235, 256, 356, 357, 390

Seaborne, 139

Seat, 19-22, 45, 49, 53-55, 57, 68, 237, 261, 303, 306

Seconds, 90

Sector, 1, 2, 60, 66, 116, 132, 139, 198, 219, 227, 244, 292, 352, 352

Semi-trailers, 97, 99, 204, 216, 380

Sensitivity, 111, 143, 197, 200, 202, 216, 285, 307, 311, 316, 320, 334

Separation, 26, 28, 30, 40, 62, 85-87, 89, 90, 137, 169, 171-176, 178, 182-189, 223, 224

Sequence, 6, 41, 43, 85, 86, 89, 90, 102, 169, 171-176, 183, 184, 203, 220-222, 226, 228, 361, 372

Serbia, 124, 125, 373,

Service, 1, 3, 5, 6, 8, 13, 14, 16, 29, 31, 33, 34, 37, 38, 43, 44-59, 62, 63, 66, 67, 71, 76, 77, 80, 84, 85, 87, 93, 95, 97-101, 103, 105, 108, 110, 111, 114, 116, 117, 119, 120, 122-129, 131-133, 139, 141-143, 145-147, 150, 152, 153, 155, 157-160, 164, 166, 167, 178, 188, 195, 197,

199, 200, 201, 203, 204, 208-210, 213, 214, 216, 218, 220, 222, 226-228, 232, 233, 237, 244, 249, 251, 253, 255, 256, 261, 264-267, 270, 273, 275-278, 292-294, 298-301, 303-306, 308, 311, 318, 320, 322, 323, 326, 327, 334, 353, 355-370, 373-375, 377-382, 390

SESAR (Single European Sky ATM Research), 78

Shanghai, 18, 153, 156, 158

Share, 8, 30, 31, 35, 36, 47, 48, 51, 65, 70, 71, 81, 97, 113, 114, 116, 117, 158, 163, 164, 198, 199, 202, 217, 229, 230, 241, 244, 252, 265, 270, 273-275, 277, 300, 301, 303-306, 308, 311

Shin Osaka, 37, 63

Shinkansen, 7, 12, 15, 36, 37, 39, 47, 63, 67-69

Shipment, 1, 4, 5, 6, 77, 97, 99, 100, 102-105, 111, 115, 116, 137-150, 154, 155, 158, 159, 190, 196-205, 207-216, 218, 219, 244, 264-266, 270, 275, 277, 285, 358, 363, 364, 371, 373, 374, 377, 380, 390

Shippers, 1, 32, 98, 99, 100, 104, 201, 220, 222, 228, 255, 265, 275, 322, 358-362, 366-372, 382

Shipping, 97, 153, 156, 166, 201, 356, 357

Ships, 5, 101, 137, 139, 140, 151, 152, 154-156, 158, 159, 161, 163, 164, 166, 190, 249, 252, 254, 255, 282, 390

Shunting, 205

Simulation, 170, 290, 291, 295, 303, 307, 318, 382

Singapore, 263

Size, 5, 14, 21, 29, 34, 43, 55, 82, 98, 99, 103, 104, 106, 116, 137-139, 141, 143, 145, 147, 1479, 150, 159-161, 164, 190, 198, 199, 201, 202, 205, 208, 214, 253, 261, 267, 297-300, 302, 305, 306, 311, 318, 326, 339, 351, 358-362, 364, 365, 367, 375, 378, 379

Slot, 77, 299,

Slovenia, 124, 125

Small, 60, 89, 205, 299, 332,

Social, 1, 2, 33, 59, 60, 65, 66, 70, 71, 76, 77, 80, 82, 84, 117, 119, 120, 129, 132, 133, 137, 140, 144, 151, 154, 164, 166, 168, 190, 195, 249, 255, 256, 257, 279, 290, 321, 322, 324, 332, 334, 336, 346, 350-352, 355, 357, 369, 370, 371, 381, 382, 389

Sofia, 373

Solar, 7

Solutions, 1, 116, 167, 168, 258, 280, 281, 285, 309-312, 321, 322, 334, 336

Space, 2, 5, 12, 16, 45, 95, 141, 149, 150, 152, 164, 165, 167, 175-177, 208, 257, 275, 354, 382

Spaced, 2, 137, 138, 167-172, 174, 175, 177, 179, 181, 183, 184, 186, 188, 189, 190, 336

SWIM (System Wide Information Management), 169

Spain, 10, 11, 24, 50, 51, 62, 64, 65, 69, 352

Spatial, 4, 5, 13, 16, 98, 104, 105, 141-143, 168, 201, 203 ,216, 218, 255-258, 266, 292, 295, 321, 324, 326, 343, 350, 351, 354, 357-361, 371, 374, 377, 379, 380, 381

Speed, 5, 8, 10-12, 14, 16-18, 20, 22-28, 31, 35, 37, 39, 40, 42-46, 50-54, 56-65, 67, 68, 71, 80, 86-89, 103, 104, 106, 110, 118, 122, 123, 127-129, 142, 146, 147, 149, 151, 152, 157-159, 161, 163, 164, 171, 172, 183, 197, 200, 203, 207, 208, 212, 216, 235, 236, 238, 239, 250-253, 256, 264, 282, 301, 327, 359, 362, 363, 368, 374, 377, 379, 380

Split, 100, 112, 199, 257, 258, 264-266, 268, 271, 282

Stand, 150, 151

Stansted, 294, 321, 334-337, 340-352

STAR (Standard Terminal Arrival Route), 88

Stations, 5, 8-10, 13, 14, 16, 18, 23, 29, 30, 34, 37-39, 41, 43, 45, 49, 50, 66, 256

Statistics, 29, 63, 80, 104, 262, 306

Stochastic, 93, 95, 152, 178, 212, 213, 278

Strip, 49, 116, 361, 371

Subsidies, 56, 113, 369, 370, 375, 378, 379

Substitution, 2, 76, 77, 116, 119, 120, 129, 130-133

Supplier, 5, 12, 138, 140-144, 146-148, 150, 152, 153, 156-158, 198, 201, 202, 264

Supply, 1, 2, 4, 6, 9, 13, 18-20, 23, 66, 137, 138, 140-145, 147, 179-156, 159, 161-167, 189, 190, 196, 200, 201, 249, 257, 258, 260-262, 356, 390

Sustainability, 142, 352, 375

Sustainable, 2, 116, 134, 258, 285, 334, 355, 390, 392

Svilengrad, 373

Swap-bodies, 97, 99, 204, 216, 380

System, 1-50, 56, 59, 60, 62-67, 70, 76-85, 87-98, 115, 117, 119, 132, 133, 137, 138, 140, 152, 166-172, 178, 180-182, 184-190, 195, 298, 200, 217, 219-223, 225-227, 229-231, 242, 244, 249, 250, 256-258, 275, 278, 279-281, 285, 286, 291, 292, 298, 301, 304, 320-324, 326-336, 339, 342-348, 351-353, 355-358, 361-364, 368, 370, 371, 374, 377, 380-383, 389-392

Triple E, 154, 155, 158, 166

Taking-offs, 2, 171, 181, 182, 189, 339

Taxiway, 83

Technical, 2, 7, 13, 18, 19, 27, 28, 32-34, 105, 119, 137, 140-147, 159, 160, 166, 190, 220, 255, 282, 352, 355, 357, 361-363, 370, 374, 377-382, 389

Technological, 2, 18, 19, 32, 33, 105, 137, 140-143, 159, 190, 203, 255,

352, 357, 361, 370, 374, 377, 379-382, 389

TENs [Trans-European Transport Network(s)], 280, 323

Terminals, 5, 6, 97-111, 138, 152-159, 167, 201, 204-208, 212-220, 256, 266-276, 324, 326, 334, 343, 347, 349, 350, 353-364, 368, 370-374, 377, 379

Terrorist, 63, 220, 223, 226

TEU (Twenty Foot Equivalent Unit), 108, 110, 146-149, 154-157, 163, 267, 270-277, 358-366, 369-375, 378, 379

TFDM (Terminal Flight Data Manager), 169

TFMS (Traffic Flow Management System), 169

TGV (Train à Grande Vitesse), 19, 20, 27, 44, 63, 68, 69, 256

Thirdparty, 325, 333, 346, 372

Threats, 223, 226

Time, 1, 5, 6, 13, 18, 24, 26-63, 66, 68, 70,76, 80, 84-90,93, 95, 96, 100, 105-118, 122, 123, 129, 131, 138-161, 164, 167-178, 181-189, 197-221, 226, 228-235, 237-240, 244, 249-262, 264, 266, 273, 275, 278-282, 285, 292-294, 298, 301-304, 315, 321, 324, 326, 330-335, 338-343, 347, 349, 351, 353, 358, 362-372, 375, 378, 380, 381

Tokaido, 36, 37, 63

Ton, 11, 19, 20, 21, 67, 108, 110, 111, 121, 127, 130, 131, 139, 140, 146-149, 157, 205-208, 254, 275, 325, 331, 348, 361-363, 370, 374, 377, 378

Ton-kilometer, 97, 105, 199

TOPSIS (Technique for Order Preference by Similarity to IdealSolution), 280, 285-287, 290, 291, 303-313, 316-320, 323, 333, 342, 346, 350-352, 356, 373, 376, 379-382

Tracks, 7-16, 25-30, 38, 39, 53, 63, 67, 71, 118, 204, 205, 358-363, 374, 377
Tractions, 127
Trading, 139, 153, 154, 261, 280, 286
Traffic, 3, 6, 16, 23-30, 38, 42, 55-65, 71, 77, 80-86, 95-102, 110, 117, 118, 123, 128-132, 137, 146, 151, 168, 169, 179, 189, 195, 212, 220-227, 257-259, 275, 278, 279, 282, 291, 292, 295, 300-306, 329, 339, 353, 355, 364-368, 372-378, 381
Trajectory, 181
Transit, 13, 14, 49, 200, 300, 338, 358, 362, 366, 368, 375, 378,380,
Transition, 10, 11, 199,
Transportation, 7, 56, 97, 99, 102, 106, 141, 148, 154, 157-161, 200, 201, 211, 214, 219, 227, 254, 293
Truck, 99, 102-105, 108, 117-126, 129-132, 152, 216, 354, 362, 363, 368-374, 377, 379
TU (Time Unit), 105, 146, 205, 327, 330-333, 363
Turkey, 124-126, 225, 373
TVM (Transmission Vole Machine), 25

UAE (United Arab Emirates), 190
UIC, 7, 9-13, 18, 36, 46, 47, 49-51, 65-68, 70, 71, 118
UK (United Kingdom), 31, 37, 46, 66, 69, 225, 226, 321, 342
Ultimate, 30, 34, 38, 39, 42, 43, 76, 84-88, 132, 137, 138, 140, 170, 171, 178, 179, 182, 184-190, 205, 266, 281, 326
UNCTAD (United Nations Conference on Trade and Development), 139, 140, 154, 159
Unit, 27, 105, 108, 130, 138, 156, 157, 205, 254, 267, 327, 358
United, 167, 204, 225, 263, 321
Unloading, 6, 39, 139, 141, 145-150, 155, 156, 158, 159, 161-163, 214, 267, 270, 358

Urban, 4, 5, 13, 16, 49, 66, 70, 71, 83, 102, 256, 275, 278, 297, 320, 354
Urbanization, 354
US (United States), 321
USA (United States of America), 7, 8, 9, 250
Use, 4, 5, 6, 15, 23, 28, 51, 52, 60, 62, 70, 71, 76, 82, 85, 89, 91, 96, 98, 102, 117, 132, 137, 138, 142-144, 154, 155, 164, 173, 184, 185, 188, 189, 198, 199, 201, 202, 204, 205, 218, 219, 228, 256, 265, 286, 297, 303, 304, 307, 318, 323, 325, 332, 334, 339, 345, 348, 349, 357, 358, 368, 371, 375 379, 381, 382, 391
User, 5, 13, 14, 16, 35, 37, 44, 63, 66, 71, 278
Utility, 264, 265, 267, 270, 272
Utilization, 2, 38, 54, 57, 70, 76-78, 83, 84, 87, 88, 93-96, 100, 132, 168, 198, 200, 201, 293, 298, 301, 318, 351, 359, 361-364, 374, 377, 380

Variable, 26, 105, 106, 156, 157, 173, 179, 209, 212-214, 220, 231, 238, 239, 259-262, 303
Varna, 373
Vehicle, 5, 101, 103-106, 108, 110, 123, 139, 140, 143, 145-147, 149-153, 216, 266, 275, 301, 327
Vertical, 10, 169-174, 184, 186, 187, 189
VFR (Visual Flight Rules), 223
Virgin, 294, 338

Wagons, 121, 205, 361-363, 374, 377, 379-381
Wake Turbulence, 168
Wake-vortex, 85, 86, 89, 169, 171, 175, 182-184, 187
Wardrop, 278
Waterways, 4, 5, 7, 97, 99, 116, 141, 199, 202, 213, 256, 353, 356, 357
Weather, 80-85, 89, 94, 101, 169, 171, 177, 184, 195, 196, 220, 223
Weight, 11, 12, 19-22, 67, 68, 88, 106, 108, 110, 118, 121, 126-128, 131,

139-141, 171, 202, 205, 208, 214,
215, 229-232, 275, 286-291, 296,
303, 308, 326, 349, 350, 362, 363,
374, 377, 378
Welfare, 2, 34, 60, 65, 295, 325, 329,
334, 344, 348, 369, 375, 378, 381
Wind, 7, 80, 82, 94, 171, 183, 235-239
WLU (Workload Unit), 306
Work, 13, 38, 44, 45, 55, 101, 103, 299,
365, 375, 378, 380, 381
Workload, 26, 31
WTMA (Wake Turbulence Mitigation
for Arrivals), 168, 169, 177

WTMD (Wake Turbulence Mitigation
for Departures), 168, 169, 177

Yard, 205
Yield, 260-262
Yr (year) 19, 20, 48, 51-57, 63, 156, 215,
260, 277, 303, 305, 330-333, 339,
342, 345, 346, 363

Zagreb, 373
Zone, 16, 40, 78, 99, 103-106, 108, 110

Printed and bound by CPI Group (UK) Ltd, Croydon, CR0 4YY

01/11/2024

01782622-0012